Understanding Our Environment:
An Introduction to Environmental Chemistry and Pollution

Second Edition

Understanding Our Environment: An Introduction to Environmental Chemistry and Pollution

Second Edition

Edited by
R. M. Harrison
University of Birmingham

ISBN 0-85186-233-0

A catalogue record for this book is available from the British Library

Published by The Royal Society of Chemistry,
Thomas Graham House, Science Park, Cambridge CB4 4WF

Typeset by Paston Press Ltd., Loddon, Norfolk
and printed by Woolnough Bookbinders Ltd., Irthlingborough, Northamptonshire

Preface

The field of environmental chemistry goes from strength to strength. Twenty years ago it existed in the UK in the form of a few isolated research groups in Universities, Polytechnics, and Research Institutes, but was very definitely a minority interest. It was not taught appreciably in academic institutions and few books dealt with any aspect of the subject. The awakening of environmental awareness, first in a few specialists and subsequently in the general public has led to massive changes. Environmental chemistry is now a component (optional or otherwise) of many chemistry degree courses, it is taught in environmental science courses as an element of increasing substance, and there are even a few degree courses in the subject. Research opportunities in environmental chemistry are a growth area as new programmes open up to tackle local, national, regional, or global problems of environmental chemistry at both fundamental and applied levels. Industry is facing ever tougher regulations regarding the safety and environmental acceptability of its products.

When invited to edit the second edition of 'Understanding Our Environment', I was delighted to take on the task. The first edition had sold well, but had never really met its original very difficult objective of providing an introduction to environmental science for the layman. It has however found widespread use as a textbook for both undergraduate and postgraduate-level courses and deserved further development with this in mind. I have therefore endeavoured to produce a book giving a rounded introduction to environmental chemistry and pollution, accessible to any reader with some background in the chemical sciences. Most of the book is at a level comprehensible by others such as biologists and physicians who have a modest acquaintance with basic chemistry and physics. The book is intended for those requiring a grounding in the basic concepts of environmental chemistry and pollution. It is therefore a companion volume to 'Pollution: Causes, Effects and Control' (also published by the Royal Society of Chemistry) which is both more diverse in the subjects covered, and in some aspects appreciably more advanced.

Mindful of the quality and success of the first edition, it is fortunate that many of the original authors have contributed revised chapters to this book (A. G. Clarke, Sir Hugh Fish, R. M. Harrison, C. N. Hewitt, and S. Smith). I am pleased also to welcome new authors who have either produced a new view on one of the topics covered in the earlier book (B. Alloway, R. Allott, and D. Hughes) or have contributed on a new topic (S. J. de Mora). The authors have been chosen for their deep knowledge of the subject and ability to write at the level of a teaching text and I must express my gratitude to all of them for their hard work and willingness to tolerate my editorial quibbles. The outcome of their work, I believe, is a book of great value as an introductory text which will prove of widespread appeal.

Roy M. Harrison
Birmingham

Contents

Contributors

R. A. Allott, *Safety and Reliability Directorate, AEA Technology, Wigshaw Lane, Culcheth, Warrington WA3 4NE, UK.*

B. J. Alloway, *Environmental Science Unit, Geography Department, Queen Mary and Westfield College, Mile End Road, London E1 4NS, UK.*

A. G. Clarke, *Department of Fuel and Energy, The University of Leeds, Leeds LS2 9JT, UK.*

S. J. de Mora, *Department of Chemistry, University of Auckland, Private Bag, Auckland, New Zealand.*

Sir Hugh Fish, CBE, *Red Roofs, Newbury Road, Shefford Woodlands, Newbury, Berkshire RG16 7AJ, UK.*

R. M. Harrison, *Institute of Public and Environmental Health, School of Biological Sciences, University of Birmingham, Edgbaston, Birmingham B15 2TT, UK.*

C. N. Hewitt, *Institute of Environmental and Biological Sciences, Environmental Science Division, Lancaster University, Lancaster LA1 4YQ, UK.*

D. J. Hughes, *Faculty of Law, University of Leicester, University Road, Leicester LE1 7RH, UK.*

S. Smith, *King's College London, University of London, Division of Biosphere Sciences, Campden Hill Road, London W8 7AH, UK.*

CHAPTER 1

Introduction

ROY M. HARRISON

1 THE ENVIRONMENTAL SCIENCES

It may surprise the student of today to learn that 'the environment' has not always been topical and indeed that environmental issues have become a matter of widespread public concern only over the past twenty years or so. Nonetheless, basic environmental science has existed as a facet of human scientific endeavour since the earliest days of scientific investigation. In the physical sciences, disciplines such as geology, geophysics, meteorology, oceanography, and hydrology, and in the life sciences, ecology, have a long and proud scientific tradition. These fundamental environmental sciences underpin our understanding of the natural world and its current-day counterpart perturbed by human activity in which we all live.

The environmental physical sciences have traditionally been concerned with individual environmental compartments. Thus, geology is centred primarly on the solid earth, meteorology on the atmosphere, oceanography upon the salt water basins, and hydrology upon the behaviour of freshwaters. In general (but not exclusively) it has been the *physical* behaviour of these media which has been traditionally perceived as important. Accordingly, dynamic meteorology is concerned primarily with the physical processes responsible for atmospheric motion, and climatology with temporal and spatial patterns in physical properties of the atmosphere (temperature, rainfall, *etc.*). It is only more recently that *chemical* behaviour has been perceived as being important in many of these areas. Thus, while atmospheric chemical processes are at least as important as physical processes in many environmental problems such as stratospheric ozone depletion, the lack of chemical knowledge has been extremely acute as atmospheric chemistry (beyond major component ratios) only became a matter of serious scientific study in the 1950s.

There are two major reasons why environmental chemistry has flourished as a discipline only rather recently. Firstly, it was not previously perceived as important. If environmental chemical composition is relatively invariant in time, as it was believed to be, there is little obvious relevance to continuing research. Once, however, it is perceived that composition is changing, (*e.g.* CO_2 in the

Table 1 *Halocarbon concentrations and trends (1990) (from IPCC*)*

Halocarbon		*Mixing ratio* p.p.t./v	*Annual rate of increase* p.p.t./v	%	*Lifetime* Years
CCl_3F	(CFC-11)	280	9.5	4	65
CCl_2F_2	(CFC-12)	484	16.5	4	130
$CClF_3$	(CFC-13)	5			400
$C_2Cl_3F_3$	(CFC-113)	60	4–5	10	90
$C_2Cl_2F_4$	(CFC-114)	15			200
C_2ClF_5	(CFC-115)	5			400
CCl_4		146	2.0	1.5	50
$CHClF_2$	(HCFC-22)	122	7	7	15
CH_3Cl		600			1.5
CH_3CCl_3		158	6.0	4	7
$CBrClF_2$	(halon 1211)	1.7	0.2	12	25
$CBrF_3$	(halon 1301)	2.0	0.3	15	110
CH_3Br		10–15			1.5

*Intergovernmental Panel on Climate Change, 'Climate Change—The IPCC Scientific Assessment', (eds) J. T. Houghton, G. J. Jenkins, and J. J. Ephraums, Cambridge University Press, Cambridge, 1990

atmosphere; ^{137}Cs in the Irish Sea) and that such changes may have consequences for humankind, the relevance becomes obvious. The idea that using an aerosol spray in your home might damage the stratosphere, although obvious to us today, would stretch the credibility of someone unaccustomed to the concept. Secondly, the rate of advance has in many instances been limited by the available technology. Thus, for example, it was only in the 1960s that sensitive reliable instrumentation became widely available for measurement of trace concentrations of metals in the environment. This led to a massive expansion in research in this field and a substantial *downward* revision of agreed typical concentration levels due to improved methodology in analysis. It was only as a result of James Lovelock's invention of the electron capture detector that CFCs were recognized as minor atmospheric constituents and it became possible to monitor increases in their concentrations (see Table 1).

2 THE CHEMICALS OF INTEREST

A very wide range of chemical substances are considered in this book. They fall into three main categories:

(i) Chemicals of concern because of their human toxicity. Thus, for example, the metals, lead and mercury are well known for their adverse effects at high levels of exposure. Chemical carcinogens are particularly topical, despite the miniscule risks associated with many of them at typical levels of exposure. Examples are benzene (largely from vehicle emissions) and polynuclear aromatic hydrocarbons (generated by combustion of fossil fuels)

(ii) Chemicals which cause damage to non-human biota, but are not believed to harm humans at current levels of exposure. PCBs are an example, believed to be having a serious effect on some animal populations, but not proven to be harmful to humans.

(iii) Chemicals not directly toxic to humans or other biota, but capable of causing environmental damage, *e.g.* CFCs.

3 THE ENVIRONMENT AS A WHOLE

A facet of the chemically-centred study of the environment is a greater integration of the treatment of environmental media. Traditional boundaries between atmosphere and waters, for example, are not a deterrent to the transfer of chemicals (in either direction), and indeed many important and interesting processes occur at these phase boundaries.

In this book, the treatment first follows traditional compartments (Chapters 2, 3, 4, and 5) although some exchanges with other compartments are considered. Fundamental aspects of the science of the atmosphere, waters, and soils are described together with current environmental questions, exemplified by case studies. Subsequently, quantitative aspects of transfer across phase boundaries are described and examples given of biogeochemical cycles (Chapter 6). Monitoring considerations are covered in Chapter 7, with the effects of chemical pollution in Chapter 8, and finally the economic and regulatory aspects in Chapter 9.

CHAPTER 2

The Atmosphere

A. G. CLARKE

1 THE GLOBAL ATMOSPHERE

1.1 The Structure of the Atmosphere

1.1.1 Troposphere and Stratosphere. The vertical structure of the atmosphere, showing the features that are most relevant to the problems covered in this chapter, is illustrated in Figure 1. The depth of the troposphere is 8–15 km, the lowest values occurring at the poles and the highest at the equator. There is also a variation with season of the year. Within this layer occurs most of the variability of conditions which leads to 'the weather' as the layman experiences it. Above the troposphere lies the stratosphere which is relatively cloud-free and considerably less turbulent—hence long distance passenger jets fly at altitudes corresponding to the top of the troposphere. The distinction between these two layers is based on a change in the temperature variation with height (Figure 1). Within the troposphere temperature decreases with height but as we enter the stratosphere the temperature starts to increase again. This point is termed the tropopause. The situation with a layer of warmer, less dense air over a layer of cooler, denser air is quite stable. Consequently air is exchanged between the troposphere and stratosphere only very slowly.

'Air pollution' normally means pollution of the troposphere within which most pollutants have a fairly limited lifetime before they are washed out by rain, removed by reaction, or deposited to the ground. However, if pollutants are injected directly into the stratosphere they can remain there for long periods resulting in noticeable effects over the whole globe. Thus major volcanic eruptions injecting fine dust into the stratosphere can lead to a reduction in the amount of solar energy reaching the ground for more than a year after the event. This could also be the result of a major nuclear conflict and has led to fears of a 'nuclear winter' which could make life extremely difficult for survivors.[1] The possibility of damage to the stratospheric ozone layer is discussed in section 1.3.

[1] 'Environmental Consequences of Nuclear War, Vol. 1 Physical and Atmospheric Effects', (SCOPE 28), 2nd Edn., ed. A. B. Pittock, John Wiley and Sons, Chichester, 1989.

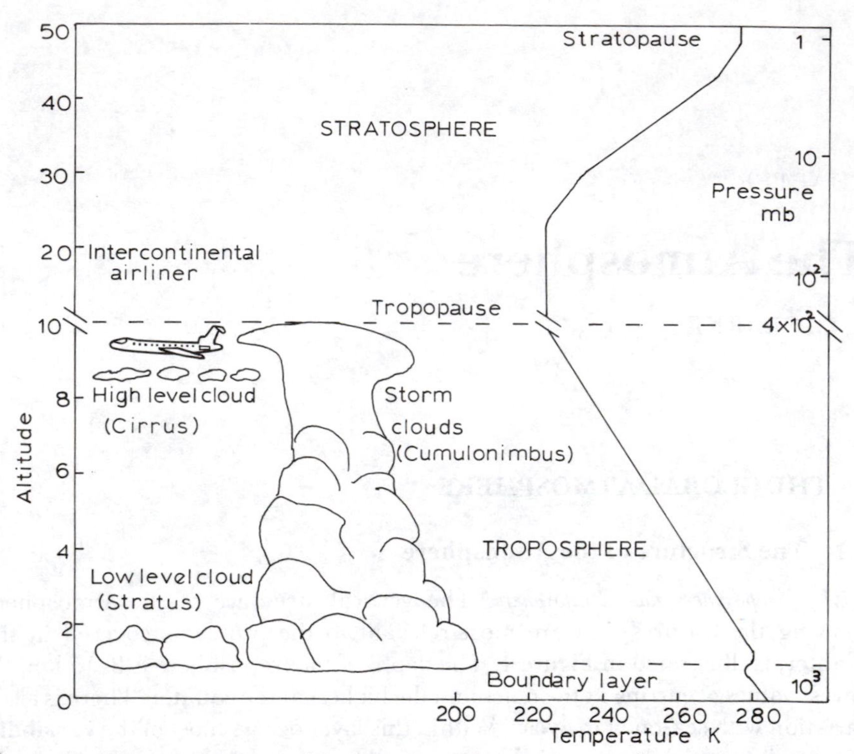

Figure 1 *The vertical structure of the atmosphere. The temperature profile would be typical for latitude 60 °N in summer. Note the change of scale used for the upper half of the figure*

1.1.2 Atmospheric Circulation. The main driving force for the circulation of the atmosphere is the incident solar radiation. The solar energy falling on a given area varies with latitude so that the poles are cold and the equatorial regions warm. The proportions of the incident energy reflected back to space, absorbed by the land or sea, and re-radiated at a longer wavelength all vary from place to place and affect the temperature distribution and circulation patterns. This energy balance is crucial to the determination of the global climate and is considered in more detail in the next section. The rotation of the earth affects the circulation patterns in a fundamental way resulting in the tendency of air to circulate in large-scale eddies around the 'low' and 'high' pressure regions on synoptic weather charts. The processes of evaporation of water, cloud formation, and precipitation also affect the energy balance and circulation patterns.

The presence of the ground has only a small effect on the overall pattern of atmospheric circulation and at most altitudes air movements approximate to those of a non-viscous fluid. The theoretical wind speed—the so-called geostrophic wind speed—can be calculated from the pressure gradient and the rotational velocity of the earth. The pressure gradient is reflected on a weather chart by the closeness of the isobars, lines of constant pressure. If the isobars are close together the wind speed will be high.

1.1.3 The Boundary Layer. Near to the ground the situation is more complicated. Turbulence is generated by mechanical forces as the air flows over uneven ground features such as hills, buildings, or trees and also by buoyancy forces. The ground may warm or cool the air next to it resulting in up-currents and down-currents. In the language of fluid mechanics we have turbulent transport of momentum and energy with corresponding velocity and temperature gradients in the vertical direction. In less technical language there is a frictional effect of the ground. Consider the variation of wind speed with height over the lowest few hundred metres of the atmosphere. This variation is greatest over rough surfaces (*e.g.* a city) when the effect could be a reduction of 40% of the wind speed aloft, that is, the geostrophic wind. Over smooth surfaces (*e.g.* sea, ice sheets) the effect is less and the reduction may be only 20%. Frictional effects also result in a variation of wind direction with height—'wind shear'. A plume from a tall chimney may appear to be travelling at an angle to the ground level wind.

Within the troposphere we therefore define a boundary layer within which surface effects are important. This is of the order of 1 km in depth but varies significantly with meteorological conditions (Figure 1). Vertical mixing of pollutants within the boundary layer is largely determined by the atmospheric stability which relates to the intensity of the buoyancy effects previously mentioned. This is the subject of a later section. As a generalization, mixing within the boundary layer is relatively rapid whereas mixing through the remainder of the troposphere is slower. This gives rise to the idea of a mixing depth within which pollutants are retained and may be transported long distances. So, for example, models of pollutant transport from the UK to the rest of Europe involve a distance scale of about 1000 km and often assume vertical mixing depth of perhaps 1 km with the pollutants uniformly distributed within this layer.

Table 1 indicates the time and distance scales involved in the dispersion of pollutants emitted from the ground. No account is taken in this table of the rates of removal of any pollutant by reaction, deposition to the ground, *etc.*

Table 1 *Time and distance scales for atmospheric dispersion of emissions*

Time of travel	*Typical distance*	*Area affected*
Hours	Tens of km	Throughout boundary layer.
Days	Thousands of km	Pollutant escaping from boundary layer into free troposphere.
Weeks	Round the earth	The whole troposphere in one hemisphere. Transport to other hemisphere beginning.
Months	Round the earth	Whole global troposphere. Some penetration into lower stratosphere.

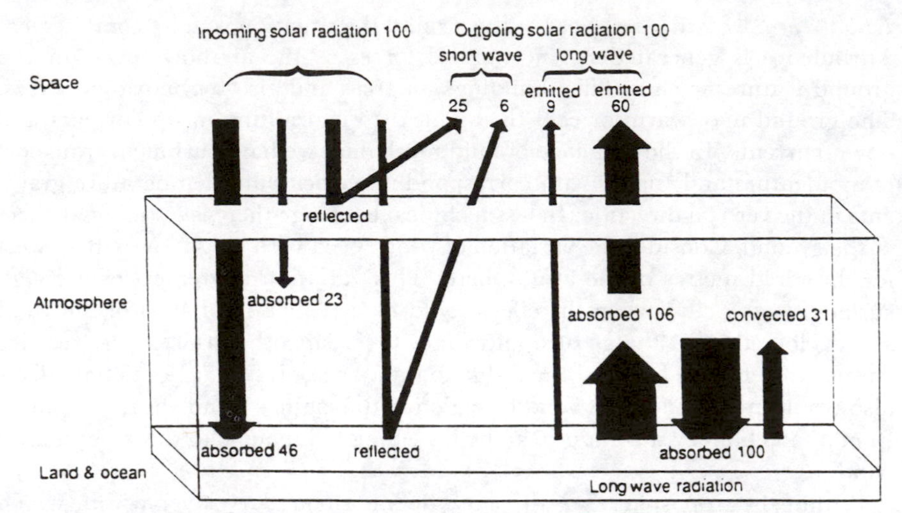

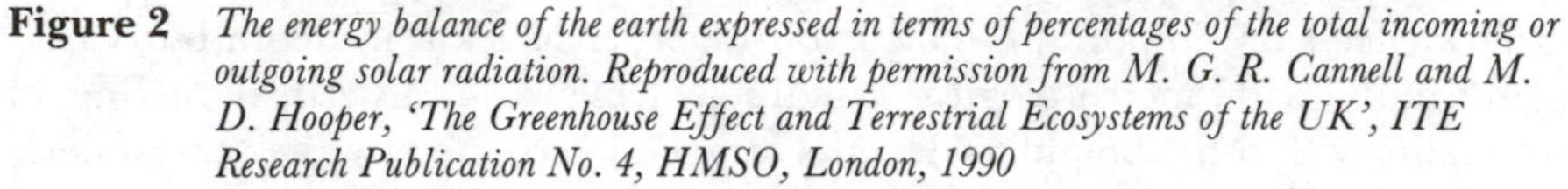
Figure 2 *The energy balance of the earth expressed in terms of percentages of the total incoming or outgoing solar radiation. Reproduced with permission from M. G. R. Cannell and M. D. Hooper, 'The Greenhouse Effect and Terrestrial Ecosystems of the UK', ITE Research Publication No. 4, HMSO, London, 1990*

1.2 Carbon Dioxide and the Global Climate[2,3,4]

1.2.1 The Global Energy Balance. If the earth had no atmosphere, the mean surface temperature would be 255 K, well below the freezing point of water. The atmosphere serves to retain heat near the surface and the earth is thereby made habitable. Most of the radiant energy from the sun lies in or near the visible region of the spectrum (*i.e.* at short wavelength *ca.* 0.6 μm). Some light is reflected unchanged from clouds or from the ground (especially by snow or ice). The fraction of reflected light is termed the albedo and is over 0.5 for clouds but below 0.1 for the oceans. The global average albedo is about 0.3. The stratosphere absorbs ultra-violet radiation primarily due to the ozone present and this results in the warming shown in Figure 1. The lower atmosphere is transparent to visible light so it gains relatively little energy from incoming radiation. The transmitted radiant energy in the visible region penetrates to the ground and is absorbed. Figure 2 shows the percentages of the radiation for different components of the overall energy balance. The radiation emitted from the ground lies in the infra-red region of the spectrum (long wavelength, *ca.* 10–15 μm) and several atmospheric constituents absorb in this region. Carbon dioxide, water vapour,

[2] 'The Greenhouse Effect, Climate Change, and Ecosystems', (SCOPE 29), ed. B. Bolin, B. R. Doos, J. Jager, and R. A. Warwick, John Wiley and Sons, Chichester, 1986.
[3] 'The Greenhouse Effect and Terrestrial Ecosystems of the UK', ITE Research Publication No. 4, ed. M. G. R. Cannell and M. D. Hooper, HMSO, London, 1990.
[4] 'Climate Change, The IPCC Scientific Assessment', ed. J. T. Houghton, G. J. Jenkins, and J. J. Ephraums, Cambridge University Press, Cambridge, 1990.

and ozone are the most important of these. Methane, nitrous oxide, and chlorofluorocarbons (CFCs) are also significant. Some of the absorbed energy will still be re-radiated back to space but a part will be returned to the ground or retained in the atmosphere. The final factor that results in surface to atmosphere transfer of energy is direct warming of the air nearest the ground together with evaporation/condensation processes. The net effect is that more energy is retained near the surface of the earth and the mean temperature is therefore higher (global average 288 K). This is sometimes described as the 'greenhouse effect' by analogy with the properties of glass. Glass is largely transparent to solar radiation while absorbing completely radiation in the infra-red at wavelengths greater than 3 μm. In fact the most important function of a greenhouse is to prevent the circulation of air, inhibiting the normal cooling processes, but the term 'greenhouse effect' has nonetheless been retained.

1.2.2 The Carbon Dioxide Cycle. There is no doubt that man's activities are leading to a gradual increase in the atmospheric carbon dioxide level and this leads to the prospect that we may eventually modify the global climate. Fossil fuel burning is the main contributor to the global annual emissions which have increased by a factor of about 10 since 1900 to an enormous 5.3×10^9 tonnes in 1980. Deforestation adds about another 1×10^9 tonnes per annum. This must be considered in relation to the total atmospheric content of CO_2 which is about 720×10^9 tonnes. The various components of the overall global balance of carbon dioxide are not well quantified. CO_2 is removed from the atmosphere by photosynthesis in plants and is released in the natural processes of respiration and decay. These processes are roughly in balance. The oceans contain vast amounts of CO_2 in inorganic form as well as in association with living organisms. Exchange of gas between the atmosphere and the upper layers of the ocean is rapid. In some areas there may be net release of CO_2 and in other areas net removal. Overall the oceans represent a net sink for CO_2. If all the anthropogenic emissions remained in the atmosphere we should expect to be seeing an increase in the CO_2 level by about 0.75% of its value each year. The measured CO_2 levels are increasing at about half this rate although the upward trend is accelerating with increasing emissions. Before the start of the industrial revolution it is estimated that the mean atmospheric concentration was below 300 p.p.m. From 1958 to 1988 the records at Mauna Loa Observatory in Hawaii show an increase from 315 to 350 p.p.m. in the annual average concentration.

1.2.3 Global Warming. The rates of growth in the atmospheric levels of the other greenhouse gases and their contribution to warming relative to CO_2 are shown in Table 2 and Figure 3.[5] Molecule for molecule, changes in CH_4, N_2O, and the CFCs have more effect than changes in CO_2. Figure 3 indicates the theoretical global warming at equilibrium for given rates of growth of the individual gases. The calculation takes no account of the years that would be necessary for the earth to make the transition to the new equilibrium state. Warming of the oceans, for example, occurs relatively slowly. Although CO_2 is the most important

[5] V. Ramanathan, R. J. Cicerone, H. B. Singh, and J. T. Kiehl, *J. Geophys. Res.* 1985, **90,D3**, 5547.

Table 2 *Atmospheric levels and rates of change of the greenhouse gases used to calculate global warming*

Gas	*Global average mixing ratio* (p.p.b.) *1980*	*2030*	*Rate of rise*
Carbon dioxide	339 000	450 000	2.4% yr^{-1} in emissions
Methane	1650	2340	0.7% yr^{-1}
Nitrous oxide	300	375	0.45% yr^{-1}
CFC 11 ($CFCl_3$)	0.18	1.1	3% yr^{-1}
CFC 12 (CF_2Cl_2)	0.28	1.8	3% yr^{-1}
Ozone (tropospheric)			0.23% yr^{-1}
Ozone (stratospheric) assumed to be +3.8% at 10 km and −38% at 40 km			

Source: 'Stratospheric Ozone', UK Stratospheric Ozone Review Group, HMSO, London, 1987

contributor the other gases taken together contribute about half the overall temperature rise. The situation with ozone is complex. There appears to be a gradual increase in the level of tropospheric ozone due to emissions of NO_x, hydrocarbons, *etc.*, at the same time as a reduction in stratospheric ozone. The surface warming effect depends on the vertical distribution of changes in ozone concentration (Table 2).

Although there is general agreement that we are committed to an increase in mean global temperature the climatological consequences of this are less well understood. Modelling the effect of increased greenhouse gas levels on the global climate is an enormously complex problem requiring the most advanced computers available. Three-dimensional global circulation models describe the vertical, latitudinal, and longitudinal variations in conditions and attempt to compare the present situation with that when for example the CO_2 level has

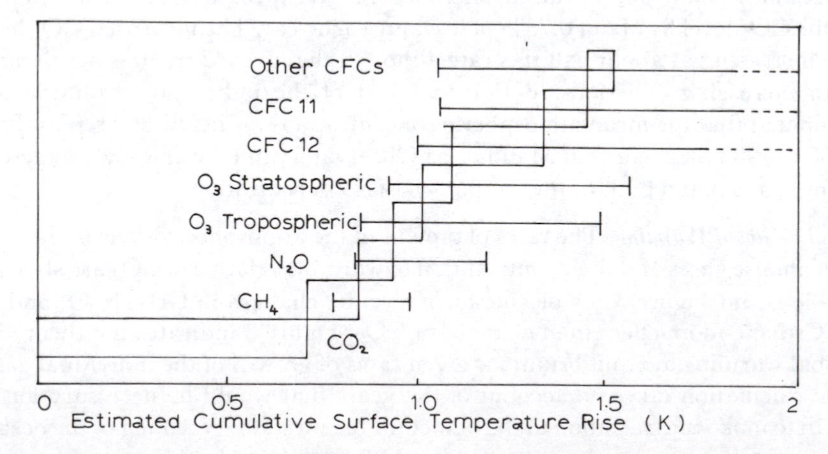

Figure 3 *Equilibrium surface warming by 2030 AD calculated by Ramanathan* et al.[5] *on the basis of the assumed concentrations and rates of increase of trace gases listed in Table 2. The bars indicate uncertainties associated with the projections of release rates*

doubled. Changes in surface and atmospheric temperatures, cloud cover, evaporation, and precipitation, *etc.* all follow the changed radiation balance but the effects are not equally distributed over the globe. For CO_2 doubling most models now suggest an increase of 2–3 °C in mean global temperatures with the largest increases of 8–10 °C over N. Europe and N. Asia (>50° N) in winter and increases of up to 4 °C in Antarctica. Although the global precipitation might increase by 5–10%, the studies suggest that the tropics and areas bordering the eastern coasts of continents would become generally wetter and the sub-tropical regions become drier. This might be critical for some central African regions which already suffer severe drought conditions. Major changes in sea level arising from the melting of the Antarctic ice-cap seem possible only on a time scale of several centuries. A significant thinning of the Arctic ice seems to have occurred already and is currently being studied. The last decade has been the hottest since accurate records began to be taken in the mid-nineteenth century. As yet there is no proof that this is due to the greenhouse effect and it may be many years before this can be established unequivocally.

International discussions are already taking place with a view to limiting the emissions of greenhouse gases. The second World Climate Conference met in Geneva in 1990. It had as its technical basis a report[4] from the UN Intergovernmental Panel on Climate Change, an international body of 300 scientists. Britain is aiming to stabilize its CO_2 emissions at 1990 levels by 2005 although elsewhere in Europe a target date of 2000 was preferred. Germany is aiming at a 30% reduction in emissions by that date. So far the USA is reluctant to commit itself to a target for economic reasons.

Reduction of global CO_2 emissions can be achieved by a number of means including the elimination of large scale deforestation by burning and encouragement of reforestation to promote CO_2 absorption. However for the majority of industrialized countries the general aim must be to reduce the amount of fossil fuels burnt for energy production. Major improvements must be made in energy efficiency and conservation. Wider use of natural gas as a substitute for coal will result in some benefit since the mass of CO_2 emitted per unit of heat released is less: gas 0.43, oil 0.62, coal 0.75 ktonne $MWyr^{-1}$. This trend will occur in the UK as natural gas begins to be used for power production and the amount of coal used is reduced. The other alternatives are the use of renewable energy sources such as wind, solar, wave, and tidal power or the further development of nuclear energy. The last of these options seems unlikely in the short term for both economic and environmental reasons.

1.3 Depletion of Stratospheric Ozone[6,7,8]

1.3.1 The Ozone Layer. Although ozone occurs in the troposphere and plays an important role in air pollution chemistry, about 90% of the total ozone content of

[6] UK Review Group on Stratospheric Ozone, 'Stratospheric Ozone', HMSO, London, 1987.
[7] UK Review Group on Stratospheric Ozone, 'Stratospheric Ozone', HMSO, London, 1988.
[8] UK Review Group on Stratospheric Ozone, 'Stratospheric Ozone', HMSO, London, 1990.

the atmosphere occurs in the stratosphere at altitudes between 15 and 50 km. The ozone layer acts as a filter for ultra-violet radiation from the sun, removing most of the radiation below 300 nm. This serves to protect humans from the adverse effects of UV which become significant below 320 nm—decreasing wavelength corresponds to higher energy photons which can cause sunburn and types of skin cancer. Any depletion of stratospheric ozone would therefore lead to a larger amount of UV radiation incident on the earth's surface and an increased risk of the induction of cancers.

Concern was first expressed about this risk in the early 1970s in connection with emissions of nitrogen oxides from supersonic aircraft such as Concorde, which fly in the lower stratosphere. The nitrogen oxides are potential catalysts for the destruction of ozone. This particular effect is now thought to be relatively minor and attention has been switched to halogen compounds, especially the CFCs or freons. The freons are a group of chlorofluorocarbons used as aerosol propellants, refrigerants, and as gases for the production of foamed plastics. Their attraction lies in the fact that they are non-toxic, non-flammable, and chemically inert. Global production of the two commonest gases, CFC 11 ($CFCl_3$) and CFC 12 (CF_2Cl_2), rose rapidly from below 50 000 tonnes per annum in 1950 to 725 000 tonnes per annum by 1976 decreasing slightly to 650 000 tonnes in 1985. About 90% of this is released directly to the atmosphere while the remainder, representing the refrigerant use, will be released when the equipment containing it is eventually discarded.

The levels of CFCs in the atmosphere are extremely small, (less than 1p.p.b., see Table 2), but are rising at a rate which correlates well with the known emissions. They are resistant to attack by molecules, radicals, or the UV radiation present in the troposphere and not subject to significant dry deposition or rain-out. The higher energy UV radiation in the stratosphere can however lead to photodissociation forming chlorine atoms which can in turn lead to the destruction of ozone. Despite the slow exchange of air between the troposphere and the stratosphere this effect is now known to be significant.

1.3.2 Ozone Depletion. The chemistry of ozone depletion is complex but a basic outline of the important processes is as follow. The ozone is formed from the dissociation of molecular oxygen by short wavelength UV radiation in the upper stratosphere:

$$O_2 \xrightarrow{UV} O^{\bullet} + O^{\bullet} \tag{1}$$

$$O^{\bullet} + O_2 + M \longrightarrow O_3 + M \quad (M = \text{inert third body}) \tag{2}$$

However, ozone itself is rapidly photodissociated:

$$O_3 \xrightarrow{UV} O_2 + O^{\bullet} \tag{3}$$

so the so-called 'odd oxygen' species, O_3 and O, may interconvert many times before they destroy one another by:

$$O^{\bullet} + O_3 \rightarrow O_2 + O_2 \tag{4}$$

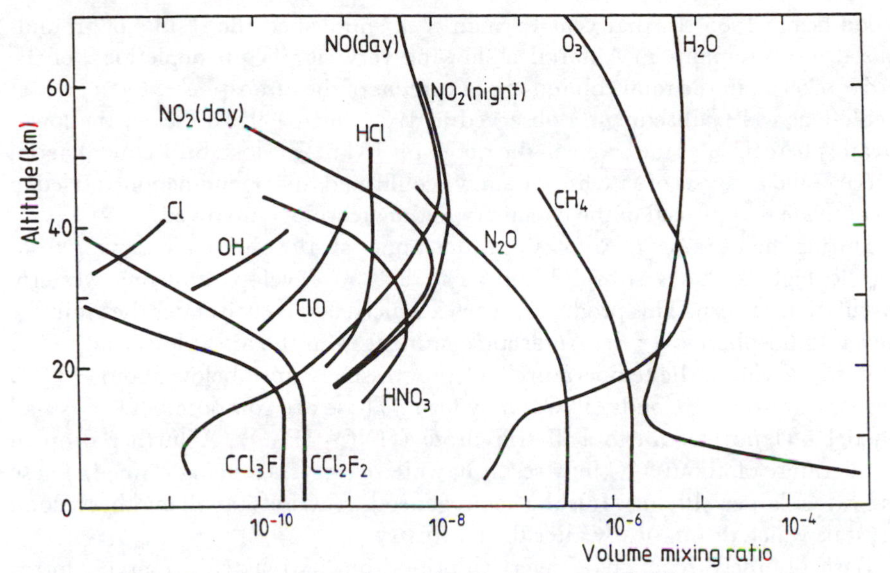

Figure 4 *Approximate concentration ranges of species in the stratosphere important to ozone chemistry. Reproduced with permission from UK Review Group on Stratospheric Ozone, 'Stratospheric Ozone', HMSO, London, 1987*

In fact, measurements of the ozone profile in the atmosphere suggest that ozone destruction must be considerably faster than could be achieved by reaction (4) alone. These other mechanisms can be represented by

$$X + O_3 \rightarrow XO + O_2 \quad (5)$$

$$XO + O^{\bullet} \rightarrow X + O_2 \quad (6)$$

Net effect $$O^{\bullet} + O_3 \rightarrow O_2 + O_2$$

If $X = NO$, the reactions form and destroy NO_2; if $X = Cl^{\bullet}$, the reactions form and destroy ClO. $X = OH^{\bullet}$ gives a third ozone destruction cycle. Other sets of reactions involving NO and $Cl^{\bullet}$ simply achieve the interconversion of O_3 and $O^{\bullet}$ and therefore have no effect on the net ozone levels. The reactive NO_x and $Cl^{\bullet}$ species can be removed by the formation of the relatively stable 'reservoir' molecules HNO_3 and HCl or the somewhat shorter lived chlorine nitrate $ClONO_2$. About half the stratospheric content of NO_x is stored as HNO_3 and about 70% of the chlorine as HCl. Although these may be reactivated by conversion back to NO_x and $Cl^{\bullet}$, they may eventually be transferred back to the troposphere and removed to the ground by rain-out. The typical concentration ranges for the species involved in these reactions are indicated in Figure 4.[6] N_2O is significant as it photolyses in the stratosphere and forms the major precursor of NO_x. Methane reactions produce hydrogen containing species including water vapour and HCl.

1.3.3 The Antarctic Ozone 'Hole'[7,8]. This was the general picture (although a highly simplified account) of the homogeneous stratospheric chemistry as understood before 1985. In that year Farman *et al.*[9] published the results of ground-based measurements in Antarctica showing very significant depletions, of the order of 50%, in the total column ozone content of the atmosphere. Even greater depletions were subsequently observed in 1987 and 1989. In fact in the lower stratosphere 97% of the ozone in the range 14–18 km was lost. Subsequent aerial surveys and analysis of satellite data have confirmed this phenomenon and led to a complete reappraisal of the chemistry and meteorology involved.

During the dark, cold Antarctic winter upper stratospheric air moving from low to high latitudes subsides and as it does so develops a strong westerly circulation pattern. This produces a vortex which effectively isolates the air in the lower stratosphere over the Antarctic continent from the air at lower latitudes. Within the vortex the temperature falls progressively until below about −80 °C polar stratospheric clouds (PSCs) may form. These are composed of very small particles (1 μm) of nitric acid trihydrate ($HNO_3.3H_2O$). A further drop in temperature of about 5 °C may result in water ice crystals being formed. These are rather larger (10 μm). It is the heterogeneous reactions involving these cloud crystals which dramatically alter the chemistry.

Basically these reactions convert chlorine from its inactive, reservoir forms (HCl, $ClONO_2$) into forms which are active ozone depletors ($Cl^\bullet$, ClO). HCl is readily incorporated into ice crystals and can undergo reaction with $ClONO_2$:

$$\underset{\text{ice}}{HCl} + \underset{\text{gas}}{ClONO_2} \rightarrow \underset{\text{gas}}{Cl_2} + \underset{\text{ice}}{HNO_3} \tag{7}$$

and

$$H_2O + ClONO_2 \rightarrow HOCl + HNO_3 \tag{8}$$

The nitric acid is left in the ice phase. The chlorine remains in the gas phase until the polar spring when the sun reappears and photodissociates it to chlorine atoms:

$$Cl_2 + h\nu \rightarrow Cl^\bullet + Cl^\bullet \tag{9}$$

The $Cl^\bullet$ atoms rapidly react with ozone generating ClO:

$$Cl^\bullet + O_3 \rightarrow ClO + O_2 \tag{10}$$

In the winter the stratosphere is thus chemically 'preconditioned' by heterogeneous reactions so that in the spring very rapid ozone depletion occurs. In addition to the ozone destruction cycle represented by equations (5) and (6) with $X = Cl^\bullet$ it is now recognized that chlorine monoxide dimers are also important. This was realized because the oxygen atom concentrations in the lower stratosphere are too low to account for the observed ozone destruction rates:

$$ClO + ClO + M \longrightarrow Cl_2O_2 + M \tag{11}$$

$$Cl_2O_2 \xrightarrow{UV} Cl^\bullet + ClOO^\bullet \tag{12}$$

$$ClOO^\bullet + M \longrightarrow Cl^\bullet + O_2 + M \tag{13}$$

These reactions by-pass the $ClO + O^\bullet$ reaction as a route for reconversion of ClO back to $Cl^\bullet$. The importance of ClO within the Antarctic stratosphere is

[9] J. C. Farman, B. J. Gardiner, and J. D. Shanklin, *Nature (London)*, 1985, **315**, 207.

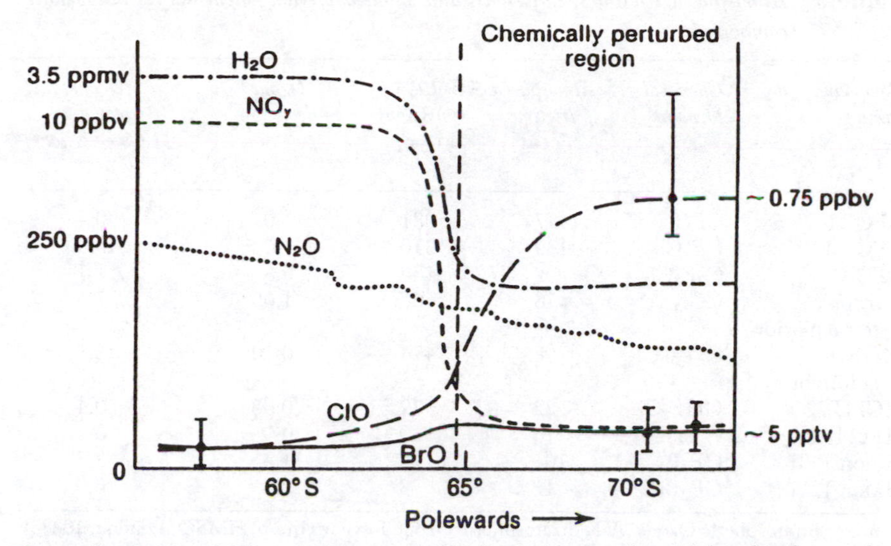

Figure 5 *Profiles of ClO and other species in the Antarctic stratosphere, 18 km altitude, near the boundary of the chemically perturbed region. NO_y refers to forms of oxidized nitrogen other than NO and NO_2, such as HNO_3. The decreases in water vapour and NO_y are due to condensation of water and nitric acid in polar stratospheric clouds followed by gravitational settling. Reproduced with permission from UK Review Group on Stratospheric Ozone, 'Stratospheric Ozone', HMSO, London,* 1988

illustrated in Figure 5. Reactions of bromine atoms in addition to those of chlorine atoms are now thought to account for about 20% of the ozone depletion. Bromine emissions occur in the form of methyl bromide which has natural and man-made sources plus another family of halocarbons, the Halons—Halon 1301 is CF_3Br, Halon 1211 is CF_2BrCl (Table 3).

But is the 'hole' in the Antarctic ozone layer significant for the rest of the globe? Several facts suggest that it is. As the Antarctic spring progresses the vortex breaks up and the ozone depleted air can then be transported to lower latitudes. Such an event has been observed over Australia. In the Northern Hemisphere airborne studies of the Arctic winter were carried out in 1988/9. Although the temperatures are not as low as in the Antarctic and the occurrence of stratospheric clouds not as common, nevertheless they are formed and the existence of a similar chemistry with high ClO concentrations has been demonstrated. The extent of ozone depletion is less marked—perhaps 10–25% in the range 20–25 km altitude representing a reduction of some 3% in total column ozone. Data from the network of ground level monitors for column ozone show a clearly decreasing trend in the Northern Hemisphere since 1970 although the considerable variability in the data makes the precise percentage decrease sensitive to the start date assumed. For Europe and North America the decrease is 2.5 to 3.5% per decade with an indication that the trend has accelerated in the last decade, in parallel with the worsening conditions in the Antarctic.

1.3.4 International Control Measures. The UN Convention on the Protection of the Ozone Layer (the 'Vienna Convention') was agreed in 1985 and subsequently

Table 3 *Atmospheric lifetimes, emissions, and ozone depletion potentials for halogenated compounds*

Compound name	*Chemical formula*	*Atmospheric lifetime* Year	*1985 emissions* kt yr^{-1}	*Ozone depletion potential (CFC 11 = 1)*	*Ozone removal % (1985 emissions)*
CFC 11	$CFCl_3$	77	281	1.0	30.4
CFC 12	CF_2Cl_2	139	370	1.0	40.0
CFC 13	$C_2F_3Cl_3$	92	138	0.8	11.7
Carbon tetrachloride	CCl_4	76	66	1.06	7.6
Methyl chloroform	CH_3CCl_3	8.3	474	0.10	5.1
HCFC 22	CHF_2Cl	22	72	0.05	0.4
HFC 134a	$C_2H_2F_4$	10	0	0	0
Halon 1301	CF_3Br	101	3	11.4	3.7
Halon 1211	CF_2BrCl	12.5	3	2.7	0.9

Source: 'Stratospheric Ozone', UK Stratospheric Ozone Review Group, HMSO, London, 1988

measures to reduce the emissions of various halocarbons were incorporated into the 'Montreal Protocol' in September 1987. A meeting in London in June 1990 further tightened the restrictions under the protocol. More than 70 countries are now signatories. The use of both CFCs and Halons will be phased out by the year 2000 as will the use of carbon tetrachloride, CCl_4. Methyl chloroform will be phased out by 2005. In the immediate future replacement chemicals such as the HCFCs will be used. Because these contain hydrogen atoms as well as halogens they are more reactive in the troposphere and have shorter lifetimes. Their potential impact on the stratosphere is therefore much reduced. Even so the expectation is that the total stratospheric chlorine loading will peak after 2000 at about 4 p.p.b. and only decline slowly through the remainder of the twenty first century.[8] How long it takes to fall to the pre-war level of below 1 p.p.b. depends on the extent of global participation in implementing restrictions and the extent of future use of alternative chlorine containing compounds such as the HCFCs (Table 3).

2 EMISSIONS TO ATMOSPHERE

The primary components of pure dry air are nitrogen, N_2 (78.1%), oxygen, O_2 (20.9%), argon, Ar (0.9%), and carbon dioxide, CO_2 (0.035% or 350 p.p.m.). Water vapour is present in amounts which typically range from 0.5–3% at ground level, depending on temperature and relative humidity. Analysis of air samples reveals the presence of hundreds of other substances in trace amounts. Some of these can be explained directly in terms of their emissions either from natural sources or human activity. Others must have arisen indirectly by chemical processes in the atmosphere. We distinguish these as *primary* and *secondary* pollutants. The secondary pollutants, including gases like ozone and particulate compounds such as sulfates, are dealt with in section 5. Here we concentrate on the primary pollutants which arise from sources we can usually identify.

2.1 Natural Emissions

2.1.1 Sulfur Species. Within a local area the levels of most pollutants are usually dominated by the contributions for which man himself is responsible. However nature is also a generator of what we call 'pollutants' and on a global scale the natural emissions may far outweigh human emissions. This is illustrated in Table 4. Sulfur in the form of SO_2 and some H_2S is emitted most dramatically from volcanoes. However much larger amounts are emitted within sea spray (as sulfate) and from biological processes. In the absence of air, biological decay results in emissions of hydrogen sulfide, H_2S, and organic compounds such as dimethyl sulfide. Both terrestrial and oceanic emissions are significant. The grouping together of all sulfur compounds is appropriate since H_2S and organic sulfur compounds are converted to SO_2 in the atmosphere. Various estimates place the total emissions at 150–250 Tg sulfur per year (1 Tg = 1 million tonne). Since the global SO_2 from combustion emissions is 70–100 Tg S yr^{-1}, the natural sulfur emissions are greater than the anthropogenic emissions.

2.1.2 Nitrogen Species. Biological processes in soil lead to the release of all of the common nitrogen oxides, NO, NO_2, and N_2O. The amounts involved are very uncertain but for NO and NO_2 are of the order of 10 Tg N yr^{-1}. Lightning and biomass burning are other major sources. Oxidation of NH_3 to NO occurs in the troposphere and some nitrogen in the form of HNO_3 is transferred to the troposphere from the stratosphere. These sources total 20–30 Tg N yr^{-1}. In comparison the anthropogenic emissions of NO + NO_2 from combustion are about 20 Tg N yr^{-1}. The sources of nitrous oxide, N_2O are not as well understood as the main loss mechanism which is decomposition in the stratosphere (6–10 Tg N_2O yr^{-1}).[2,3] The major source is the release from soil especially in situations

Table 4 *Natural emissions of S[a] and N[b] compounds*

Source	*Tg S yr^{-1}*	*Source*	*Tg N yr^{-1}*
Volcanoes	<2	Lightning	8
Biogenic gases from land	35	NH_3 oxidation	1–10
Biogenic gases from water	35	From stratosphere	0.5
Sea spray	171	Biogenic production	8
		Biomass burning	12
Natural total	243	Natural total	33
Anthropogenic	75	Anthropogenic	21

[a]D. Möller, *Atmos. Environ.* 1984, **18**, 19 and 29.
[b]J. R. Logan, *J. Geophys. Res.* 1983, **88**, 10785.

where fertilizer has been added. Smaller releases occur from the oceans and from combustion processes.

Anthropogenic emissions of ammonia, NH_3, the other nitrogen-containing gas, are only a few Tg yr^{-1}, mainly from waste treatment. Natural emissions from biological decay and animal excrement exceed 100 Tg yr^{-1} although the exact amount is highly uncertain.

2.1.3 Hydrocarbons. The largest natural sources of methane are anaerobic fermentation of organic material in rice paddies and in northern wetlands and tundra, plus enteric fermentation in the digestive systems of ruminants (*e.g.* cows).[2,3] Methane is also released from insects, from coal mining, gas extraction, and biomass burning. Total emissions are 300–550 Tg yr^{-1} and appear to be increasing at a rate of 50 Tg yr^{-1}. Within the UK, animals, mining, landfill gas, and gas leakage account for most of the 3.5 Tg yr^{-1} emissions.[10]

Heavier hydrocarbons, such as terpenes, are released directly to the atmosphere from trees. Global emissions considerably exceed anthropogenic emissions.

2.2 Stationary Combustion Sources

2.2.1 Carbon Dioxide. The main elemental components of fossil fuels are carbon and hydrogen. When burnt with the oxygen present in air, carbon is converted to carbon dioxide, CO_2, and hydrogen to water vapour, H_2O. It is these overall reactions which convert the chemical energy of the fuel into useful heat. CO_2 is harmless to man at concentrations below about 1% being a natural product of our metabolic processes which is exhaled in our breath. The concern over its wider role in the atmosphere and its possible effect on climate has already been discussed. UK emissions were estimated to be 157 Tg (as carbon) in 1989 representing 2.3% of global man-made emissions.[10]

2.2.2 Carbon Monoxide.[11] In a flame the hydrocarbon fuel molecules are progressively broken down to smaller fragments and combined with oxygen. The end product of this stage of the reaction is carbon monoxide, CO. Subsequent further oxidation converts the carbon monoxide to CO_2. Providing there is sufficient oxygen present, all the CO should be converted to CO_2. However, the formation of carbon monoxide is relatively fast whereas its removal by the reaction:

$$CO + OH^{\bullet} \rightarrow CO_2 + H^{\bullet} \qquad (14)$$

is relatively slow. For its removal to be complete requires high temperatures to maintain rapid reaction and sufficient time before the gases are cooled. These

[10] Department of the Environment, 'Digest of Environmental Protection and Water Statistics', No. 13, HMSO, London, 1990.

[11] R. C. Flagan and J. H. Seinfeld, 'Fundamentals of Air Pollution Engineering', Prentice Hall, Englewood Cliffs, 1988.

conditions are met fairly readily in most combustion appliances providing they are properly adjusted. Faulty or improperly adjusted appliances can produce dangerous amounts of CO (several percent in the flue gas) usually due to some abnormal limitations of the air supply. CO emissions from internal combustion engines are more of a problem and are discussed below.

2.2.3 Soot.[11] Formation of soot commonly accompanies carbon monoxide formation and is generally due to inadequate air supply. In these circumstances partially degraded fuel molecules can polymerize to produce carbon nuclei which can accumulate further material by surface reactions and coagulation. The resulting soot particles are commonly sub-micron ($<1\ \mu m$) and because they are of comparable size to the wavelength of light they are effective both at light absorption and light scattering. A relatively small mass concentration will therefore render the exhaust or flue gases opaque.

Combustion of fuel oils is achieved after spray atomization. Droplets of lighter oils (*e.g.* diesel, kerosine, gas oil) will normally evaporate completely and any soot that is formed will arise from the gas-to-particle route just described. With heavy fuel oils only the lighter components are rapidly vaporized, leaving a tarry residue which is thermally cracked by the heat of the flame. The end product is a hollow, spherical, carbonaceous particle not a lot smaller than the original oil droplet. Given sufficient time and high enough temperatures, these will burn away but some may survive and be deposited within the combustion system or emitted from the chimney. This type of soot has a particle size up to 50–100 μm. Since heavy fuel oils contain sulfur, some of which may be converted to sulfuric acid, an additional problem can be the formation of acid smuts. Soot deposited on the inside of the chimney wall can become saturated with condensing acid. Flakes of this material later blown out of the chimney will eat holes in clothing and corrode the paintwork of cars. This is a very local problem restricted to a few hundred metres distance from the chimney.

Combustion of coal proceeds in a rather similar way to oil droplets. Volatile material is released and burns as a gas, the remaining char burns more slowly and a final residue of ash is left. Combustion of pulverized coal suspended in air can normally be controlled to avoid smoke formation. However burning lumps of coal on a grate is more of a problem because the mixing of air with the volatile matter released by the heat is rather inefficient. Hence the original emphasis in the UK on smoke control by the substitution of smokeless fuels for bituminous coal. Anthracite, coke, and the other manufactured smokeless fuels contain low levels of volatile material and so avoid the problem. Incomplete combustion of volatile matter is also the major problem with wood smoke and smoke from bonfires or stubble burning.

A breakdown of the total UK emissions of black smoke, SO_2 NO_x, CO and hydrocarbons is given in Table 5.[10] From this table it is possible to see the relative importance of transport, domestic, and industrial emissions on a national scale.

2.2.4 Hydrocarbons. Most boilers and central heating units burning fossil fuels result in very low emissions of gaseous hydrocarbons or oxygenated hydrocarbons such as aldehydes. The term Volatile Organic Compounds (VOCs) is

Table 5 *Estimated UK emission of primary pollutants by source type for 1989*

Source	*Black smoke*	*Thousand tonnes* SO_2	NO_x [1]	*CO*	*VOC* [2]
Domestic	191	135	68	339	50
Commercial/industrial	92	683	337	342	43
Power stations	25	2644	785	47	12
Refineries	2	109	36	1	—
Processes + solvents[3]	—	—	—	—	1059
Road vehicles petrol	15	22	702	5649	590
diesel	182	30	596	102	172
Railways	—	3	32	12	8
Forests[4]	—	—	—	—	80
Gas leakage[5]	—	—	—	—	34
Other	5	65	144	31	18
Total emission	**512**	**3699**	**2700**	**6522**	**2067**

[1] NO_x is expressed as NO_2 equivalent; [2] Volatile organic compounds excluding methane; [3] Includes evaporation of motor spirit during production, storage, and distribution; [4] Estimate of natural emissions; [5] Gas leakage is an estimate of losses during transmission along the distribution system.
Source: 'Digest of Environmental Protection and Water Statistics', No. 13, 1990, HMSO, London

used to describe organic material in the vapour phase excluding methane. These compounds are of importance in internal combustion engines as can be seen in Table 5 and will be discussed later.

High molecular weight hydrocarbons are often found in association with soot. Because of the relatively low temperatures on a domestic grate, the smoke contains not only soot but also tarry material that has essentially been distilled from the coal and recondensed in the flue. Such material contains potentially toxic compounds. Workers in the old types of coal gasification plant before the advent of modern standards of environmental control were susceptible to a type of cancer that was attributed to polynuclear aromatic hydrocarbons, referred to as PNAs or PAHs of which benzo(*a*)pyrene is one example:

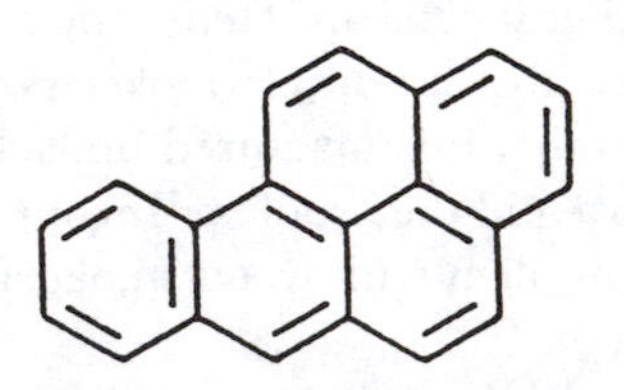

Consequently there has been a long-standing interest in the presence of such carcinogens in the general environment and in the levels emitted from combustion processes. Attention has been switched from coal smoke to diesel engine exhaust smoke as a source of these compounds but it remains to be established

Table 6 *The approximate elemental compositions of fossil fuels (% by wt) used in the UK*

Fuel	*Carbon*	*Hydrogen*	*Oxygen*	*Sulfur*	*Nitrogen*
Wood	50	6	43	0.5	0.5
Bituminous coal	82–92	4.0–5.5	2.5–8.0	0.7–1.0*	1.5–1.9
Anthracite	93.7	3.3	1.1	0.7	1.2
Coke	96.5	0.5	1	1	1
Natural gas	74.8	24.5	0.4 (as CO_2)	—	0.2 (as N_2)
Gasoline	85–88	12–15	—	<0.1	—
Diesel fuel	86.5	13.2	—	0.3	—
Fuel oils	85–86	11.4–13.2	0.1–0.4	0.5–3.0	0.1–0.4

*The compositions for solid fuels are on a dry, mineral-matter free basis. Mineral matter, such as pyrites, is excluded from the figure given. The total sulfur contents of coals, including mineral sulfur, is in the range 1.0–1.8% by weight

whether there is a significant health hazard at the very low concentrations to which the general public are exposed.[12]

2.2.5 Ash. The levels of ash for UK solid fuels range from 5–20% by weight. Bituminous coals burnt in power stations tend to be of lower quality; a typical ash content might be 16%. Coke and anthracite are rather cleaner fuels, with ash contents below 10%. Emission of pulverized fuel ash from current power stations is limited by HM Inspectorate of Pollution to 115 mg m^{-3} and this requires the removal of over 99% of the ash originally present in the flue gases. The limit for new stations has been reduced to 50 mg m^{-3} in line with European requirements. Electrostatic precipitators are the usual means of achieving the high particle removal efficiency.

2.2.6 Sulfur Oxides. Sulfur dioxide, SO_2 arises from the sulfur present in most fuels. Table 6 shows the typical elemental compositions of a range of fossil fuels.[13] In coal the sulfur takes the form partly of mineral matter (pyrites, FeS_2) and partly of organic sulfur compounds. In the table it is the levels of 'dry, mineral matter free' sulfur which are shown, the pyrites contributing additional sulfur. Much of the sulfur in the lighter petroleum fractions is removed during refining but that concentrated in the heavier fuel oils is difficult to remove being strongly bound in aromatic ring compounds. During combustion conversion of sulfur dioxide is virtually complete although about 10% may be retained by coal ash. For the highest sulfur fuel oils (3% S) the flue gas concentrations can reach 2000 p.p.m., while for a typical power station coal of 1.6% S the concentration is about 1200 p.p.m. No control over SO_2 emissions can be achieved by modification of combustion conditions and reductions must be sought by pre-treatment of the fuel or desulfurization of the flue gases after combustion.

[12] K. Steenland, *Am. J. Ind. Med.*, 1986, **19**, 177.
[13] 'Technical Data on Fuel', 7th Edn., ed. J. W. Rose and J. R. Cooper, Scottish Academic Press, Edinburgh, 1977.

A small fraction of the SO_2, usually less than 1%, may be further oxidized to sulfur trioxide, SO_3 which combines with water vapour to form sulfuric acid, H_2SO_4 in the flue gas. This can have serious implications for in-plant corrosion especially on any low-temperature surfaces in the air heaters, ducts, or chimney where the acid condenses. The problems are most acute with heavy fuel oils for which there is the additional difficulty of acid smuts.

2.2.7 Nitrogen Oxides.[11,14] There are two main routes to the formation of the nitrogen oxides NO and NO_2 together described as NO_x. One involves the combination of nitrogen and oxygen in the air at the peak flame temperatures to form nitric oxide, NO. This is termed 'thermal NO_x'. The other starts with the nitrogen originally present in the fuel. Depending on the conditions, some of the fuel-nitrogen will be converted to NO and some to nitrogen gas, N_2. Factors such as the burner design, the intensity of combustion, the overall shape and size of the furnace, and the amount of excess air all influence NO formation and can be modified to achieve a certain measure of control. However this falls considerably short of eliminating the emissions. Typical flue gas concentrations for coal-fired power stations are about 550 p.p.m., but with new designs of low-NO_x burners currently being installed this will be reduced by about 40%. Nitrogen dioxide, NO_2 forms only a small fraction of the waste gases from combustion (usually less than 10%) so the description 'NO_x emissions' actually refers mainly to NO emissions. Once in the atmosphere, oxidation of NO to NO_2 occurs and the relative proportions of the two oxides may then be comparable.

2.2.8 Hydrogen Chloride.[15] In addition to SO_2 and NO, the combustion of coal produces hydrogen chloride, HCl originating from chlorides in the fuel at a level of about 0.2% by weight. Although not generally considered to be as important a contributor to pollution problems, it adds to the acidity of precipitation. Incineration of waste containing chlorinated plastics such as PVC also produces HCl. Petroleum products have no naturally occurring chlorine but leaded petrol contains both chlorine and bromine added in the form of ethylene dichloride and ethylene dibromide in conjunction with the lead anti-knock additives. The chlorine and bromine scavenges the lead oxide formed during combustion and converts it to more volatile lead halides which are emitted from the exhaust. The remainder will be emitted as HCl and HBr but the total amounts involved are small compared to the HCl from coal combustion.

2.3 Emissions from Engines[16]

2.3.1 Gases. Internal combustion engines, whether of the spark ignition (petrol) or compression ignition (diesel) variety, have particular features which warrant discussion beyond that already given. The combustion takes place under high pressure instead of atmospheric pressure and the peak temperatures are higher

[14] J. A. Miller and C. T. Bowman, *Prog. Energy Combust. Sci.*, 1989, **15**, 287.
[15] P. J. Lightowlers and J. N. Cape, *Atmos. Environ.* 1988, **22**, 7.
[16] J. B. Heywood, 'Internal Combustion Engine Fundamentals', McGraw Hill, London, 1988.

Table 7 *Emissions factors and emissions limits for motor-vehicles*

A. Emission factors (g kg^{-1} fuel)

	CO	*VOC*	*NO_x*	*SO_2*	*Black smoke*
Motor spirit	236	25	29	0.9	0.6
DERV	10	17	59	3.8	18

Based on Warren Spring Laboratory estimate of UK national emissions divided by mass of fuel consumed in 1989

B. Emission limits (g test^{-1})

Engine capacity	*CO*	*HC + NO_x*	*NO_x*	*Pm (diesel)*
<2000 cc	30	8		1.1
>2000 cc	25	6.5	3.5	1.1

These are from directives 88/76/EEC[17] (gases) and 88/436/EEC[18] (particles). They have now been replaced by a 'consolidated' directive with one limit for all passenger cars, diesel (<3.5 te), and gasoline, directive 91/441/EEC:

CO	*HC + NO_x*	*NO_x*	*Pm (diesel)*
2.72	0.97	—	0.14 g km^{-1}

than in a normal boiler. The time available for combustion is limited by the engine's cycle to a few milliseconds instead of a second or more.

In petrol engines, lacking any control devices, incomplete burnout of the fuel leads to high carbon monoxide and significant hydrocarbon emissions, especially during idling and deceleration. Nitrogen oxide emissions are also high due to the high temperatures but are at a maximum during acceleration. Diesel engines have much lower carbon monoxide and hydrocarbon emissions but comparable NO_x emissions to a petrol engine (Table 7). The importance of vehicle emissions of these gases relative to other sources is shown in Table 5.

Emissions from petrol engines are legally controlled in Europe under UNECE Regulation 15 and corresponding EEC directives.[17] These require type approval for new automobiles and set limits for CO, hydrocarbons, and NO_x based on the mass emissions during a simulated urban driving cycle. The limitations of the latest directive depend on the capacity of the engine as indicated in Table 7. An additional extra-urban cycle at higher speeds has been adopted and emissions from diesel engines in cars and light commercial vehicles are now covered.[18] From November 1991 CO and hydrocarbon emissions are checked as part of the annual M.o.T. test in the UK.

[17] Directive 88/76/EEC, 'Emissions standards for vehicles up to 3.5 tonnes', *Off. J. Eur. Comm.*, 1988, **L36**, 1.

[18] Directive 88/436/EEC, 'Limiting gaseous pollutants from diesel vehicles up to 3.5 tonnes', *Off. J. Eur. Comm.*, 1988, **L214**, 1.

2.3.2 Particulates. Smoke formation is not a significant problem with spark ignition engines because the fuel and air are well mixed before entering the cylinders. Diesel engines suffer from the disadvantage of producing more smoke especially under heavy load or acceleration conditions. The reason is the relatively poor mixing of air with the fuel spray injected into the cylinder. This produces regions which are too rich in fuel for complete combustion, leading to soot formation. Cities with large numbers of diesel vehicles such as buses suffer from elevated roadside levels of smoke compared to the general urban environment. The odours associated with diesel emissions arise from partially burnt, oxygenated hydrocarbons such as aldehydes.

Diesel engine smoke has been controlled in the EC by opacity limitations.[19] However new regulations have specified mass emission limits both for light duty and heavy duty vehicles.[18,20]

2.3.3 Lead. Emissions of lead from automobile exhausts arise from the use of lead tetra-alkyl anti-knock additives to improve the octane rating of the petrol. About 70% of the lead is emitted from the tail-pipe, mainly as mixed halides (see 2.2.8 above). Increasing concern over the potentially harmful effects of lead on health, particularly the health of children, resulted in a gradual decrease in the maximum permitted amount which can be added to petrol. This is currently 0.15 g Pb l^{-1} [21] having been reduced from 0.4 g Pb l^{-1} in 1986. All new cars must now be able to run on unleaded gasoline. The emissions of lead in 1990 were about 30% of what they were in 1975 and 40% of total gasoline sales in the UK are of unleaded petrol.[10]

2.4 Non-combustion Sources

2.4.1 Gases. For the primary pollutants discussed so far, the emissions from combustion processes generally exceed those from other industrial sources although in local areas emissions from specific industries may be significant. Sulfur dioxide is emitted from sulfuric acid plants and from the roasting of sulfide ores during non-ferrous smelting. NO_2 is emitted from nitric acid plants and hydrocarbons or hydrogen sulfide from refineries. However the largest source of VOCs is the use of solvents, including those released from paints, together with the evaporative losses of gasoline during storage and distribution (Table 5).

Other pollutants not encountered in combustion can be significant in the process industries. For example, hydrogen fluoride is emitted from brick kilns where fluoride-containing clays are used, as in Bedfordshire. It is also emitted in primary aluminium smelting due to the fluoride in the molten electrolyte, from

[19] Directive 72/306/EEC, 'Measures to be taken against emissions of pollutants from diesel engines for use in motor vehicles', *Off. J. Eur. Comm.*, 1972, **L190**, 1.

[20] Directive 88/77/EEC, 'Emission of gaseous pollutants from diesel lorries and buses', *Off. J. Eur. Comm.*, 1988, **L36**, 33.

[21] Directive 85/210/EEC, 'Lead content of petrol and the introduction of lead-free petrol', *Off. J. Eur. Comm.*, 1985, **L96**, 25.

fertilizer works employing fluorapatite—a mixed calcium fluoride/phosphate ore, and from some glass-making processes. Fluoride pollution can cause fluorosis in cattle grazing in the vicinity of the source leading to bone damage and ultimately to death.

2.4.2 Dusts. Probably the most prevalent form of air pollution not necessarily connected with combustion is dust. Particulate matter other than soot can be emitted in many forms—pulverized fuel ash, metal oxide fumes, silica, *etc.* It is convenient to distinguish between different sizes of particles and in the UK the following terminology is used:

Grit —large particles ($>76\ \mu m$ diameter) which will rapidly settle out of the air due to gravity and are just visible to the naked eye.
Fume—very small particles ($<1\ \mu m$ diameter) which can remain suspended in air for long periods and which are visible only by electron microscopy.
Dust —the intermediate size range ($1\ \mu m <$ diameter $< 76\ \mu m$) which can be seen under an ordinary optical microscope.

Emissions of grit, dust, and fume from various processes are controlled under the Clean Air Acts and the Environmental Protection Act 1990.

In addition to process emissions, dust can obviously arise by the disturbing action of outdoor industrial activity on the ground or on raw materials. Quarrying, open cast mining, tipping, digging, the action of heavy lorries, or simply a strong wind acting on stock piles can lead to grit and dust blowing beyond the site and causing a nuisance to others in the area. Such 'fugitive' emissions are difficult to quantify, and control depends on careful industrial practice with an awareness of the possible problems. Water can often be used to good effect to minimize the amounts of dust raised.

2.4.3 Odours.[22] The last air pollutant which will be considered in this survey is the most difficult to deal with scientifically because, to some extent, it is a subjective problem. The unpleasantness of an odour will be judged differently by different people and the sensitivity of individuals to the same level of odour is quite variable. Odours are frequently complex mixtures of airborne substances and may be near the limits of detection of current measurement techniques. One method of quantification is to determine the number of times a sample of odorous air has to be diluted with clean odour-free air so that 50 percent of a group of panellists can no longer detect it.

The most commonly encountered sources of odour arise from agricultural practices—intensive livestock production, disposal of slurry to land, *etc.* Of more immediate concern to town-dwellers are sewage works and landfill sites. Better control of tipping and coverage of the waste with inert material has reduced the problems at such sites. Perhaps the most serious smells arise with the rendering of animal wastes. This provides a disposal route for the remains of animal carcasses from abattoirs and a valuable source of edible fats, tallow, feeding meals, and

[22] 'Odours Control—a Concise Guide', Warren Spring Laboratory, Stevenage, 1980.

gelatine. These processes will have to be registered under the Environmental Protection Act and odorous emissions will have to be strictly controlled.

3 TRENDS IN EMISSIONS AND AIR QUALITY

3.1 Smoke and Sulfur Dioxide

Following the outcry over the London smog of 1952 and the Clean Air Act of 1956, a concerted effort was made to reduce levels of smoke and sulfur dioxide in the air. The main components of this effort were control of visible smoke emissions from industrial chimneys, control of chimney heights to ensure adequate dispersion of SO_2, and the introduction of 'smoke control zones' by local authorities. Within such zones the burning of bituminous coal in domestic grates is forbidden and 'smokeless fuels', *viz.* anthracite, coke, or other manufactured smokeless fuels, must be used in approved types of grates or stoves. The alternative to solid fuels is to switch to electricity, oil, or gas, all of which have negligible smoke emissions compared to coal. In practice, because of the widespread availability of natural gas from the North Sea in the years following 1970, gas has captured most of the domestic heating market and shares the commercial market with oil. Domestic coal is limited and the major users are now the electricity supply industry and steel making. Even this pattern is changing with the privatized generating companies switching to gas for new power stations for economic and environmental reasons.

The trends in emissions of smoke and sulfur dioxide are illustrated in Figure 6.[10] The smoke emissions in 1960 were about double those in 1970 where the figure commences. There has been a very marked reduction in smoke concentrations in parallel with the decreased emissions. However some caution is needed in interpreting the data because of the method of measurement. 'Black Smoke' is related to the darkness of the stain on a filter paper through which air has been drawn. The result is reported as a mass concentration ($\mu g\ m^{-3}$) based on a calibration devised at a time when smoke was primarily coal smoke. Different smokes have different degrees of blackness per unit mass. For example diesel engine soot is 3 times blacker than coal smoke and this is incorporated into the estimation of the 'black smoke' emissions. It is estimated that over 50% of smoke readings in city centre areas may be contributed by diesel engine emissions although the overall national average is 36%. Another problem in relation to the changing composition of the particles is that originally the smoke readings equated to the total mass of particles in the atmosphere. This is not the case today—only one-half to one-third of the mass of particles in the urban atmosphere is black soot. The smoke readings therefore considerably underestimate the total mass concentration of particles.

For sulfur dioxide the overall upward trend in emissions was not reversed until much later than in the case of smoke. 1970 was the peak year. There was however a significant shift from low level emissions (domestic and commercial) to high level emissions (industrial and power generation). In 1960 power generation accounted for only 30% of the SO_2 emissions but by 1989 this had risen to 71%.

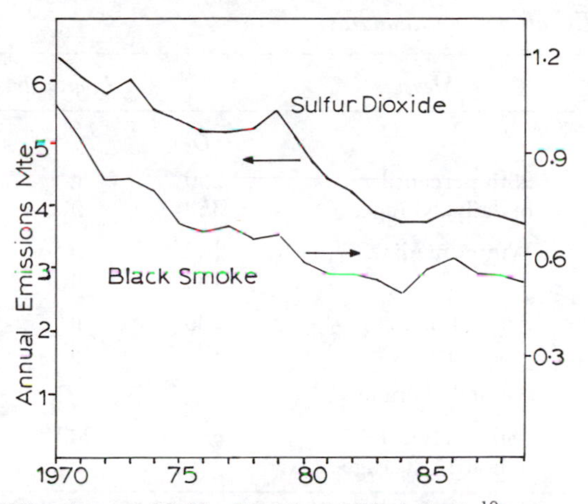

Figure 6 *Trends in UK emissions of Sulfur Dioxide and Black Smoke*[10]

The trend in SO_2 is illustrated in Figure 6 and shows that there has been more than a 30% decrease since 1970. The result on ambient ground level concentrations has been great improvements in the major cities but less marked changes in rural areas.

The EC has agreed health protection standards for smoke and sulfur dioxide,[23] for lead,[24] and for nitrogen dioxide, NO_2[25] (which is more toxic than NO). These are summarized in Table 8. The smoke and sulfur dioxide values are complicated by the criteria for SO_2 being made dependent on the levels of smoke. Most areas of the UK meet the EC criteria. In 1988/9 only one site in mainland Britain exceeded the levels plus two sites in Northern Ireland.[10]

3.2 Nitrogen Oxides[26]

Emissions of nitrogen oxides have gradually increased due to increased traffic density (Figure 7) and will only come down slowly as the new emissions regulations begin to take effect. Since nitrogen dioxide is the main health hazard, most attention has been paid to monitoring urban levels of NO_2 as opposed to total NO_x. The EC air quality standard for NO_2 (Table 8) is 200 $\mu g\ m^{-3}$ or 104.6 p.p.b. which must not be exceeded by more than 2% of the hourly averages. Two sites in London for which automatic monitoring data is available marginally failed this criterion in 1989.[10] The highest recorded level in recent years was 1817 p.p.b. in 1981. Annual average concentrations in urban areas of the UK range from 25 to 45 p.p.b.

[23] Directive 80/779/EEC, 'Air quality limit values and guide values for sulfur dioxide and suspended particulates', *Off. J. Eur. Comm.* 1980, **L229**, 30.
[24] Directive 82/844/EEC, 'Limit value for lead in air', *Off. J. Eur. Comm.*, 1982, **L378**, 15.
[25] Directive 85/203/EEC, 'Air quality standards for nitrogen dioxide', *Off. J. Eur. Comm.*, 1982, **L87**, 1.
[26] UK Photo-chemical Oxidants Review Group, 'Oxides of Nitrogen in the United Kingdom', Harwell Laboratory, 1990.

Table 8 *EEC air quality standards*

Pollutant	*Measure*	*Limit values (μg m⁻³)*			
		SO_2	...	*Smoke*	*Smoke*
SO_2/smoke[a]	98th percentile of daily averages	250	if	>150	250
		350	if	<150	
	Winter median	130	if	>60	130
		180	if	<60	
	Annual median	80	if	>40	80
		120	if	<40	
Lead[b]	Annual average		2		
NO_2[c]	98th percentile of hourly averages		200		

Directive numbers: [a]80/779/EEC[23], [b]82/844/EEC[24], [c]85/203/EEC[25]

NB The EEC smoke levels refer to a calibration curve proposed by OECD which is slightly different from the normal British Standard calibration. The corresponding BS levels may be calculated from BS Smoke = 0.85 × OECD

3.3 Carbon Monoxide

Emissions of carbon monoxide are predominantly from road vehicles and have increased as traffic density has increased (Figure 7). Carbon monoxide levels beside busy roads can under adverse conditions reach approximately half the occupational exposure level of 50 p.p.m. but general urban levels are usually below 10 p.p.m. and annual averages only 1–2 p.p.m. The 98% point in the hourly data for Cromwell Road in London was 10 p.p.m. in 1989.[10] 10 p.p.m. is in fact the WHO guideline for 8 hour averages not one hour averages. Apart from roadside locations under low windspeed conditions, CO is most likely to be a problem in road tunnels and underground car parks where there may be limited ventilation.

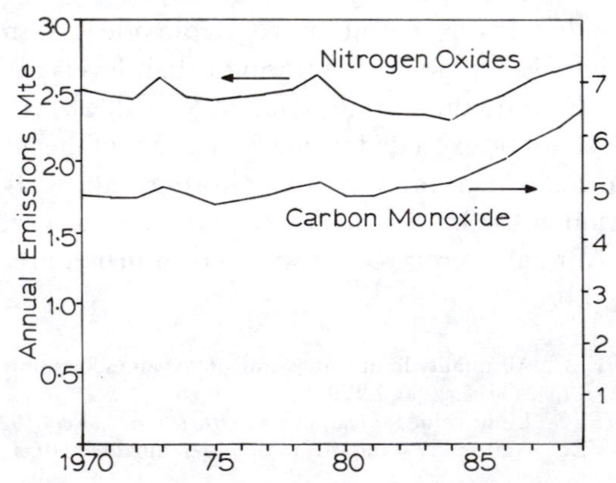

Figure 7 *Trends in UK emissions of Nitrogen Oxides (expressed as NO_2) and Carbon Monoxide*[10]

Table 9 *Department of the Environment air quality classification*

	Concentration range p.p.b.		
	NO_2	SO_2	O_3
Very good	0–50	0–60	0–50[1]
Good	50–100[2]	60–125[3]	50–100
Poor	100–300	125–500	100–200
Very poor	300+	500+	200+

The upper limits approximate to:
[1] the draft EC directive for 98% of hourly averages,
[2] the existing EC directive for 98% of hourly averages,
[3] the existing EC directive for 98% of daily averages

3.4 Ozone[27]

Ozone is a secondary pollutant formed in the atmosphere by photochemical reactions. However there is also a significant background level in the troposphere of about 30 p.p.b. which arises partly from transfer from the stratosphere. The chemistry involved is discussed in section 5. In the UK elevated levels generally occur in summer, anticyclonic conditions. Concentrations tend to be higher in southern England where polluted air masses from continental Europe have a significant effect. The highest recorded concentration in the UK was 258 p.p.b. at Harwell in 1976. Annual averages are near or below 30 p.p.b. with 98% of the hourly averages being in the range 50–80 p.p.b.[10]

The Department of the Environment has established a classification for air quality which is used in public statements and air quality bulletins.[28] The bands for ozone, NO_2, and SO_2 are shown in Table 9.

3.5 Metals

Average lead levels in London, are currently around 0.2 $\mu g\ m^{-3}$, (Table 10)[10] having fallen significantly with the reduction of the maximum permitted level of lead in petrol in 1986 and the gradual introduction of lead-free petrol. The EC air quality standard is 2 $\mu g\ m^{-3}$ annual average.

Occupational exposure limits are valuable as a general indicator of relative toxicity when considering atmospheric levels of pollutants. However, tolerable levels of exposure appropriate for the general public are obviously much lower than those for healthy workers. Adopting a factor of 10 for safety and a further factor of 4 to allow for a working week (40 h) only being a quarter of the total hours in a week, we arrive at the following rule of thumb—ambient levels of pollutants should be kept below 1/40 of the occupational exposure limit. The

[27] UK Photo-chemical Oxidants Review Group, 'Ozone in the United Kingdom', Harwell Laboratory, 1987.

[28] Department of the Environment, 'Air Quality Data Bulletins: Explanatory Briefing', 15th November, 1990.

Table 10 *Levels of trace metals in the atmosphere of central London*

Metal	*Annual mean* ng m^{-3} *1979/80*	*1988/89*	*OEL* mg m^{-3}	*Annual mean as* % *of* OEL/*40*
Zinc	160	95	5	0.08
Iron	1260	710	5[1]	0.57
Vanadium	22	10	0.5[2]	0.08
Lead	730	200	0.15	5.3

[1]as oxides, [2]inorganic lead
Source: Digest of Environmental Protection and Water Statistics No. 13, 1990, HMSO, London

application of this rule of thumb to data on urban heavy metal levels in London is shown in Table 10. It can be seen that amongst the toxic metals listed lead is the only one to which the general public is exposed at a significant fraction of OEL/40 and this is now only at the level of 5%.

4 ATMOSPHERIC TRANSPORT AND DISPERSION OF POLLUTANTS

4.1 Wind Speed and Direction

In general low wind speeds result in high pollutant concentrations and vice versa. If we imagine the wind blowing across the top of a chimney emitting smoke at a constant rate, the *volume* of air into which the smoke is emitted is directly proportional to the wind speed. The *concentration* of smoke in the air is thus inversely proportional to the wind speed. A similar description can be applied to a source distributed uniformly over a wide area (*e.g.* domestic emissions from a city), Figure 8. The concentration of pollutants in the hypothetical box of air into which the pollutants are mixed is proportional to the emissions rate and inversely proportional to the wind speed. In practice this picture is grossly oversimplified.

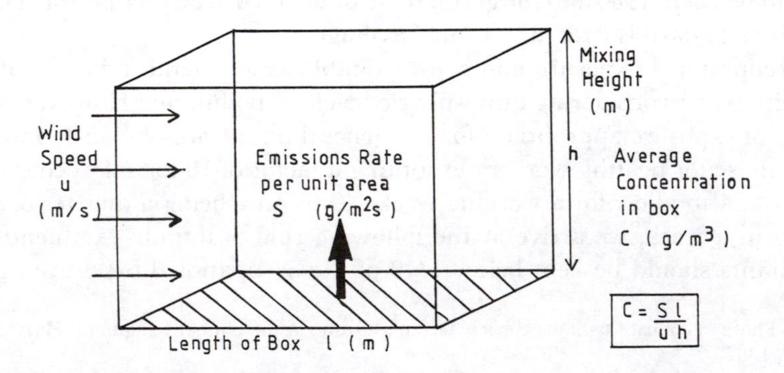

Figure 8 *A simple box model applied to an area source such as a town*

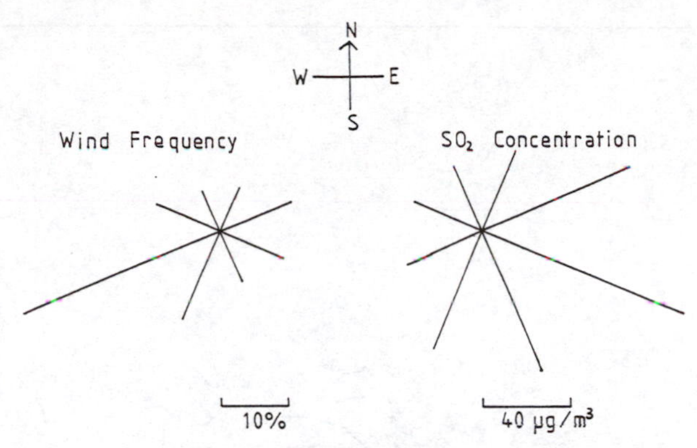

Figure 9 *Wind roses illustrating the directional dependence of (a) wind frequency and (b) sulfur dioxide concentration. Data for Leeds 1982/3*[29]

The concentration of pollutants measured in urban areas rarely decreases with wind speed as rapidly as predicted.

For reasons which will be discussed later, the wind speed at ground level tends to drop overnight and rise again during the morning, especially during cloud-free conditions. Of course, emissions also tend to drop overnight—fewer fires, boilers, and furnaces are alight, fewer cars are on the roads. So some of the highest pollution levels occur in the morning when emissions increase rapidly before the wind speed picks up and dispersion conditions improve.

The people most affected by air pollution are those who are situated downwind of the major sources. A knowledge of the prevailing wind direction is therefore important in predicting the likely impact of these sources. Within the UK the most common winds are westerly or south-westerly. A more detailed presentation of the information can be made using a *wind rose* in which the length of the line is proportional to the frequency of occurrence of the wind in each sector of 45° or 30° (Figure 9). A similar approach can be used to show which wind directions are associated with the highest pollution levels. Figure 9 shows the average sulfur dioxide levels in Leeds for each sector.[29] Wind roses are valuable in identifying the directions of important source regions but need interpreting with considerable care. In Leeds the sulfur dioxide *concentrations* appear lower from W or SW than from E to SE but W or SW winds are stronger than those from the east and south. The total *mass* of SO_2 blowing over Leeds is as great with W or SW winds as with E or SE and this is consistent with the directions of the most important source regions. Variations of atmospheric stability with wind direction also complicate the interpretation.

The wind direction at a point is not sufficient to identify high pollution levels with the effect of distant sources. Over the scale of hundreds to thousands of kilometres the overall path of the particular mass of air must be computed for periods of perhaps 1 to 3 days. Such trajectories can be quite curved as illustrated

[29] A. G. Clarke, M. J. Willison, and E. M. Zeki, 'Physico-chemical Behaviour of Atmospheric Pollutants: Proceedings of the 3rd European Symposium', ed. B. Versino and G. Angeletti, D. Reidel Publishing Company, Dordrecht, 1984.

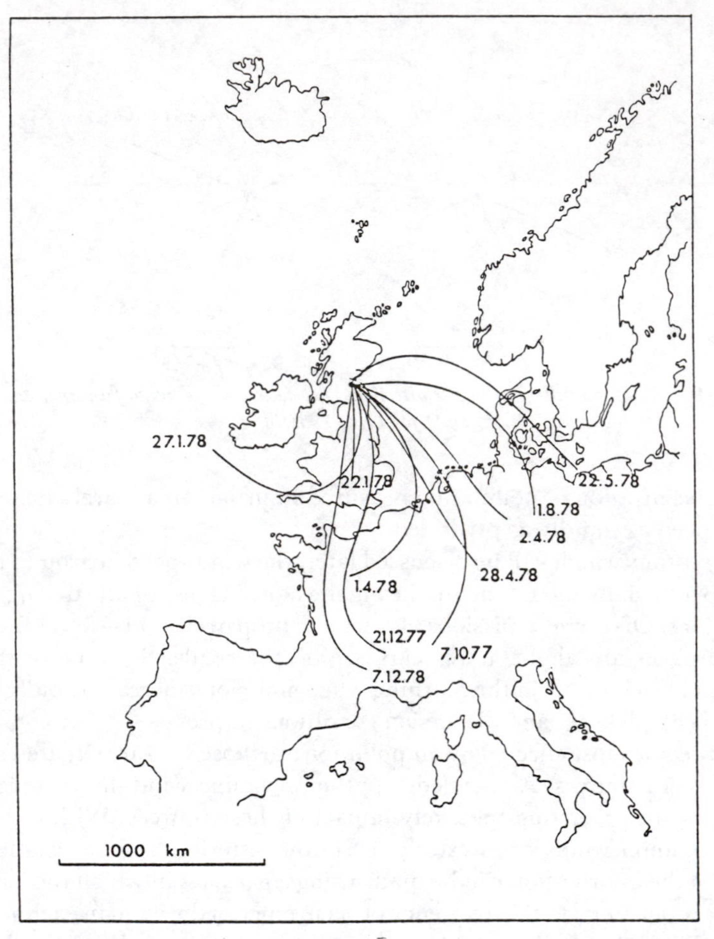

Figure 10 *Back trajectories over 48 h leading to rainfall at Bush, Scotland with pH < 4. Reproduced with permission from* Atmos. Environ., *1984,* **18**, *1863*

in Figure 10, which relates to incidents of high rainfall acidity (pH < 4) at Bush in Scotland. The origin of the acidity is clearly the pollutant emissions from the southern UK and/or continental Europe. Corresponding trajectories from the Atlantic generally bring much less acidic rain.

4.2 Atmospheric Stability

4.2.1 The Lapse Rate. The roughness of the ground produces a certain amount of turbulence in the lowest layer of the atmosphere which promotes the mixing and dispersion of pollutants. This effect increases with the scale of the surface roughness and is greater for a city with large buildings than for open ground with few obstructions. However the major factor affecting atmospheric stability and turbulence is thermal buoyancy. The pressure in the atmosphere decreases exponentially with height. Ascending air expands as the pressure decreases and

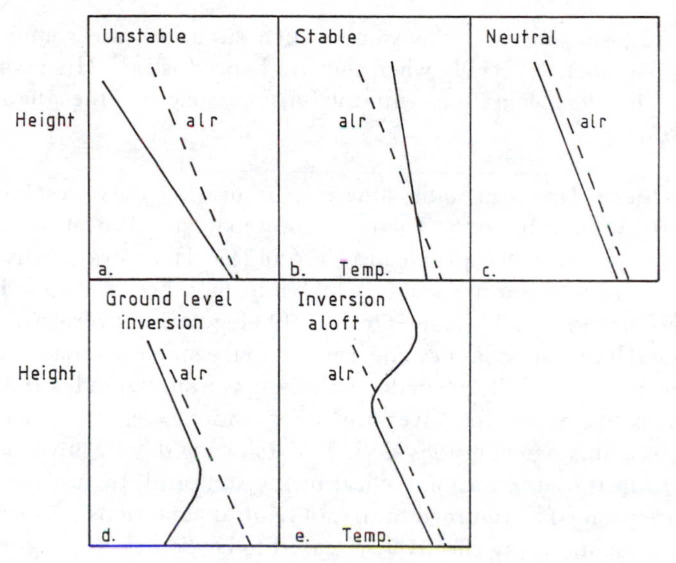

Figure 11 *Schematic illustration of the atmospheric lapse rate for various stability categories. Full lines: actual temperature profile; dashed lines: adiabatic lapse rate (a.l.r.)*

as it expands it cools. A simple calculation based on the properties of gases leads to the conclusion that we should expect a decrease in temperature with height of 9.86 °C km^{-1} or about 1 °C for every 100 m of dry air. The figure is somewhat lower for moist air. The variation of temperature with height is called the lapse rate and the calculation for the ideal case leads to what is known as the adiabatic lapse rate (a.l.r.).

In the real atmosphere the lapse rate can be greater than, smaller than, or close to the adiabatic lapse rate. This fundamentally affects the extent of vertical mixing of air as the following two examples show:

Case (a). Temperature decreases with height more rapidly than the a.l.r., Figure 11a. Air that is slightly warmer than its surroundings starts to rise and to cool at the a.l.r. The temperature difference between the rising parcel of air and its surroundings increases with height, thus the upward movement due to thermal buoyancy continues and is magnified. The atmosphere is *unstable*. Upcurrents in one location are balanced by downcurrents elsewhere and rapid vertical mixing of air occurs, promoting rapid dispersion of pollutants.

Case (b). Temperature decreases with height less rapidly than the a.l.r. or actually increases with height, Figure 11b. Air that is slightly warmer than its surroundings starts to rise and cool at the a.l.r. and the temperature difference between the rising parcel of air and its surroundings soon decreases to zero. Upward movement due to thermal buoyancy ceases. The atmosphere is *stable*. Any vertical movement of air tends to be damped out. Lower polluted layers stay near the ground and pollutant concentrations will be high.

Unstable situations occur with bright sunlight warming the ground to a temperature above that of the air. The air adjacent to the ground is subsequently warmed and rises due to its buoyancy. Such situations are common during daytime in summer, especially when the wind speed is low. High wind speeds tend to lead to *neutral* conditions with the lapse rate close to the adiabatic value (Figure 11c).

4.2.2 Temperature Inversions. Stable situations occur when the lowest layer of air is cooled by the ground beneath. The most common cause is overnight radiation cooling of the ground which often leads to ground level temperature inversions on clear nights. The temperature profile, which may have been shown in Figure 11c in late afternoon, gradually changes into profile Figure 11d overnight. The effect is that ground level emissions become trapped in the stable inversion layer which may not be more than 100–200 m deep. Emissions from tall industrial chimneys however may be above the inversion layer and be vertically dispersed by relatively good mixing conditions aloft. The following day the inversion layer is gradually eroded by the warming effect of the sun until, by mid-morning, the temperature profile has returned to that of neutral conditions (Figure 11c) and the trapped pollutants are effectively released to be dispersed to higher levels.

Another factor contributing to high pollution levels during inversion conditions is the lowered wind speed. Since high level, faster moving air does not mix with low level air, there is no mechanism for the downward transport of momentum. The lowest level air therefore becomes stagnant. In the low temperature conditions, dew, frost, or fog formation may occur. Fog adds to the problem by slowing down the break-up of the overnight inversion layer because the sun's energy is reflected by its upper surface and does not reach the ground. The ground therefore stays cool rather than warming. In extreme cases fog may persist for several days as happened in the London 1952 smog. Polluted fogs are more persistent than clean fogs because the chemicals dissolved in the water droplets prevent complete evaporation even when the relative humidity drops well below 100%.

The other main type of inversion occurs during anticyclonic conditions and is described as a subsidence inversion. Within an anticyclone air is subsiding from a high level to lower levels and progressively warming with decreasing height. The result is the development of an elevated inversion layer as illustrated in Figure 11e. Below the inversion layer the air may be neutral or even unstable so that good mixing occurs but only up to the inversion height. Local urban pollution levels are rarely as great under such conditions as during ground level inversions. However subsidence inversions are often associated with warm, dry weather and they provide the ideal conditions for a long-range transport of pollution. Summer haze conditions in which the UK receives already polluted air from Europe before the addition of our own emissions can lead to particulate pollution levels as high as those on the most polluted winter days. At the same time the levels of photochemical oxidants such as ozone are also high.

In the discussion so far, no reference has been made to the geographical situation in which air pollution levels are being considered. Towns situated in valleys are particularly susceptible to pollution problems. Cool air will tend to flow downhill into the valley so aggravating the problem of low level inversions.

Mixing between the air in the valley and the air above is reduced. Fogs will persist longer. Often in winter a layer of polluted air over the town with cleaner air above can be clearly seen from the higher ground. Towns situated by the sea may be subject to sea breezes. The proximity of relatively warm ground to the cool water surface results in a circulation of air from sea to land. The sea breeze will be cooler than the overlying air, another example of inversion conditions, so in polluted areas conditions will be worse than for a corresponding inland site. Sea mists may blow inland aggravating the general discomfort.

4.3 Dispersion from Chimneys[30]

4.3.1 Plume Rise. After the waste gases leave the top of a chimney the plume may rise a considerable height before achieving more or less horizontal travel. The major factor is the thermal buoyancy of the plume. The hotter the flue gases, the greater the plume rise. A lesser factor is the vertical momentum of the gases due to their efflux velocity out of the chimney top. Plume rise can mean that the effective chimney height is double the physical chimney height under low wind speed conditions and so is a considerable asset in achieving effective dispersion. Conversely any control technology for pollutant reduction which also reduces the flue gas temperature (such as a scrubber) results in poorer dispersion of the plume.

4.3.2 Ground Level Concentrations. The effluent gases leaving a chimney gradually entrain more and more air and the plume width both in the vertical and in the horizontal directions increases with downwind distance. In the horizontal direction there is, of course, a rapid fall-off in concentration as we move away from the plume centre line defined by the average wind direction. The ground level concentration very close to the chimney will be zero because this point is well below the base of the plume. Some distance downwind the dispersing plume reaches ground level and the concentration rises rapidly to a maximum value. For example, for Eggborough Power Station this is at 8 km from the 200 m chimney. Thereafter the concentration falls off with increasing distance as the plume becomes more and more dilute, Figure 12. Unstable conditions will bring the plume down to ground more rapidly, that is closer to the chimney, but the rate of dilution is large so the concentration falls rapidly from its maximum value as we go to greater distances. Stable conditions result in the plume dispersing very slowly. It may remain visible for a considerable downwind distance. The point of maximum ground level concentration is also a long way from the chimney.

The basic models of plume dispersion suggest that the maximum ground level concentration will depend on the square of the chimney height. A 40% increase in chimney height should roughly halve the ground level impact. However this strictly refers to the effective chimney height including the plume rise so the actual degree of improvement may not be quite as great.

[30] 'Recommended Guide for the Prediction of the Dispersion of Airborne Effluents', ed. J. R. Martin, American Society Mechanical Engineers, New York, 3rd Edn., 1979.

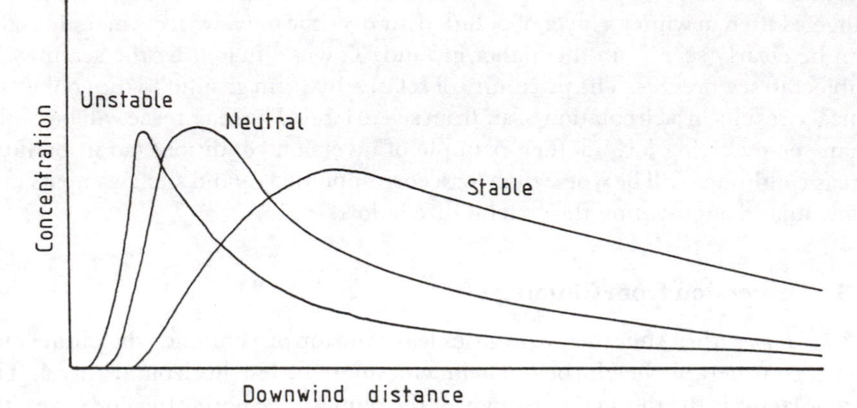

Figure 12 *The variation in ground-level concentration along the plume centre line downwind of a chimney for various stability classes*

Above the chimney top, the vertical dispersion of the plume may be hindered by an inversion layer. The pollutants are then trapped between the inversion height and the ground. At sufficient distance downwind, the concentration may be virtually constant at all heights within the mixing layer.

4.3.3 Time Dependence of Average Concentrations. Atmospheric turbulence makes the dispersion process irregular and the visible plume will often look ragged. From the point of view of an observer with a measuring instrument on the ground, one minute he may measure zero concentration and the next a very high concentration. The longer averaging time he takes, the less the variability of his results. This can be illustrated by the results of measurements made by the Central Electricity Generating Board near Eggborough, a 2000 MW power station in Yorkshire.[31] At the radius of maximum effect, which is about 8 km from the chimney, the peak 3 minute concentrations of sulfur dioxide due to the power station were approaching 1000 $\mu g\ m^{-3}$. Such peaks were extremely rare and 95% of the 3 minute averages were below 10 $\mu g\ m^{-3}$. The highest daily averages were around 100 $\mu g\ m^{-3}$ and the highest monthly averages around 10 $\mu g\ m^{-3}$. The overall annual average contribution of the power station to the ambient SO_2 concentration was only 2–3 $\mu g\ m^{-3}$ in an area where the prevailing concentration from other sources was about 40 $\mu g\ m^{-3}$.

4.4 Mathematical Modelling of Dispersion

Quantitative calculations of the dispersion of a pollutant from a point source such as a large chimney are usually based on the Gaussian Plume Model[30] which will

[31] A. J. Clarke, 'Environmental Effects of Utilizing More Coal', ed. F. A. Robinson, Royal Society of Chemistry, London, 1980.

be discussed in Chapter 7. This treats steady state emissions and situations which can be described as a superposition of a number of different steady states. For example long term average distributions around a source can be calculated by using the statistically averaged meteorological data. The model can be adapted to treat area sources such as wind-blown dust from a stockpile or odours from a sewage works. Urban areas can be modelled as a sum of area sources representing domestic and commercial emissions plus larger point sources such as factories or power stations. Another application is dispersion from sources distributed along a line such as a motorway. Space does not permit a detailed description of these techniques, however, in all cases the following information is required:

(a) Emissions data, *i.e.* the rate of emission of all contributing sources; the heights of the discharge points; flue gas temperatures, efflux velocities, and stack diameters (for plume rise calculations). If particles are involved their particle size range and corresponding sedimentation velocities are needed.
(b) Meteorological data, *i.e.* wind speed and direction, stability class, mixing height. If long-term averages are required, these data must be available in the form of joint frequency distributions. Typically these might involve 5 wind speed ranges, 6 stability classes, and 16 wind direction sectors.

Detailed modelling of urban pollution requires an emission inventory on a grid basis, for example 1 km squares. Long term concentrations can be estimated using software based on the Gaussian Plume principles. However the modelling of diurnal variations of concentration with time dependent emissions and meteorology is much more complex. This involves numerical solution of the diffusion equations. Similarly, modelling the coupled dispersion and atmospheric chemistry of NO_x, hydrocarbons, and ozone is highly complex.

Emissions inventories for the most common pollutants are now available for the whole of Europe on a national basis,[32] on the 150 km × 150 km grid used by the European Monitoring and Evaluation Programme (EMEP),[33] and on grid sizes down to 10 km × 10 km for the UK.[34] Models of regional transport of SO_2 are often based on large numbers of back-trajectories (see Figure 10) for specified receptor points taking account of emissions and loss processes along the path. These are described as Lagrangian models. The alternative, Eulerian, approach is to estimate concentrations and depositions at each grid point as a function of time.[35,36]

[32] C. Veldt, *Atmos. Environ.*, 1991, **25A**, 2683.
[33] J. Saltbone and H. Dovland, 'Emissions of Sulfur Dioxide in Europe in 1980 and 1983', EMEP/CCC Report 1/86, Norwegian Institute for Air Research, Lillestrom, Norway, 1986.
[34] H. S. Eggleston and G. McInnes, 'Methods for the compilation of UK air pollutant emission inventories', Report LR 634(AP), Warren Spring Laboratory, Stevenage, 1987.
[35] UK Review Group on Acid Rain, 'Acid Deposition in the United Kingdom, 1986–1988', Warren Spring Laboratory, Stevenage, 1990.
[36] F. Pasquill and F. B. Smith, 'Atmospheric Diffusion', 3rd Edn., Ellis Horwood, Chichester, 1983.

5 BEHAVIOUR OF POLLUTANTS IN THE LOWER ATMOSPHERE

5.1 Gases

5.1.1 Atmospheric Photochemistry.[27] Figure 13 illustrates the processes which need to be considered between the emission of pollutants and their eventual destruction or removal to the ground. The chemical processes involved are complex and changes both in the gas phase and in the aqueous droplet phase are important. Not only do we have to consider the transformation of the primary pollutants but also the formation of secondary pollutants which can have adverse effects on the environment and on human health.

Of basic importance to an understanding of the gas phase chemistry is the effect of the sun. Photons of ultra-violet light provide a means of initiating chemical reactions which would otherwise not take place. In addition to stable molecules, photochemical reactions involve free radicals such as hydroxyl $OH^{\bullet}$, hydroperoxy $HO_2^{\bullet}$, and methyl $CH_3^{\bullet}$. Free radicals are extremely reactive and have very short lifetimes. Concentrations in the atmosphere are small but nonetheless significant. For example, $OH^{\bullet}$ concentrations in polluted atmospheres may be in the range 10^6–10^7 radicals cm^{-3}, *i.e.* one radical for every 10^{13} nitrogen molecules.

The overall process we have to describe is one of oxidation, that is, the combination of atmospheric oxygen with the primary pollutants. For the three commonest inorganic pollutants the overall results are:

carbon monoxide	$CO \rightarrow CO_2$	carbon dioxide
nitrogen oxides	$NO, NO_2 \rightarrow HNO_3$	nitric acid
sulfur dioxide	$SO_2 \rightarrow H_2SO_4$	sulfuric acid.

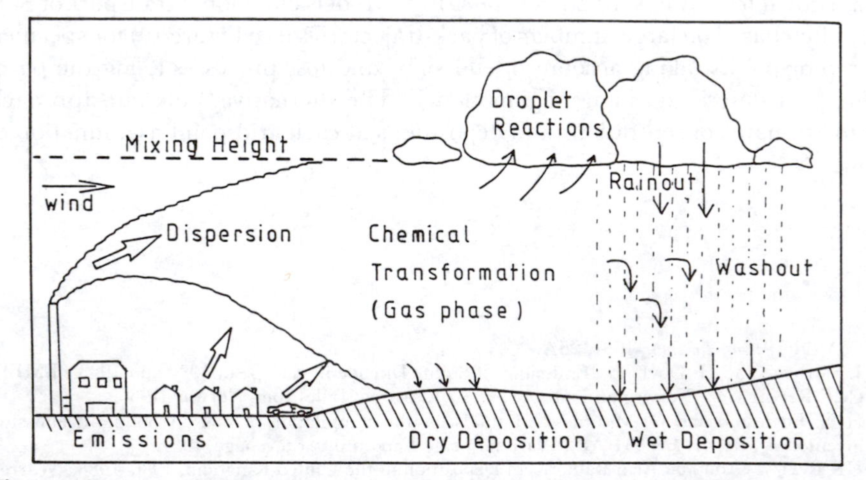

Figure 13 *Processes which may be involved between the emission of an air pollutant and its ultimate deposition to the ground*

Table 11 *Chemical reactions for the atmospheric oxidation of CO, SO_2, and CH_4*

Carbon monoxide

$$CO + OH^{\bullet} \rightarrow CO_2 + H^{\bullet}$$
$$H^{\bullet} + O_2 + M \rightarrow HO_2^{\bullet} + M \quad (M = \text{an inert third body})$$

Sulfur dioxide

$$SO_2 + OH^{\bullet} \rightarrow HSO_3^{\bullet}$$
$$HSO_3^{\bullet} + O_2 \rightarrow SO_3 + HO_2^{\bullet}$$
$$SO_3 + H_2O \rightarrow H_2SO_4$$

Methane

$$CH_4 + OH^{\bullet} \rightarrow CH_3^{\bullet} + H_2O$$
$$CH_3^{\bullet} + O_2 + M \rightarrow CH_3O_2^{\bullet} + M$$
$$CH_3O_2^{\bullet} + NO \rightarrow CH_3O^{\bullet} + NO_2$$
$$CH_3O^{\bullet} + O_2 \rightarrow HCHO + HO_2^{\bullet}$$

$HO_2^{\bullet}$ to $OH^{\bullet}$ conversion

$$HO_2^{\bullet} + NO \rightarrow OH^{\bullet} + NO_2$$

Later we shall consider the involvement of the two acids in particles and droplets but as far as the gas phase chemistry is concerned these species mark the end point of the process. For organic hydrocarbon species there may be a number of intermediate stable molecules formed but the overall process is rather like combustion with the end product being carbon monoxide; for example:

$$\text{methane } CH_4 \rightarrow \text{formaldehyde } HCHO \rightarrow \text{carbon monoxide}$$

In all these cases the main species initiating the sequence of reactions is the hydroxyl radical:

$$CO + OH^{\bullet} \rightarrow CO_2 + H^{\bullet} \tag{14}$$

$$NO_2 + OH^{\bullet} \rightarrow HNO_3 \tag{15}$$

$$SO_2 + OH^{\bullet} \rightarrow HSO_3^{\bullet} \tag{16}$$

$$CH_4 + OH^{\bullet} \rightarrow CH_3^{\bullet} + H_2O \tag{17}$$

In the case of nitric acid formation, there are no remaining free radicals to continue the chain of reactions. In the other cases the hydroxyl radical is eventually regenerated but only after several further steps which interlink the chemistries of the various species (Table 11). The carbon monoxide oxidation is a slow process and the lifetime of CO in the atmosphere is several years. The oxidation rate of SO_2 can be around 2% h^{-1} in urban air or a factor of 10 lower in clean air resulting in an overall lifetime of a few days. Radicals other than $OH^{\bullet}$ such as $HO_2^{\bullet}$, $CH_3O_2^{\bullet}$, or other hydrocarbon peroxy radicals will also attack SO_2 but at a slower rate and their contribution to the overall oxidation is thought to be relatively small.

We must now address the question of where the hydroxyl radicals come from in the first place. One source present even in non-polluted atmospheres at a background level of 20–40 p.p.b. is ozone, O_3 (see section 3.4). Ultra-violet light of wavelength below 310 nm can dissociate ozone producing electronically excited oxygen atoms, O* which rapidly split molecules of water vapour:

$$O_3 \xrightarrow{\text{UV light}} O_2 + O^* \tag{18}$$

$$H_2O + O^* \longrightarrow 2OH^{\bullet} \tag{19}$$

Aldehydes (R.CHO, including formaldehyde H.CHO) can also be photolysed producing hydrogen atoms which eventually result in $OH^{\bullet}$ radicals via reactions already shown in Table 11.

Of basic importance to the understanding of polluted urban atmospheres is the photolysis of NO_2 and the subsequent formation of ozone above the background levels. It was noted earlier that only a small proportion of NO_x emissions are in the form of NO_2, the rest being NO. NO emitted into the atmosphere can be slowly oxidized to NO_2 by the reaction with molecular oxygen:

$$O_2 + 2NO \rightarrow 2NO_2 \tag{20}$$

Photolysis of NO_2 by UV light below 395 nm produces oxygen atoms and subsequently ozone:

$$NO_2 \xrightarrow{\text{UV}} NO + O^{\bullet} \tag{21}$$

$$O_2 + O^{\bullet} \longrightarrow O_3 \tag{22}$$

However the process is reversed by the reaction

$$O_3 + NO \rightarrow O_2 + NO_2 \tag{23}$$

so the net result is an ozone level in equilibrium with NO and NO_2 and dependent on the intensity of the solar radiation. The observed levels of ozone are higher than would be predicted on the basis of this limited scheme. High ozone levels imply a high NO_2/NO ratio or rapid NO → NO_2 conversion which cannot be achieved by the molecular reaction (20). The types of reactions responsible have already been shown in Table 11; they are the transfer of an O-atom from $HO_2^{\bullet}$, $CH_3O_2^{\bullet}$, and other peroxy radicals. Crudely we can say that the photochemical reaction mixture catalyses the NO to NO_2 conversion resulting in the build-up of ozone. Hydrocarbon molecules differ in their ozone production potential. The most important species include the substituted benzenes, such as toluene and the three xylene isomers, and light unsaturated hydrocarbons such as ethene, propene, and but-2-ene.[37] The aromatics are present in high concentration in gasoline and the unsaturated compounds are typical products of engine combustion.

The diurnal variations of NO, NO_2, and O_3 typically detected in a photochemical smog situation are illustrated in Figure 14. The morning rush hour peak in the NO and hydrocarbon emissions is followed by the gradual conversion to NO_2 and subsequent rise of O_3 which decays as the sun goes down in late

[37] R. G. Derwent and M. E. Jenkin, 'Hydrocarbon involvement in photochemical ozone formation in Europe', AERE—R13736, UK Atomic Energy Authority, Harwell.

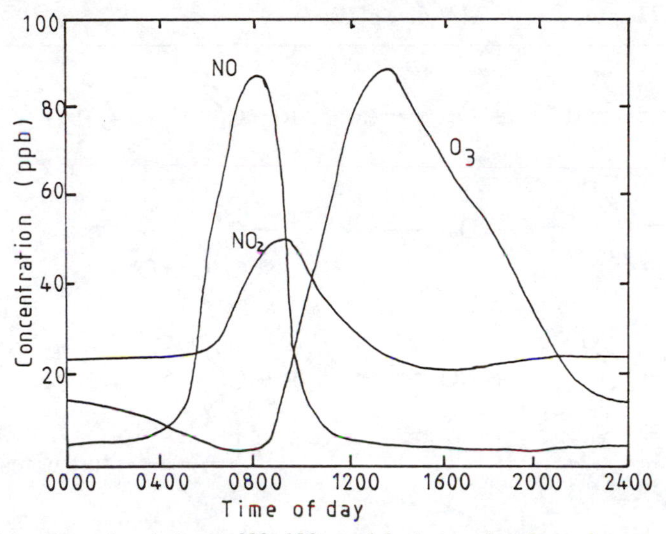

Figure 14 *Diurnal variations in NO, NO_2, and O_3 during photochemical smog conditions*

afternoon. In an air mass moving downwind from a city, the ozone peak may be worse in the surrounding countryside than in the city centre. During the night most of the reactions that have been described die down but there is an additional route for conversion of NO_2 to nitric acid via the nitrate radical NO_3 which is formed from reaction of NO_2 with O_3. NO_3 is photolytically unstable in daylight.

Space does not permit a detailed discussion of the hydrocarbon chemistry in the atmosphere which is extremely complex. In addition to reactions with $OH^{\bullet}$ radicals and molecular oxygen similar to those for methane shown in Table 11, hydrocarbon species are attacked by the oxygen atoms released in reaction (21) and by ozone. One result of the interaction of the hydrocarbon and NO_x chemistry is the formation of a group of lachrymatory substances including peroxyacetyl nitrate (PAN) and peroxybenzoyl nitrate (PBzN). The thresholds for eye irritation for these compounds are only 700 and 5 p.p.b. respectively. They are formed through the reaction of NO_2 with oxidation products of aldehydes as illustrated in Table 12.

5.2 Particles

5.2.1 Particle Formation. The nitric and sulfuric acids formed in the gas reactions described earlier generally undergo further changes. They are both water soluble and will be rapidly absorbed into water droplets if these are present. They may react with solid particles forming sulfates and nitrates. For example, limestone particles (calcium carbonate, $CaCO_3$) can be converted to calcium sulfate, salt particles (NaCl), of marine origin, can be converted to sodium sulfate, Na_2SO_4 or sodium nitrate, $NaNO_3$ with the displacement of hydrogen chloride gas, HCl. However the most common reactions are those involving ammonia:

$$NH_3 + HCl \rightleftarrows NH_4Cl \qquad (24)$$

Table 12 *The formation of PAN from aldehydes*

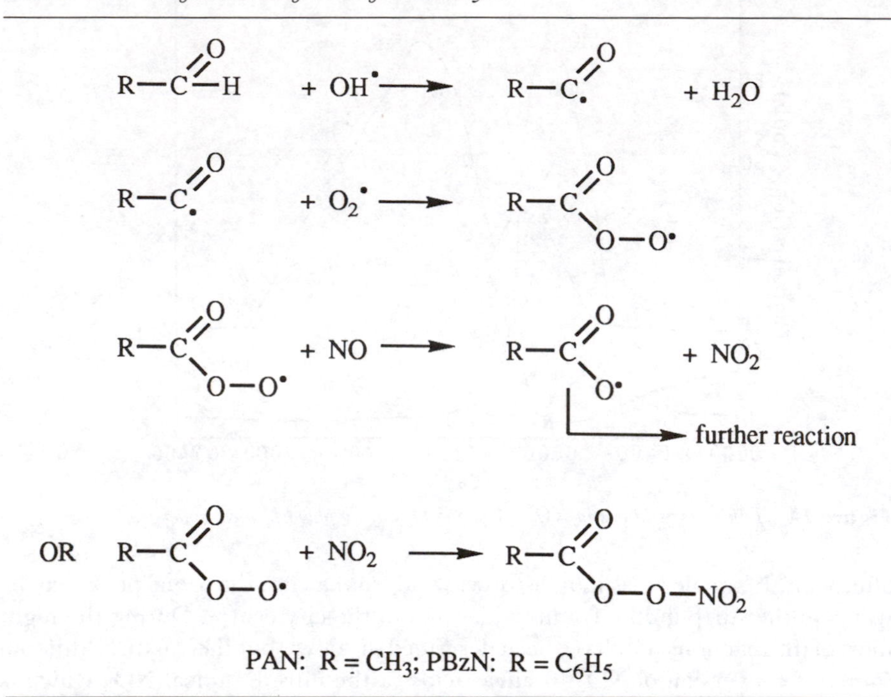

$$NH_3 + HNO_3 \rightleftarrows NH_4NO_3 \quad (25)$$

$$NH_3 + H_2SO_4 \rightarrow NH_4HSO_4 \quad (26)$$

$$NH_3 + NH_4HSO_4 \rightarrow (NH_4)_2SO_4 \quad (27)$$

Reactions 24 and 25 are actually reversible and the position of equilibrium depends on the concentrations and the temperature. Significant dissociation of NH_4Cl and NH_4NO_3 occurs during warm summer weather. One end product of the oxidation of SO_2 in the atmosphere is ammonium sulphate, a substance better known as a fertilizer. The degree of neutralization will depend on the supply of ammonia and in the UK sulphate particles are found to be well neutralized on average.[29] The largest source of ammonia is thought to be animal urine.[38]

Particles formed by gaseous reactions are initially very small (<0.1 μm) but grow rapidly either by surface accumulation of material from the gas or by particle–particle coagulation. Once in the size range 0.1 to 2.0 μm, they become relatively stable towards further growth and can remain airborne for periods of days. Most smoke emissions are in this category along with the sulfates and nitrates. At the other end of the size spectrum (2–50 μm) is coarse dust either emitted from industrial processes or raised by the wind from the ground. The

[38] H. M. ApSimon, M. Kruse, and J. N. B. Bell, *Atmos. Environ.*, 1987, **21**, 1939.

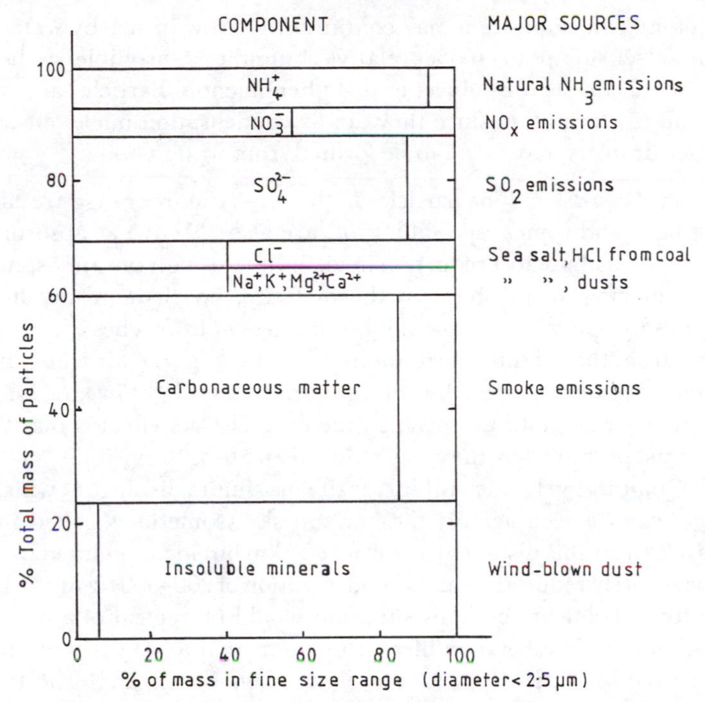

Figure 15 *Composition of atmospheric particles, their origins, and size ranges. Fine particles* $<2.5\ \mu m$ *diameter, coarse particles* $2.5\ \mu m <$ *diameter* $< 15\ \mu m$. *Data for Leeds 1982/3*

action of road vehicles is another mechanism for raising such dust and sea salt left from the evaporation of spray is also fairly coarse.

5.2.2 Particle Composition. Figure 15 presents a typical breakdown of the components of the total suspended particulate matter for an urban area based partly on results from a survey in Leeds[39] where the annual average concentration is 40–50 $\mu g\ m^{-3}$. The fine particles are dominated by the ammonium sulfate and nitrates plus carbonaceous material. About one third of this is elemental carbon and the other two thirds organic carbon (*i.e.* high molecular weight solid hydrocarbons) which together comprise 'smoke'. The coarser particles are dominated by wind-blown dusts (clays, silica, limestone, *etc.*) and include the sea salt component but has smaller amounts of sulfates, nitrates, and carbon with little ammonium.

5.2.3 Deliquescent Behaviour. The particles of water soluble compounds such as sulfates, nitrates, and chlorides will exist in the atmosphere either as solid particles or droplets depending on the relative humidity. For pure compounds the transition from solid to liquid is a sharp one. For example, it takes place with salt, NaCl at 75% relative humidity and with ammonium sulphate at 86%.

[39] A. G. Clarke, M. J. Willison, and E. M. Zeki, *Atmos. Environ.*, 1984, **18**, 1767.

Particles of mixed composition may continuously grow in size by water absorption from 60–70% up to 100% relative humidity. Insoluble carbonaceous particles are, of course, not subject to this phenomenon. Particles are important in cloud and fog formation since they act as condensation nuclei on which the much larger droplets may begin to be formed from water vapour.

5.2.4 Optical Properties.[40] Fine particles in the 0.1–2 μm size range are effective at scattering light and some, especially soot, are also effective at absorbing light. These effects contribute to a reduction in visibility through the atmosphere. If an observer is looking at an object in the distance, he distinguishes it from its surroundings by optical contrast—if it is dark, less light reaches the eye from the object than from the brighter surroundings. Particles in the air will scatter light from brighter regions *out* of the line of sight and scatter light *into* the line of sight that otherwise would not have reached the eye. The net effect is that the dark/bright contrast perceived by the eye is reduced and thus the visibility of the object impaired. Quantitatively, the visibility is the maximum distance at which a large dark object can be seen against the horizon sky (sometimes called the visual range). In clean air this distance can be over 50 km but in air polluted by particles this can be severely reduced. A mass concentration of 200–300 $\mu g\ m^{-3}$ will reduce the visibility to below 5 km. This situation would be typical of a mid-summer haze on a hot day. It is totally different phenomenon to an early morning mist which is caused by water droplets at a relative humidity near 100%. In the UK such highly polluted summer days are usually associated with air transported from Europe by winds from the east or south. Wintertime visibility has been improved and the frequency of fogs reduced in urban areas by the measures taken to control smoke and SO_2 emissions. Any additional measures to reduce pollutant emissions, especially of NO_x and SO_2, will bring corresponding benefit in terms of improved visibility throughout the year.

5.3 Droplets[35]

Liquid water occurs in the atmosphere as clouds, mists, and fog within which the concentration of water can be up to 1 g m^{-3}. Smaller amounts of water are also present in association with deliquescent particles as discussed in the previous section. At relative humidities below 95%, this secondary amount will normally be less than 1 mg m^{-3} and will not be considered further.

Water droplets can accumulate pollutants by adsorption of either gases or particles and within the droplets chemical reactions can proceed changing the nature of the adsorbed species. There are several mechanisms for the oxidation of SO_2 to sulfuric acid in the aqueous phase. In a cloud situation the most important oxidants appear to be ozone and hydrogen peroxide which is formed from the reaction of two $HO_2^{\bullet}$ radicals. The oxidants must first be adsorbed into water from the gas phase and then result in the reactions:

$$O_3 + SO_2 + H_2O \rightarrow 2H^+ + SO_4^{2-} + O_2 \qquad (28)$$

[40] A. P. Waggoner *et al.*, *Atmos. Environ.*, 1981, **15**, 1891.

$$H_2O_2 + SO_2 \rightarrow 2H^+ + SO_4^{2-} \quad (29)$$

The sulfuric acid formed will be completely dissociated to ions H^+ and SO_4^{2-}. The ozone reaction is inhibited by low pH whereas the H_2O_2 reaction is not. In acidic droplets the oxidation by H_2O_2 is therefore dominant whereas at high pH the O_3 becomes more significant. In a cloud or fog situation where there has been a significant input of particulate pollution the oxidation of SO_2 by atmospheric oxygen catalysed by metal ions (iron Fe^{3+} or manganese Mn^{2+}) or by soot can also be important. Although there is partial neutralization by ammonia, the water in polluted fogs and clouds can be much more acidic than in collected rainfall. pH values down to 2 have been measured in urban fogs—more acid than vinegar. Similarly low values have been measured in the plumes of large power stations tracked over the North Sea by instrumented aircraft.

It is difficult to distinguish experimentally between the photochemical reaction mechanism leading to H_2SO_4 with subsequent absorption of the acid into water and the aqueous phase oxidation of SO_2. However it appears that both routes are important. In the case of nitrogen oxides it is likely that the dominant process is the gas phase oxidation to nitric acid followed by absorption into water. This arises because of the poor solubility of NO_2 and especially NO in water.

5.4 Deposition Mechanisms[35]

5.4.1 Dry Deposition of Gases. As illustrated in Figure 13, the life cycle of an air pollutant normally involves emission, dispersion, and transport, chemical transformation, and finally deposition to the ground. Understanding the rates and mechanisms of deposition is important to the assessment of the environmental impact of many pollutants.

Experimentally we can measure the concentration of the pollutant (μg m^{-3}) and the total rate of deposition (μg m^{-2} s^{-1}). The higher the ground level concentration, the more rapid the deposition but the ratio of these two quantities gives a useful measure of the efficiency of the deposition process. It is called the deposition velocity.

$$\text{Deposition velocity, V} = \frac{\text{deposition rate}}{\text{air concentration}} \quad (\text{units: m s}^{-1})$$

Crudely we can envisage the process of surface adsorption and downward mixing of the SO_2 by turbulent diffusion as a drift of SO_2 to the ground with the calculated deposition velocity. Different surfaces (water, soil, ice, *etc.*) will have correspondingly different deposition velocities. The description 'dry deposition' is used in all cases even if the removal is to a wet surface.

Deposition to vegetation is rather more complex than deposition to a plane surface. The pollutant's progress may be retarded either by the slowness of diffusion through the air within the canopy of vegetation (*i.e.* between the leaves, *etc.*) or by the rate of transfer from air to leaf surface (the cuticle) or to the interior of the leaf via stomata. Moisture makes a considerable difference in that transfer to a wet surface is generally faster than to a dry surface. Overall the deposition velocities measured experimentally for SO_2 range from below 5×10^{-3} m s^{-1} to nearly 2×10^{-2} m s^{-1} and a typical value of 1×10^{-2} m s^{-1} is often assumed. Similar considerations apply to other pollutants which are subject to a significant rate of adsorption at the ground or on to vegetation (NO, NO_2, HNO_3, *etc.*). This is explored further in Chapter 6.

5.4.2 Wet Deposition. The term 'wet deposition' is used to describe pollutants brought to ground either by rainfall or by snow. This mechanism can be further sub-divided depending on the point at which the pollutant was absorbed into the water droplets. In-cloud absorption followed by precipitation is termed 'rainout'; below cloud absorption, *i.e.* pollutants collected as the raindrops fall, is termed 'washout'. The rate of removal of a pollutant by washout will increase in proportion to the rainfall rate. For SO_2 about half of the gas below the clouds will be removed in two hours of heavy rain. The rates for the nitrogen oxides are lower due to their reduced solubility in water. In remote areas the majority of the wet deposition of sulfur appears to be due to rainout. Washout becomes relatively more significant near the sources of pollution where the gas concentrations are high.

Dry deposition and wet deposition are of comparable importance in the case of SO_2. Dry deposition is most significant where ground level concentrations are high, in other words close to the sources. More than half the total deposition in central and eastern England is due to dry deposition whereas in the north and west of Scotland wet deposition dominates (see section 5.5.3). There is a commonly held misunderstanding about the role of chimney height in determining the amount of sulfur deposited in other countries from UK sources. 'We have solved our local pollution problems by building tall chimneys but the gases now travel further and are causing problems elsewhere'. The truth is that the height of the emission makes only a very small difference (perhaps 10%) to the amount of long-range transport. This arises because a slightly higher proportion of low level emissions are lost by dry deposition to the ground near the sources than in the case of high level emissions.

Another mechanism of deposition is when fog or cloud droplets are removed directly to the ground or to vegetation. This is termed 'occult deposition'. It becomes significant at elevated locations such as mountains or hill tops. There is the potential for more severe damage to foliage than with acid rain since, as was previously mentioned, polluted fog or cloud droplets can contain much higher concentrations of acidic pollutants than raindrops.

5.4.3 Deposition of Particles. The mechanism of deposition of particles depends on the particle size. Large particles with diameter greater than 10 μm fall slowly by gravitational settlement. The larger the particles, the more rapidly they fall. The sedimentation velocities for particles of density 2 g cm^{-3} are as follows:

Diameter, μm	**Velocity, m s^{-1}**	**Diameter, μm**	**Velocity m s^{-1}**
5	1.5×10^{-3}	50	1.4×10^{-1}
10	6.1×10^{-3}	100	4.6×10^{-1}
20	2.4×10^{-2}	150	8.0×10^{-1}

Particles larger than 150 μm diameter, falling at over 1 m s^{-1} remain airborne for such a short time that they do not need to concern us as air pollutants. Particles less than 5 μm have sedimentation velocities which are so low that their movement is determined by the natural turbulence of the air, just as for gases.

Intermediate particles, between 1 and 10 μm diameter, can be removed by impaction onto leaves and other obstacles. Particles in the 0.1 to 1 μm range which include most of the nitrates and sulfates, are only removed very slowly by dry deposition. The deposition velocities are of the order of 10^{-3} m s^{-1} a factor of 10 lower than for SO_2. The most likely route for their removal is rain-out following water vapour condensation and droplet growth in clouds. Wash-out is not very efficient for these fine particles although it becomes more significant for larger particles such as coarse dust.

5.5 Acid Rain

5.5.1 The Effects. The phrase 'acid rain' has come to be used very loosely to mean almost anything to do with air pollution whether or not rain is actually involved. Three particular effects have received most attention. Historically the first of these was the increased acidification of lakes and streams in Scandinavia leading to loss of fish and other aquatic organisms. This was attributed to the acidity of rain polluted by sulfur and nitrogen oxides and the UK was blamed as being the chief culprit. More recently the acidification of fresh waters in our own country has been demonstrated, especially in South-west Scotland.[41]

The second type of environmental damage is the die-back of forests in Central Europe. The effects became very noticeable in Germany in the early 1980s and are worst in the south-west (the Black Forest) and on the eastern border with Czechoslovakia, a country which shares the same problem. Since then other countries have reported similar phenomena. Although the reasons for the damage are still being debated, it seems likely to be a combination of factors. Predisposing factors include drought and high altitude. Some forests in the Eastern European countries suffer from the effects of very high SO_2 levels but this is not the case further west. Possible mechanisms include the effect of ozone initiating the attack and subsequent further deterioration being due to acid rain or acid mists and fogs. Acidification of the ground with consequent effects on the soil chemistry may also contribute. If we accept that both ozone and acidic precipitation are important for forest damage then attention must be given not only to SO_2 emissions but also to NO_x and hydrocarbon emissions as the precursors of ozone.

Surveys have been carried out in the UK which indicate that forest damage exists here but the importance of pollution among the various possible factors is not well established.[42] It is clear that the ozone levels are lower than in Central Europe and the occasional high values tend not to persist due to our maritime

[41] UK Acid Waters Review Group, 'Acidity in United Kingdom Fresh Waters', HMSO, London, 1986 and Second Report, 1988.

[42] UK Terrestrial Effects Review Group, 'The Effects of Acid Deposition on the Terrestrial Environment in the United Kingdom', HMSO, London, 1988.

climate. Few forests exist at the high elevations common in Germany. Direct comparison between the two countries is therefore difficult.

The third problem is the attack on stonework and the decay of many of our most famous cathedrals and other buildings constructed of limestone.[43] Both wet and dry deposition of sulfur dioxide are involved. Under moist conditions SO_2 or sulfuric acid will convert calcium carbonate to gypsum, $CaSO_4$. Since the sulfate is more soluble than the carbonate, the reacted stone can be removed by dissolution. The solid gypsum also occupies a larger volume than the original carbonate and this leads to spalling of material from the surface. A combination of these factors leads to a rate of loss of stone which depends on the deposition rate of SO_2 to the surface. Deposition to a moist surface is more rapid than to a dry surface so the fraction of time the surface is wetted as well as the SO_2 concentration is important. It is generally assumed that in urban areas the dry deposition of SO_2 gas is the major factor. Control of urban SO_2 levels requires a completely different approach to the control of national emissions, which is the objective of the wider European movement for the reduction of acid rain. Reduction in sulfur content of fuel oils and further conversion from coal and solid fuels to natural gas or electricity are all relevant. Limitations on power station emissions may help in some areas but in other areas will make no difference. For example, in London itself there are no longer any operational stations and emissions from more distant stations contribute less than 10% of the city's SO_2 levels.

5.5.2 Rainwater Composition. Even in the absence of air pollutants, rain-water is slightly acidic (pH 5.6) due to atmospheric carbon dioxide. 'Acid rain' therefore refers to rain with a pH below about 5. Annual average acidities over much of the UK correspond to pH values between 4 and 4.5, whilst the highest recorded acidities are below pH 3.[35]

A typical ionic composition in units of micro-equivalents per litre of a fairly polluted rain-water with high ionic concentrations and a pH of 4.1 might be as follows:

Cations H^+ 80, NH_4^+ 120, Na^+ 80, K^+ 10, Ca^{2+} 60, Mg^{2+} 25

Anions SO_4^{2-} 170, NO_3^- 90, Cl^- 120

Qualitatively we can see that the acidity originates from SO_2 and NO_x which have been oxidized to H_2SO_4 and HNO_3. Overall about two-thirds of the acid is associated with SO_2 emissions and one-third with NO_x emissions. This is consistent with the relative emissions (see Table 5). The acid has been partially neutralized by ammonia and other ions such as Ca^{2+} which may have originated as calcium carbonate, $CaCO_3$. Taking the UK as a whole about one fifth of the neutralization is due to calcium so the effect of the ammonia is dominant. In the ground NH_4^+ can release H^+ during the process of being oxidized to NO_3^- by bacteria. The H^+ can then be leached out into streams. From the point of view of

[43] UK Building Effects Review Group, 'The Effects of Acid Deposition on Buildings and Building Materials', HMSO, London, 1989.

freshwater acidification, ammonia should therefore not necessarily be regarded as an ally even though it reduces the rain-water acidity.

Most of the sodium, chlorine, and magnesium in precipitation originate from sea salt. Non-marine chloride levels are only significant near major sources such as coal-fired power stations. The East Midlands and North-East England are areas of enhanced chloride deposition for this reason. On a national basis the potential acidity from HCl is only about 4% of that from SO_2 and NO_x.

5.5.3 Patterns of Deposition.[35] The most acidic rain (pH < 4.2) falls in the east of England in an area covering the East Midlands, South Yorkshire, and Humberside. This is the downwind side of the country and has major emitters such as power stations within it. However the west receives the largest volume of rain. Consequently the largest values of the total deposited acidity are in places such as western Scotland, the Lake District, and western Wales. It is unfortunate that these are also areas which have rather poor, acidic moorland soils with little lime to buffer the acid. Thus the potential for harm is significant. Figure 16 shows the pattern of annual wet deposition of acidity side-by-side with the distribution of areas geologically susceptible to acidification.

As is shown in Figure 10 individual acidic rainfall events may be significantly influenced by European emissions. Table 13 shows the sulfur and nitrogen budgets for the UK. As far as sulfur is concerned, 80% of the dry deposition and a half to two thirds of the wet deposition derives from UK emissions. The total measured deposition is only 20% of the emissions — the remainder is transported

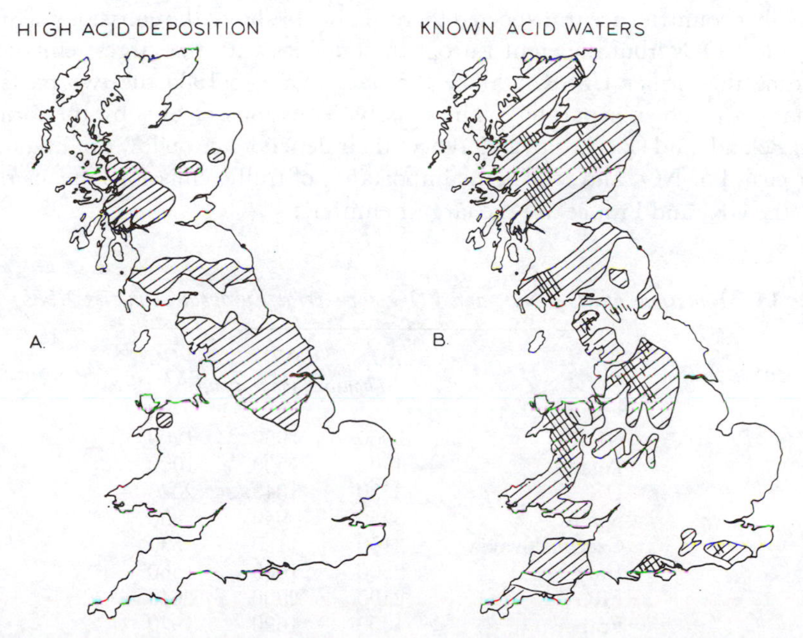

Figure 16 *(A) Regions of high acid deposition* (>0.04 g H^+ m^{-2} yr^{-1}) *based on reference 35. (B) Geological regions susceptible to acidification (lined) and regions of known acid waters (hatched) based on reference 41*

Table 13 *Sulfur and oxidized nitrogen budgets for the United Kingdom in 1987 (k tonnes S or N)*

	Sulfur		*Nitrogen*	
Emissions	1935		700	
Deposition:	Computed	Measured	Computed	Measured
Dry — UK	254		46	
— Elsewhere	70		17	
— Total	324	160	63	53
Wet — UK	98		38	
— Elsewhere	57		30	
— Total	155	232	68	98
Total	**479**	**392**	**131**	**151**

Reference: 'Acid Deposition in the United Kingdom 1986–1988', UK Review Group on Acid Rain, HMSO, London, 1990

out of the country. A similar picture applies to nitrogen deposition. As far as wet deposition is concerned there appears to be a significant European component.

The UK is a net exporter of air pollutants while other countries of Europe are net importers, for example Norway and Sweden. The UK is the largest single contributor to sulfur deposition in southern Norway although emissions from many other countries are transported there. Table 14 shows the emissions of SO_2, NO_x, and VOCs from the eight European countries with the largest emissions excluding the former USSR.[32] Since the data relate to 1985 the two parts of Germany are shown separately. The massive emissions of SO_2 by the former GDR, Poland, and Czechoslovakia reflect their dependence on low quality, high sulfur coal. For NO_x and VOCs the importance of traffic emissions means that FRG, the UK, and France are the largest emitters.

Table 14 *Emissions of SO_2, NO_x and VOC, for several European countries (1985)*

	SO_2	*NO_2*	*VOC*
	Thousand tonne annum^{-1}		
GDR	5000	950	1200
Poland	4300	1500	1000
UK	3550	1840	2020
Spain	3250	950	850
Czechoslovakia	3150	1130	530
Italy	2500	1600	1560
FRG	2400	2900	2640
France	1850	1690	1970

Reference: C. Veldt, *Atmos. Environ.* 1991, **25A**, 2683

5.5.4 Control Measures. In 1988 the Commission of the European Communities passed a directive[44] that aimed to effect reductions of 60% in the SO_2 emissions from large stationary sources (over 50 MW thermal input) and 40% in the NO_x emissions. The reference year is 1980 and the target years are 2003 for SO_2 and 1998 for NO_x. Some variations in the required reductions were allowed between member states and the UK has a target of only 30% for NO_x. This will be achieved by a progressive programme of burner replacement in coal fired power stations. New low NO_x burners can achieve reductions of 40–50%. The SO_2 target will be achieved by a variety of measures. These include installation of flue gas desulfurization at two or three large power stations, construction of new combined cycle gas turbine plants burning natural gas and the substitution of some British coal with imported coal of lower sulfur content. Tighter control of other industrial emissions will be effected through the 1990 Environmental Protection Act. Significant reductions in vehicle emissions of NO_x and VOCs will occur as new cars constructed to the most recent standards (section 2.3) are introduced. Taken together these measures should lead to reductions in levels of deposited acidity and in ozone.

[44] Directive 88/609/EEC, 'Limiting emissions of certain pollutants into the air from large combustion plants', *Off. J. Eur. Comm.*, 1988, **L336**.

CHAPTER 3

Freshwaters

SIR HUGH FISH, CBE

1 THE NATURE AND COMPOSITION OF FRESHWATERS

1.1 Natural Characteristics

Water, H_2O, hydrogen oxide, is an extraordinary chemical compound of absolutely fundamental environmental importance. The electronic structure of its molecules impart the very special chemical and physical properties which lead to water being described as 'the universal solvent' or 'the liquid of life'. Water in the gaseous state is made up of single molecules, the structure of which is usually represented as shown in Figure 1. Water in the liquid state is made up of groups of molecules associated together by linkages (hydrogen bonding) between each of the hydrogen atoms of one water molecule and the oxygen atom of an adjacent water molecule, as depicted in Figure 2. In ice the water molecules are associated in tetrahedral structure formed by four molecules arranged around a central

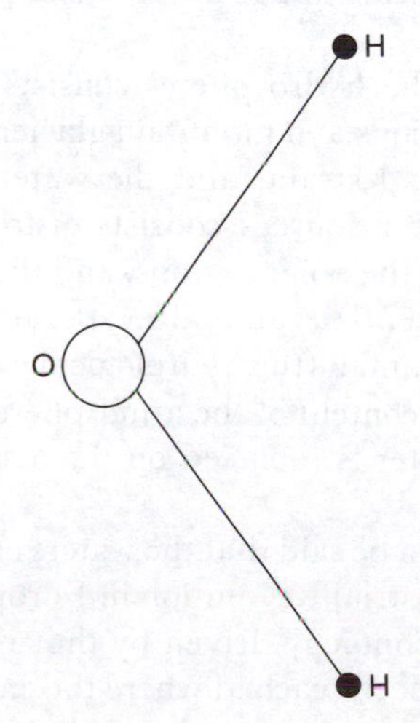

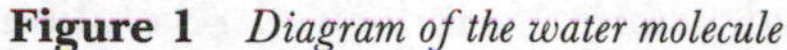

Figure 1 *Diagram of the water molecule*

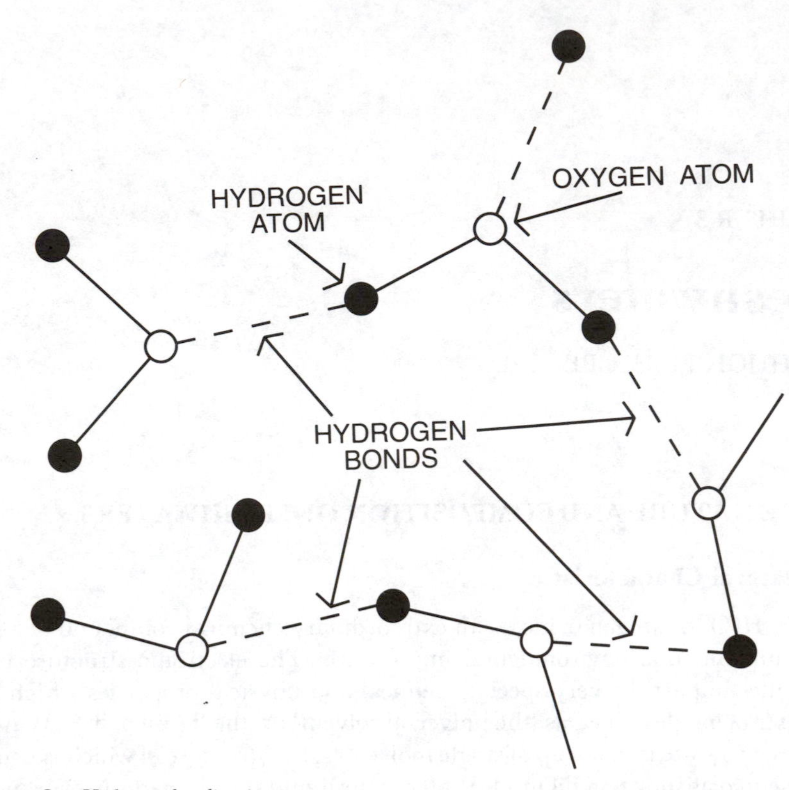

Figure 2 *Hydrogen bonding in water*

molecule. It is the nature of the molecular structure and the ways in which these molecules associate in hydrogen bonding which yield, among other chemical and physical characteristics, the general solvent power, the chemical reactant power in biological processes, and the heat storage and transfer power of water as a key environmental resource.

The Earth's water resources, the 'hydrosphere', consists of the oceans and seas, the ice and snow of the polar regions and mountain glaciers, the water contained in surface soils and underground strata, and the water in lakes, rivers, and streams. Less than 1% of these resources consists of freshwater, some 2% is freshwater ice located mainly in the polar regions, and the remaining 97% or so consists of seawater and sea ice. The annual evaporation of water from the hydrosphere, and its return as rainfall (the hydrological cycle) amounts to about 260×10^{12} m^3. The total water content of the atmosphere is about 7×10^{12} m^3, indicating that atmospheric water is replaced on the average some 37 times a year.

For all practical purposes it can be said that the waters of the hydrosphere were of natural quality until the Industrial Revolution in Europe and North America initiated the development of technology driven by the energy of the fossil fuels coal and oil. Now the stage has been reached where the entire hydrosphere, save that deep in the polar ice which was formed before the industrial revolution began, is contaminated by the polluting activities of man. Prior to the 'Environmental Revolution', which began to occur a mere 30 years ago and focussed world-wide attention on the growing pollution of air, land, and water, general interest in water pollution in the developed world was very limited. Now of course there is world-wide concern on pollution and all environmental matters, exemplified in the UK in 1988 by the introduction of environmental politics by the Prime Minister in her speech to the Royal Society.

Nevertheless the bulk of freshwater resources is not yet grossly polluted—that is, sufficiently contaminated by impurities derived from the activities of man that the waters are unsuitable for the legitimate uses that man wishes to make of them, or are unable to support a reasonable diversity of the species of flora and fauna typical of the individual waters. Such pollution is nevertheless fairly commonplace world-wide in built-up areas, and it is not at all an exaggeration to believe that the overall situation is steadily depreciating as demands for freshwater use and for the disposal of polluted water after use increase. However the pollution of water, though this is the major and most pernicious destroyer of the water environment, is not the only cause of man-made degradation of that environment. Water abstraction for domestic, industrial, and agricultural supply, and the lowering of near-surface groundwater levels to increase agricultural output, can bring about major depreciations of the aquatic environment. A typical example of this exists currently in the UK. The low rainfall over South-East England over the years 1989 to 1991, coupled with excessive groundwater abstractions, has led to the complete drying out or gross flow reduction on many reaches of normally beautiful streams draining from downland chalk areas. Table 1 lists some National Rivers Authority schemes for improvement of artificially low river flows.

The quality of waters is defined in physical, chemical, and biological (including microbiological) terms. This definition may be lengthy and detailed, or short

Table 1 *Some National Rivers Authority proposals for improvement of artificially-low river flows*[a]

River and NRA region	*Likely improvement action*
River Allen, Wessex	Reduction of public supply abstractions. Approximate cost—£5.5 × 10^6
Battlefield Brook, Severn Trent	Augmentation of flows from a borehole and possible lining of river channel
River Darent, Southern	Relocation of public supply abstractions, partial revocation of licences, river augmentation scheme. Approximate cost—£10 × 10^6
River Misbourne, Thames	Reductions in existing abstractions and lining of lake and river bed. Approximate cost—£5 × 10^6
Upper Waveney, Anglian	Relocation of existing abstractions and/or augmentation of flow from boreholes (cost unknown)

[a] National Rivers Authority, London, NRA(91)INF28, May 1991

and broadly indicative, according to the purpose of the definition. Where the purpose is to decide whether or not the water is fit for a particular use, for example as a source of potable water supply, chemical parameters of quality will predominate in the definition. In contrast, if the purpose of the definition is to decide the extent to which a particular reach of river is subject to pollution, the occurrence of sporadic toxic chemical pollution is more likely to be deduced from observation of the biology and the physical characteristics of the river regime than by other means. Of course the proof of occurrence of any such sporadic chemical pollution will inevitably be expressed in chemical terms. Furthermore the parameters of water quality which are specified as limitations of the acceptability of a water for particular use purposes, or as a general environmental resource, are predominantly chemical and biochemical ones. Such specifications usually appear as 'quality standards', 'quality criteria', or 'quality guidelines' promulgated by international bodies and individual nations.

Rainfall, even when it was in its wholly natural state, was quite impure water in the scientific sense. Its physical characteristics in the liquid phase are that it would normally be naturally clear, colourless, and odourless with a temperature varying from ambient to colder depending on the nature and height of its precipitation. Rainfall naturally contains materials dissolved from the atmosphere. Firstly it contains the atmospheric gases—mainly nitrogen, oxygen, and carbon dioxide dissolved from the atmosphere. Secondly it contains matter dissolved from impurities in the air derived from the earth's surface—such as sea spray, volcanic emissions, wind-borne dusts, hydrogen sulfide and methane from anaerobic decomposition, and volatile organic compounds derived from land and aquatic plants. Thirdly it contains traces of ozone and other gases derived from chemical reactions, triggered by solar and cosmic radiation, of naturally occurring materials in the atmosphere. While it is essential to understand this natural chemistry of rainfall, it is the chemical consequences of the man-made pollutions of the atmosphere, as referred to in Chapter 2, which cause the greatest concern. Thus, apart from the effects of atmospheric pollution, the natural quality of rainfall is determined mainly by seawater and dust take-up, varying according to weather, locality, and major volcanic events. Rainfall naturally contains around 10 to 20 mg l^{-1} of dissolved solids according to circumstances. It is naturally somewhat acidic in character, with a pH value of around 5.5, because of its solution of atmospheric acid gases, mainly carbon dioxide.

On reaching the ground, rainfall picks up more impurities naturally from vegetation and the ground surface. If the ground is permeable, all or some of the rainfall will pass underground according to circumstances. Some will be evaporated and the balance will run off to streams and rivers and via underground strata *en route* to the sea. On the average some 70% of the rainfall evaporates directly, or indirectly via plant transpiration. Of the remaining 30%, more travels through underground strata than along rivers and streams, and about 90% of the total run-off reaches the oceans. Natural run-off from bare, substantially impermeable, hard rock gathers material from gradual weathering of the rock to yield low concentrations of dissolved calcium, magnesium, sodium, and potassium. Traces of other materials will be present according to the chemistry of the rock. In

Table 2 *Analysis of run-off from a boulder clay catchment*

	$mg\ l^{-1}$ *(unless otherwise stated)*
Ammonia nitrogen	0.2
Chemical Oxygen Demand (dichromate value)	4.0
pH Value	7.8 units
Chloride (Cl)	50
Nitrate (N)	6.0
Total hardness ($CaCO_3$)	410
Non-carbonate hardness ($CaCO_3$)	250
Magnesium (Mg)	8.0
Phosphate (PO_4)	0.6
Sulfate (SO_4)	140
Silicate (SiO_2)	6.0
Iron (Fe)	0.05

contrast, where the rock underlies a cover of peat and associated vegetation the take up of stable organic matter will be high. The run-off from peat areas is usually acid in the pH range 5 to 6, and has a pronounced yellow-brown colour, highly so in times of heavy rainfall. A similar situation applies to the run-off from woodland areas where ground cover of leaf mould produces a similar effect. Research[1] has shown that trees, particularly certain conifers, can themselves naturally contribute to the acidity of run-off.

The run-off from clay soil is characterized by its content of calcium sulfate and calcium carbonate, compounds which are naturally abundant in clays. Turbidity in the water is often evident because of the presence of colloidal clay particles. An analysis of a typical water running off a clay area in South-east England is shown in Table 2. The run-off from clay soils during prolonged heavy rainfall, particularly from ground with little plant cover, is heavily loaded with suspended particles of clay and fine sand derived from the clay. While during heavy rainfall the water referred to in Table 2 would contain up to 0.1% suspended soil particles, the larger rivers of the world subject to tropical rainfall intensity carry relatively immense concentrations of solids in suspension. The most striking example of this is on the Huang He (Yellow River) in China, where heavy rainfall run-off from the immense area of loess clay in that river catchment causes the river water to contain some 2% of very fine suspended clay particles.

1.2 Groundwaters

Rainfall which passes from the surface soil into deeper permeable stratum (an aquifer) continues to move downwards through the unsaturated zone of the aquifer until it reaches the saturated zone of accumulated water lying over the impermeable layer of rock marking the base of the aquifer. The major aquifers in the UK are in the chalk regions which are located in East Yorkshire and East

[1] R. Harriman and B. R. S. Morrison, *Hydrobiologia*, 1982, **88**, 251.

Lincolnshire, but mainly South and East of a line drawn from the Wash to the Dorset coast, in the downlands of the Home Counties, in East Anglia, and along the South Coast. Other important aquifers lie in the bunter sandstone areas of Nottinghamshire, Cheshire, and Lancashire and the oolitic limestone areas of the Cotswolds and Yorkshire. Magnesian limestone in Durham, Yorkshire, and Lincolnshire provides valuable aquifers. In other areas useful aquifers lie in the greensand, the coal measures, and river valley gravels. There are very many locations where small but useful aquifers exist, many in glacial–gravel deposits. When rainfall percolates through the surface soil into permeable strata, it is filtered free of suspended matter and picks up matter into solution. In its continuing passage downwards the water undergoes a major change in quality, in a variety of ways according to its rate of travel, the nature of the stratum, and the chemical reactions proceeding. In unconfined chalk, that is chalk which lies at, or very close to, the ground surface, the water is characterized by its content of calcium and bicarbonate ion. The source of the bicarbonate ions is the carbon dioxide dissolved from the atmosphere and particularly from the surface soil. The source of the calcium ions is of course the chalk itself. Table 3 shows a typical analysis of the water in an unconfined chalk aquifer.

Solid chalk has a low permeability and its aquifer properties depend essentially on the fissuring of the chalk. Such fissuring or cracking was initially produced by

Table 3 *Typical analysis of water from unconfined chalk*

	$mg\ l^{-1}$ *(unless otherwise stated)*
Ammonia	Trace
Chemical oxygen demand	0.05[b]
pH Value	7.2 units
Colour	Nil
Turbidity	Trace
Chloride (Cl)	8.5
Nitrate (N)	6.2[a]
Free CO_2	38
Total hardness ($CaCO_3$)	303
Non-carbonate hardness ($CaCO_3$)	28
Magnesium (Mg)	1.2
Calcium (Ca)	119
Sodium (Na)	6.2
Potassium (K)	1.2
Phosphate (PO_4)	0.10[a]
Sulfate (SO_4)	2.5
Silicate (SiO_2)	11[a]
Iron (Fe)	Trace

[a] This level of contamination is probably partly due to the effects of agricultural practices; [b]Chemical oxygen demand is a laboratory determination of the take up of oxygen from acid dichromate by organic matter present in water. This gives an indirect indication of the total organic carbon content of the water

massive folding of the chalk during past natural geological upheavals, and since modified by the passage of the water through the fissures. A similar situation, but usually more pronounced, exists in the limestone aquifers. The sandstone and gravel aquifers are porous, some more so than others, and the water percolating through these aquifers dissolves calcareous and other mineral matter held in the interstices between the sandstone grains or between the packed gravel in the aquifer. In the confined sector of an aquifer, that is where the outcropping permeable stratum passes underneath an overlying layer of clay, further changes in the chemistry of the water occur. These are very varied and complex, involving natural ion-exchange and often biochemical change and generally result in changes in the extent and nature of mineralization of the water. For this reason the quality of waters from confined aquifers varies greatly. The limestone aquifers produce waters of essentially similar quality to those from the chalk, but often containing more magnesium and other mineral salts. The waters from gravel aquifers are very variable in quality because of the wide variety of mineral matter associated with the gravel deposits. Waters from riverside gravels often contain sufficient concentrations of iron and manganese as to render them unfit for use for water supply without specific treatment to reduce these concentrations. Glacial sands and gravels also yield waters of varying quality according to circumstances. Usually these waters contain much calcium sulfate rather like the run-off from clay catchment areas. The waters passing through aquifers eventually discharge to rivers or lakes via springs or offshore to the sea bed where the stratum of each aquifer, down slope, outcrops.

1.3 Freshwater Rivers, Streams, and Lakes

By and large the natural quality of rivers and streams (henceforth collectively referred to as rivers) at any point reflects the quality of the upstream contributions of surface run-off and groundwater discharge. Similarly, but to a lesser degree according to circumstances, the natural quality of a lake reflects the quality of the inflows of water that maintain the lake level. However these waters, open to the energy of sunlight, the solution of oxygen from the air, and containing the mineral nutrients sufficient to support plant growth, naturally become the media for the growth of aquatic biota. These biota can be divided into four categories, namely:

- the phytoplankton, being plants mainly algae, floating within the water;
- the zooplankton, which include bacteria and protozoa, floating within the water;
- the rooted plants, mostly growing from the river or lake bed, but some floating on the water;
- the larger animals which are either free swimming or are attached to the bed or the larger plants.

The nature, variety, and abundance of these biota—the biology of the water—is essentially determined by the physics and chemistry of the water. However there is a feed-back loop in this system which results in the growth of the biota exerting

Table 4 *Variation over 24 h of the oxygenation of a stream*

Time of day	*Concentration of dissolved oxygen,* mg l^{-1}
00.00	9.5
04.00	6.0
08.00	8.5
12.00	16.5
16.00	18.0
20.00	15.0
24.00	8.5

an impact on the physics and chemistry of the water—for example the effects of phytoplankton growth in increasing the oxygenation and alkalinity of the water, and in transferring some of the calcium, carbonate, and phosphate content of the water to the river or lake bed. Table 4 refers to an extreme case of super-oxygenation of water largely caused by algal photosynthesis.

The natural shape of the channel of a river at any point is determined by the rate of flow (discharge) of water and the topology and geology of the terrain through which the river passes. These factors determine the gradient of the river, its velocity of flow, its depth and width, its shoals and pools, and the nature of the river bed at various points—silt, gravel, or bare rock as the case may be. Even as the growth of river biota has a secondary effect on the physical and chemical characteristics of the water, so the growth of the biota can influence the effective shape of the river channel. A good example of this is the heavy growth of rooted water plants which occurs in the shallow, calcareous waters of chalk streams. Heavy growths of these plants, if not controlled by judicious cutting and removal from the rivers, often so reduce the flow cross-section as to cause flooding during heavy summer storms. Another major cause of quality variability is the biochemical activity which proceeds within micro-organisms in the water, particularly bacteria, as key factors in the operation of the well known carbon and nitrogen cycles and other biochemical transformations. The most important of these are summarized in Table 5. It needs to be borne in mind of course that straight-forward chemical reactions occur also—for example the precipitation of metals from solution in the water.

1.4 Water Pollution

It is part of the natural scheme of things that man, a saprophytic animal, should cause environmental pollution in almost all he does. Fortunately it is also part of the natural scheme of things that man, blessed with thinking ability, should recognize the need to control pollution and devise technological and administrative means of effecting this control. That this recognition is now apparently widespread does not necessarily mean that the required controlling action will be

Table 5 *Main biochemical processes proceeding in water*

Process	*Requirements*[a]	*Main end products*
Oxidation of organic carbon	Dissolved oxygen, $T > 0°C$	CO_2, H_2O
Oxidation of ammonia	Dissolved oxygen, $T > 4°C$	NO_3^-, little NO_2^-
Reduction of nitrate	Dissolved oxygen absent, $T > 0°C$	N_2, some N_2O
Reduction of organic carbon	Dissolved oxygen absent, nitrate absent, $T > 4°C$	CH_4, CO_2

[a] Temperature requirements relate to those below which the processes become very slow

taken at the most appropriate time. The possibility, that public concern about environmental pollution is not simply an acceptance that pollution control is a 'good thing' without the will to pay for it, has yet to be tested. Local attempts to enforce water pollution control in England began some 600 years ago, in the reign of Richard II. Over the years the legislation has been repeatedly made more comprehensive and purposeful. We now have a very comprehensive code for water pollution control, expressed mainly in the Water Act 1989, yet we still have more widespread water pollution proceeding than ever before.

Man-made pollution of water is divided into two kinds, namely into point sources and non-point sources. The most important of these are listed in Table 6. As might be expected, the point sources are mainly discharges of wastewaters from sewage works, factories, and farms, while the non-point sources arise mainly from specific categories of general land use (*e.g.* high intensity farming). As regards the types of pollution which occur, these can be listed according to the effect exerted by the polluting matter, as follows:

Table 6 *Main point and non-point sources of pollution*

Point sources	*Non-point sources*
Discharges from sewage treatment works to rivers	Run-off and underdrainage from agricultural land into rivers
Discharges of industrial wastewaters to rivers	General contamination of recharge rainfall to outcropping aquifers
Discharges of farm effluents to rivers	Septic tank soakaways into permeable strata
Discharges from small domestic sewage treatment plants to rivers	Wash-off of litter, dust and dry fallout, from urban roads to rivers
Discharges by means of well or borehole into underground strata	General entry of sporadic and widespread losses of contaminants to rivers
Discharges of collected landfill leachate to rivers	Seepage of landfill leachate to underground strata and to rivers

(a) Substances acutely toxic to man and/or aquatic flora and fauna (*e.g.* lead, mercury, cadmium, cyanide, pesticides);
(b) Substances which are hazardous to man and/or to flora and fauna in causing chronic, or long dormant cumulative, harm (*e.g.* polynuclear aromatic hydrocarbons, chlorophenols, trihalomethanes);
(c) Substances at very low concentrations which are not highly toxic but which can either be rendered acutely toxic by biochemical transformation in the water (*e.g.* methylation of inorganic mercury) or by bioconcentration (*e.g.* triphenyl tin, see Case Study 1);
(d) Substances which add to the load of biochemical oxygen demand* in the water or in benthal (bottom) deposits, (*e.g.* sewage effluent, food-industry wastewater, farm wastes);
(e) Substances which add to the eutrophication (plant-nutrient content) of the water (*e.g.* sewage, farm wastes);
(f) Substances which have a detrimental effect on the physical appearance of the water (*e.g.* oil, detergent foam, litter, suspended matter);
(g) Substances which only have a polluting effect at relatively high concentrations in water (*e.g.* mineral salts such as sodium chloride);
(h) The waterborne organisms which are pathogenic to man (*e.g. Salmonella, Cholera vibrio*).

In this list, the organic substances in categories (a), (b), and (c) exert their polluting effects at very low ($\mu g\ l^{-1}$) concentrations. The inorganic substances exert their effects at higher concentrations (in the range of $mg\ l^{-1}$). Those in categories (d) to (g) exert their effects at multi $mg\ l^{-1}$ concentrations. The hygienic safety of waters, through which man may be exposed to infection by the organisms in category (h), can only be high where bacteria of faecal origin cannot be detected in 100 ml of water (per litre in the case of a plaque forming unit of polio virus).

1.4.1 Case Study 1—Triphenyltin Pollution. On a large agricultural holding, a scheme for the storage of $0.5 \times 10^6\ m^3$ of winter surface water for use in summertime spray irrigation was carried out. The reservoir was developed as a rainbow trout fishery, and very good fishing, and deep freezers filled with trout, were enjoyed for two seasons. Around the middle of the third season, it was reported that the trout were no longer rising to take fly, and catches had dropped almost to zero. Examination of the fish showed them to be infected with the eye fluke, *Diplostomum spathaceum* and blind. Shortly after this it was found that the blind fish were dying in small numbers daily. Since the dead trout had been consuming snails and were quite plump it was suspected that some powerful toxic pollution must have occurred. Careful enquiry showed that about 3 weeks before the fish began to die, the large acreage of potatoes surrounding the reservoir had been aerially sprayed against blight with triphenyltin. Further, that having regard to the terrain it was likely that some of the spray material had entered the

*Biochemical Oxygen Demand is a measure of the rate at which the microbiological oxidation of organic matter depletes the oxygen dissolved in water (see Table 5). Severe depletion of dissolved oxygen renders the water unfit for support of normal aquatic life.

reservoir, and that heavy rainfall shortly after the spraying could have caused wash-off of the pesticide to the reservoir. Analysis of the reservoir water and of run-off thereto showed no detectable concentrations of tin or other toxic contaminant. However analysis of the muscle of recently dead trout showed up to 500 μg kg^{-1} of tin present, while the snails had 1000 μg kg^{-1} in their tissue. The explanation of the fish loss was then complete. The organo-tin contamination, although certainly of low overall concentration in the reservoir water, had been deposited, presumably by adsorption on suspended matter or taken up by algae and settlement, on the reservoir bed. From there the material had been concentrated in snails feeding on bottom algae and detritus. The trout, being blind, had fed almost exclusively on snails gathered from the reservoir bed and had thereby received a fatal dose of the pesticide.

1.5 Diffuse Pollution

Pollution from non-point sources, or diffuse pollution, in its very nature presents difficult control problems. If it is known that a river, lake, or an aquifer is polluted by a particular substance or group of substances, but the point of access of this pollution to the water cannot be located and no particular person or corporate body can be proved to have caused or knowingly permitted the pollution to occur, the normal processes of penalizing polluters and making them pay for remedial and/or preventive measures are thwarted. Special research and administrative action has then to be taken, and since this is a difficult and slow process, such action can be almost guaranteed to take place too late to avoid major damage arising. There are three main problems before us. The first is the acidification of waters caused by acid rain and the second is the high and still rising level of nitrate in many waters, particularly groundwaters. The third relates to water contamination by leachate from landfill sites and spills of chemicals.

1.5.1 Acidification of Waters. The acidification of rivers and lakes by acid deposition from the atmosphere is essentially a local problem in parts of Wales, Scotland, and Cumbria. There is now irrefutable evidence that much of the acidity is a part of the consequences of the major problem of atmospheric pollution in industrial Europe. The second report of the UK Acid Waters Review Group[2] concludes that parts of Scotland, Wales, and Northern England possess moderately severe acid waters leading, in some cases, to depletion of fish stocks. Further that in some regions of the UK, afforestation has exacerbated freshwater acidity by increasing the amount of acid deposition in the forest canopy. The effects of acid deposition varies greatly according to the type of soil on which it falls. Alkaline soils based on limestone can neutralize large amounts of acid whereas soils based on peat or granite cannot do so. The problem first came to light in Scandinavia because much of its surface soil has little buffering capacity.

There are two practicable ways in which the impact of acidification can be eased. One is to reduce the emission of the atmospheric pollutants. The second is

[2] Acidity in the United Kingdom Freshwaters, Second Report, UK Acid Waters Review Group, HMSO, 1988.

to add a neutralizing alkali (such as powdered limestone) to the acid-sensitive areas. The key link between the emissions and their ecological impact is the transfer of the acidity from deposition to run-off to rivers and lakes, and understanding of this link is essentially a matter of the chemistry of the interactions between the soil and the soil water. In essence the lightly buffered soils, thin and base poor, are the most sensitive to acid deposition and are the main generators of acid rivers and lakes from that deposition. In seeking to ameliorate the problem, the first reaction is inevitably that the phenomenon has been developing over some 150 years and it could be impossible to achieve significant reversal of the ecological damage done within reasonable time and at reasonable cost. In short the cost/benefit ratio of remedial action appeared most unpromising. Fortunately, recent work using mathematical models of acidification of ground water in catchments[3] shows the position to be more encouraging, although in the area of Wales studied, the model predictions showed that a reduction of about 60% in acid deposition would be required to give a significant recovery at most sites.

1.5.2 Nitrate in Waters. The Royal Society Report in 1983 on the Nitrogen Cycle in the United Kingdom[4] includes a comprehensive analysis of the problems of nitrate in waters. The EC directive on the quality of water intended for human consumption sets the maximum acceptable limit for nitrate in water at 50 mg l^{-1} (11.3 mg l^{-1} nitrate nitrogen). The implications of this limit are much more serious in the case of groundwater than surface water.

Nitrate problems in groundwaters arise mainly because of heavy loadings of nitrogenous matter on land through which percolating rainfall recharges the underlying aquifer. The only practicable way of reducing the rate of increase of nitrate concentration, and eventually reducing the actual concentration in the groundwater, is to reduce the intensity of agricultural activity which produces the heavy nitrogenous loading. The Government, working in collaboration with the National Rivers Authority (NRA) has designated 10 pilot Nitrate Sensitive Areas in England in which certain agricultural practices (*e.g.* cropping and use of fertilizer) can be regulated. In these areas farmers will be advised on the ways of minimizing the leaching of nitrate into water. They will also qualify for annual payments to help them reduce the intensity of their operations. Of course, if the objective of taking remedial action is simply to protect the quality of water supplies drawn from aquifers, there are ways in which abstracted waters may be denitrified to meet the EC requirements. Such action would only be wise in special circumstances. There is plenty of evidence that nitrate levels in many groundwaters have increased greatly over recent decades, in some cases to the extent that water abstractions from boreholes has had to be curtailed or even abandoned. Further, computer models indicate that this position will worsen unless remedial action is taken (Figures 3 and 4).[5]

[3] A. Jenkins, P. G. Whitehead, T. J. Musgrove, and B. J. Cosby, *J. Hydrology*, 1990, **116**, 403.
[4] The Nitrogen Cycle of the UK, 'A Study Group Report', The Royal Society, London, 1983.
[5] W. B. Wilkinson and L. A. Greene, *Philos. Trans. R. Soc. London, Ser. B*, 1982, **296**, 459.

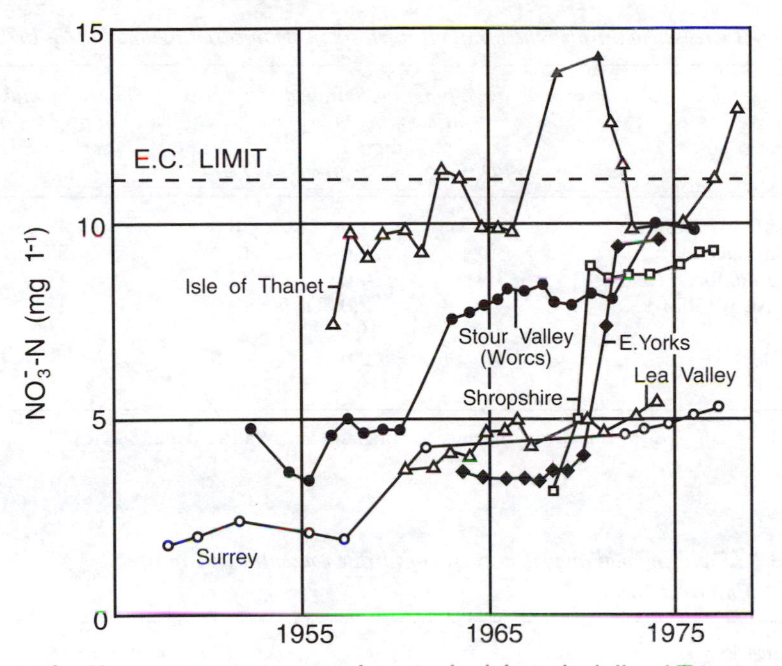

Figure 3 *Nitrate concentration in some abstraction boreholes in the chalk and Triassic sandstone*

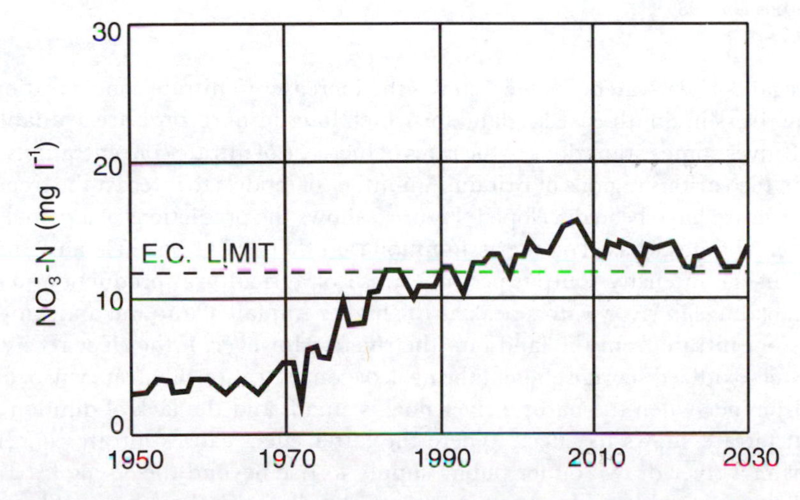

Figure 4 *Model predictions of the change in nitrate concentrations of the water in a chalk aquifer draining to the River Itchen*

Table 7 *Increases in nitrate concentrations in rivers in south east England, 1928–1976*[a]

River	*Average concentration of nitrate nitrogen in year* mg l^{-1}			*% Sewage effluent in river water at summer DWF*
	1928	*1960*	*1976*	
Thames at Hampton	2.5	4.5	8.0	49
Lee at Chingford	3.0	5.0	12	55
Stour at Langford (Essex)	3.0	4.0	7.5	—
Great Ouse at Offord	—	2.0 (1961)	8.0	45
Frome (Dorset)	—	2.0 (1964)	3.8	—

[a]After J. J. Marsh, 'Towards a Nitrate Balance for England and Wales', Water Services, 1980, pp. 601–606

Table 8 *Likely regional annual increases of nitrate concentrations in rivers in Great Britain*[a]

Region	*Annual increase in nitrate concentration* mg l^{-1}
Scotland, Wales, North West	0.1 to 0.4
North East	0.1 to 0.7
Yorkshire, Severn Trent, Thames, Southern, Wessex	0.3 to 0.8
Anglian	0.7 to 1.1

[a] DoE Central Directorate of Environmental Pollution, 'Nitrate in Water', Pollution Paper No. 26, HMSO, London, 1986

As regards river waters, Table 7 shows the increases in nitrate concentrations in some rivers in South-east England for which long-term records are available. Table 8 gives some estimations of the rates of increase of nitrate concentrations in rivers in the various regions of Britain. A number of models to forecast the trends into the future have been developed. Figure 5 shows the predictions of a model of the River Thames,[6] based on the assumption that there will be a 1.5% annum^{-1} growth in the intensity (output per unit area) of agricultural production. The nitrate loadings in river waters peak as the higher rainfall of autumn and winter washes out nitrate from the land into the rivers. However, if the river receives much well-oxidized sewage effluent the peak of concentration of nitrate will occur in midsummer when the natural river flow is small, and the lack of dilution of effluent nitrate shows its effect. Where the latter effect causes nitrate concentrations in a river drawn on for public supply to rise beyond the 50 mg l^{-1} EC limit, it is practicable and economic to partially denitrify the sewage effluents

[6]C. A. Onstad and J. Blake, 'Proceedings of the Symposium on Watershed Management', American Society of Civil Engineers, Boise, 1980, 961.

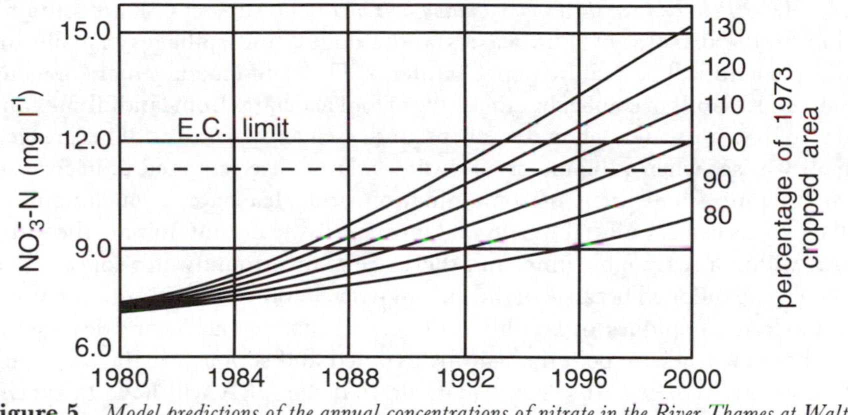

Figure 5 *Model predictions of the annual concentrations of nitrate in the River Thames at Walton*

involved to solve the problem (see Case Study 2). Where the winter nitrate position poses a similar public supply problem, but the abstracted water is given long-term storage before treatment for supply, dilution and natural removal of nitrate within the reservoir could ease the problem.

1.5.3 Case Study 2—A Nitrate Problem in the River Lee. In the early 1970s the continuing increase in the peak nitrate concentration in summer on the River Lee began to exceed 50 mg l^{-1} for several days at a time. To avoid putting treated river water into supply in contravention of the WHO* recommendations on nitrate levels in drinking water, the abstracted river water was put into one of the Lee Valley reservoirs and low nitrate water was drawn from another reservoir to meet raw water supply needs. During 1974, the first of the three dry years 1974, 1975, and 1976, this problem worsened and it was concluded by the then Thames Water Authority (TWA) that a remedy should be urgently formulated and applied. It was clear that the nitrate problem arose from the high concentrations of high-quality, but nitrate rich, effluents from Luton and Middle Lee sewage works discharged upstream of the waterworks intakes. The river, the waterworks, and the sewage works were all managed by TWA and so there were no problems about which authority should do what, as existed before TWA was formed in 1974. It was speedily decided that, with the scientific and technical assistance of the former Water Pollution Research arm of the Water Research Centre, denitrification of the Middle Lee sewage effluent should be attempted on a very large experimental scale. This was done by arranging for the activated sludge channels to be modified to give a one hour detention at the entrance to the channels without air blow, where the nitrate in the return activated sludge (30 mg l^{-1}) would be reduced to nitrogen gas in the newly created anaerobic section of the channels. This worked remarkably well, solved the problem, and reduced operating costs by 15% by saving on the costs of air blowing. This technique is now a standard practice wherever a well-oxidized effluent is required irrespective of whether denitrification is necessary, simply because of the cheaper operating mode involved.

*No European standards of water quality were in existence at the time of the events described.

1.5.4 Water Pollution via the Contamination of Land. The subject of land contamination by the disposal of solid wastes and sludges, and spillages of polluting liquids, to land will be dealt with in Chapter 5. The sound management of water resources demands of course that the entry of foul leachates from landfill sites and other sources to waters should be kept under effective control. The greatest difficulty arises when valuable groundwater sources are rendered unfit for use, often by quite small levels of contamination, from leachates from landfill or spillages. It is usually most difficult, and most costly, to decontaminate the water source within a reasonable time, and there are many groundwater sources that have been abandoned because of the virtual irreversibility of the contamination. The modern techniques of landfill control and management[7] are developing such that new leachate pollution can be avoided and serious existing contamination greatly reduced. It seems inevitable that the NRA will have to specify Water Protection Zones in the near future so that the risks of accidental spillage of hazardous chemicals into important aquifers can be minimized.

2 WATER RESOURCES DEVELOPMENT AND MANAGEMENT

2.1 Flood Defence

Defence against river flooding and tidal inundation is a very important environmental protection service. It is aimed at protecting life and property from the natural excesses of the atmospheric and aquatic sectors of the environment, as distinct from protecting the atmospheric and aquatic environments from the excesses of man. This service is provided in England and Wales by the NRA under the sponsorship of the Ministry of Agriculture, Fisheries, and Food. The natural function of any river is to convey the excess of rainfall to the sea, and of course this results from time to time in natural flooding of land. The consequences of this natural flooding are often made much worse than they otherwise would be by permitting development in the natural flood plain. Also, the frequency and extent of natural flooding is artificially increased downstream of the development, as a result of the increased run-off of rainfall from the roofs and paved surfaces of the development.

The possibility that global warming will cause changes in distribution and patterns of rainfall in Europe and other temperate zones of the Earth, leading to more severe flooding than heretofore, is a matter of concern.[8,9] That the same global warming might manifest itself in more, and more violent, storms is of even greater concern. The tidal defences of the estuaries and sea coasts protecting the low-lying land of the world are not in general in first class condition to meet present flood defence needs. More powerful storms causing greater tidal upsurges

[7] D. J. Lisk, *Sci. Total Environ.*, 1990, **116**, 403.

[8] J. T. Houghton, G. J. Jenkins, and J. J. Ephraums, "Climate Change: 'The IPCC Scientific Assessment'", ed. J. T. Houghton, G. J. Jenkins, and J. J. Ephraums, Cambridge University Press, Cambridge, 1990.

[9] UK Climate Change Impacts Review Group, First Report, 'The Potential Effects of Climatic Change in the UK', HMSO, London, 1991.

could wipe out those flood defences and the people sheltering behind them. We do not yet know, and may not know for at least ten years, what climatic change is likely to come from global warming. Nevertheless the consensus of opinion in the developed countries, that the increasing emission of greenhouse gases must be halted, seems correct.

The essence of flood defence activity embraces accurate rainfall forecasting, aided by weather radar, flood forecasting using computer models incorporating radar input and relevant hydrometric data, the preparation and management of flood warning and emergency arrangements, and the maintenance and improvement of flow channels, control structures, and protective walls and levees. However in the latter activities, much of the construction work involved can be quite detrimental in environmental terms, particularly in reducing water levels in rivers and the adjacent land. So, much care needs to be taken in works planning, design, and execution to enhance overall rather than degrade the riverain environment.[10]

2.2 Water Conservation

Conservation of the quality and quantity of freshwater resources is the key activity in managing the freshwater environment. Its objective is to secure the best value possible from the available water resources, which value embraces both environmental value and use value. Alternatively, it can be said that freshwater conservation is the only practicable means whereby sustainable development[11,12] of the freshwater environment can be achieved. Responsibility for the conservation of water resources in England and Wales now rests with the NRA. That authority was created in July 1989 when the Water Act 1989 was given the Royal Assent and a few months before the flotation of the newly privatized water services plc's. Prior to this change, all water management functions were integrated into the responsibility of ten Regional Water Authorities. In effect, the 1989 Act split up the existing public service of integrated *water* management into a new public service of integrated *river* management under the NRA, and a new privatized sewerage and water supply service under ten independent water services plc's. The integrated water management system which existed in England and Wales between 1974 and 1989 was regarded as revolutionary and quite unique world-wide. The current system is even more revolutionary and again quite unique. Figure 6 shows the areas of the water plc's and of the NRA's management regions.

The total quantity of water abstracted from surface and groundwaters in England and Wales in 1989 amounted to $33.3 \times 10^6\ m^3\ d^{-1}$. Of this some 20% was groundwater. In its regulation of the quantitative aspects of water resources

[10] J. L. Gardiner, *Reg. Rivers Res. Man*, 1988, **2**, 445.
[11] World Commission on Environment and Development, 'Our Common Future', Oxford University Press, Oxford, 1987.
[12] A Perspective by the United Kingdom on the Report of the World Commission on Environment and Development, DOE, 1988.

Figure 6 *Map of the 10 regions of the NRA*

management, the NRA licences all water abstractions (except those of not greater than 20 m^3 d^{-1}) from rivers and lesser watercourses, estuaries, and groundwaters, and also licences all impoundments of watercourses. The NRA lays down conditions regarding the operation of these licences, especially as regards how much water may be abstracted, when, and by what means. Annual charges are levied by way of licence fee and relating to the volume, timing, and source of the licensed abstraction and the nature of the use to which the water is put. The income levied on authorized abstractions must, taking one year with another, balance the costs incurred by the NRA in all its water resources work. The NRA is also responsible for operating hydrometric schemes for measuring the quantity of water resources, and for ensuring that the major licensed abstractors make and carry out proper plans for developing and conserving water resources. The NRA is specifically charged that in carrying out its water resources function it shall have regard to environmental and related considerations. Further that it shall not permit any new abstraction from surface waters or groundwaters which will cause the flow of a watercourse to fall below any specified minimum acceptable flow or the level required to meet downstream authorized abstractions, and to meet the requirements of public health, navigation, land drainage, and fisheries.

Reference has already been made to the current problems of very low flows in many rivers in the south east of England and the steps that are being taken by the NRA to ease these problems. The underlying cause of the problems is that insufficient investment has been made over recent years in properly developing water resources to meet rising demands, and now urgent action will have to be taken to develop new resources. The NRA recently completed a survey of demands for water and of the reliable yields of water resources currently available. Figure 7 compares the licenced abstractions for each region with the effective rainfall under drought conditions. The methods of further development of water resources currently in vogue are set out in Table 9. Of these, the most environmentally promising are the conjunctive use of surface and groundwaters and the artificial recharge of overdrawn aquifers, in that their environmental

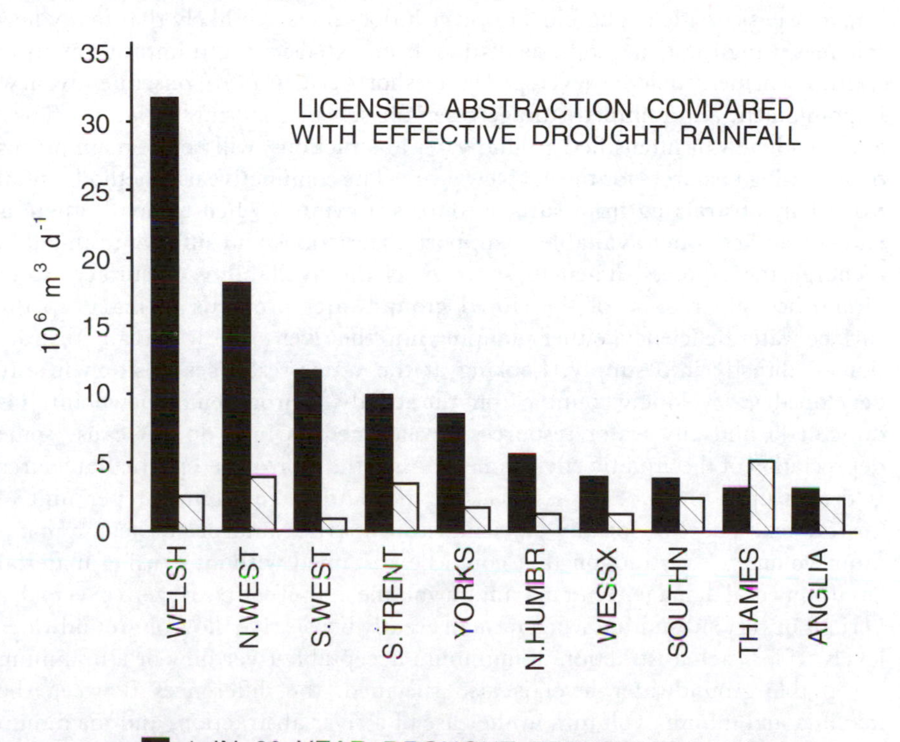

Figure 7 *Comparison of licensed abstraction with drought effective rainfall in the 10 NRA regions*

Table 9 *Methods of development of water resources*

Method of development	*Comments*
Abstraction of river or groundwater for direct supply	Often environmentally unfriendly
Draw-off from impounding or pumped-storage reservoir for direct supply	Reasonably satisfactory if only high river flows stored
Conjunctive use of surface and groundwater for direct supply	Environmentally friendly if abstractions carefully controlled
Reservoir or groundwater discharges to rivers for flow augmentation	Used to support down-river abstraction or to improve down-river environment
Artificial recharge of aquifers with surplus surface water for summer abstraction	Environmentally friendly but relatively expensive
Clean-up of lowland river water for multiple abstraction and re-use	The clean up is now starting. The most environmentally sustainable method

impact can be made negligible. However it does not seem likely that many new schemes using these methods, as distinct from extensions and improvements of existing schemes, could be developed in the short-term future. Consequently, new impoundment and pumped-storage reservoir schemes storing high river flows, along with new or intensified groundwater abstractions, will be the main means of increasing resources for the next few years. The conjunctive use methods entail essentially abstracting from surface sources in winter, when normally there is enough surface water available to support abstractions and sufficient rainfall to recharge the aquifers. Then in summer as the availability of surface water diminishes, abstraction of the stored groundwater proceeds to make up the surface water deficiency, either pumping into the river for downstream abstraction or directly into supply. Looking at the water resources position in any developed or developing country from the strictly environmental viewpoint, it is difficult to find any water resources developments which do not cause some depreciation of the aquatic environment. Also, the more that a particular source is drawn upon the worse the effect on the natural environment per unit of abstraction. There is, for any particular source (river and/or aquifer), a maximum volume of abstraction that could be sustained without causing material environmental damage. That maximum volume may of course be zero in very dry periods or very substantial when rainfall greatly boosts river flow or groundwater levels. If for each abstraction, a minimum acceptable river flow or a minimum acceptable groundwater level is also specified, the differences between the maxima and minima volumes in the case of a river abstraction, and maximum and minimum groundwater levels in the case of a groundwater abstraction, would define the sustainable environmental capacity of the water source for abstraction. Using this approach, if all licences to abstract were restricted within the prescribed sustainable environmental capacities for abstraction, the freshwater environment would be properly protected in water quantity terms.

Furthermore, this approach would provide a basis for a new scheme of charging for water abstraction according to the proportion of environmental capacity used by individual abstractors, and could provide much more powerful incentives to economize in water resources use than are applied at present. The present legislation would have to be amended to enable such a charging scheme to be introduced.

2.3 Water Pollution Control

The objective of water pollution control is to maintain and improve the aquatic environment and the utility value of water resources. Accordingly, water pollution may be described as 'any man-made alteration of the chemical, physical, or biological quality of a water which results in unacceptable depreciation of the utility or environmental value of the water'. This definition should stand the test of time in that it leaves the matter of unacceptability to be determined according to economic and social expectations and requirements at any time. That is precisely what is intended to happen in the main thrust of water pollution control carried out by the NRA in accordance with the provisions of the Water Act 1989. This main activity will involve the following principles:

(a) The quality of waters may be classified according to a system to be specified in statutory regulations. The criteria specified in these regulations are to embrace one or more of the characteristics related generally to the uses for which the waters are suitable, or related to specified chemical quality, or related to other specified characteristics of the water;
(b) Water quality objectives (WQOs) may be established by notice of the Secretary of State for the Environment. Such objectives are to be expressed by reference to the classification system;
(c) It is the duty of the Secretary of State and the NRA to ensure as far as practicable that specified water quality objectives are met at all times. It is also the duty of the Authority to monitor the extent of pollution of waters;
(d) The NRA grants consents (if it thinks fit) to persons and corporate bodies to make discharges of wastewaters to rivers and other watercourses, tidal waters, and underground. In granting consents, the NRA may attach conditions relating, among other things, to the volume and quality of the discharge. These conditions have to be set with great care, otherwise the WQOs may not be met;
(e) Where a discharge of wastewater is made without consent, or not in accordance with the conditions of consent, an offence is committed except in certain extenuating circumstances;
(f) It is possible to vary WQOs and consents and consent conditions from time to time, thereby giving flexibility to accommodate changing environmental expectations and requirements in the future.

The proposals to produce a river quality classification scheme and water quality objectives are not entirely new. Such a scheme was adopted by the

National Water Council in 1977.[13] The classification is summarized in Table 10. However that system was not applied uniformly by the then 10 water authorities, and the National Water Council had no powers to direct those authorities. It is now expected that the proposed new system, with statutory backing, will be applied uniformly across England and Wales by the NRA. A summary of the NRA's forecast of the change in river water quality between 1991 and 1994 is given in Table 11. A key factor in setting both WQOs and consent conditions for discharges, and in checking conformity with WQOs and consent conditions, is the availability of information about water discharge quality and quantity. The monitoring programmes of the NRA are the main source of this information, summaries of which are held available for public scrutiny in registers at the main offices of the NRA.

When a WQO has been set at a particular point on a river, limiting for example the Biochemical Oxygen Demand (BOD) of the water to 5 mg l^{-1}, and the river at that point is at its lowest drought flow of say 0.5×10^6 m^3 d^{-1}, the load of BOD carried by the river is 2500 kg d^{-1}. This is the maximum load of BOD that can be carried if the river is to meet its WQO at very nearly all times. In effect the environmental capacity at that point on the river to carry BOD pollution has been fixed. If there is a discharge of sewage effluent made to the river, not far upstream of the point at which the WQO applies, with a rate of discharge of 62.5 $\times 10^3$ m^3 and with a maximum BOD of 20 mg l^{-1}, the load of BOD discharged in the effluent will be 1250 kg d^{-1}. This means that the discharge will take up half of the sustainable environmental capacity of the river to accept BOD. COD figures could be used in the same way. It would seem reasonable to levy charges on the use of this sustainable environmental capacity of a river to accept and contain pollution—which after all is a public good. If this were done it could provide powerful incentives to individual dischargers to reduce the loads of pollution they discharge in much the same way as suggested above in respect of water abstractions. The NRA has published[14] for public comment a proposed scheme of levying charges in the year 1992/3 for discharges of wastewater to controlled waters. However this scheme is restricted by the requirement of the 1989 Act that the total income from these charges, taking one year with another, shall not exceed such amount as may be reasonably attributable to the expenses of the Authority in carrying out its functions in relation to discharges. Amending legislation would be required to permit the use of a charging scheme which included the levying of charges for the use of environmental capacity for assimilation of wastewater discharges.

2.4 Economy in Water Use and Management

In the pursuit of the sustainable development of the freshwater environment, it is of crucial importance that the waste of water should be minimized. Some of the

[13] 'River Water Quality; the Next Stage', National Water Council, London, 1978.

[14] Scheme of Charges in respect of Applications and Consents for Discharges to Controlled waters, National Rivers Authority, London, 1988.

Table 10 *Summary of river classification*

River quality class	*Quality criteria applicable* Dissolved O_2 % *satn.*	BOD mg l^{-1}	NH_4 *as N* mg l^{-1}	*Current and potential uses*
Class 1A	>80	3	0.4	All uses. Supporting high class fisheries. High amenity value
	also non-toxic to trout and coarse fish			
Class 1B	>60	5	0.9	Lower quality than Class 1A but usable for the same purposes
	also non-toxic to trout and coarse fish			
Class 2	>40	9	—	Suitable for potable supply after advanced treatment. Moderate amenity value
Class 3	>10	17	—	Low grade industrial use. Fish absent or rarely present
Class 4	Anaerobic at times	—	—	Grossly polluted. Completely fishless and nuisance likely

Table 11 *Forecast change in river water quality 1991–1994*[a]

Class	*km in Class 1991*	*km Change 1991–94*	*% Difference*
1A	12 544	+15	0.0
1B	13 389	+346.1	+0.8
2	10 426	+129.8	+0.4
3	3485	−384.2	−0.9
4	526	−142.6	−0.3
Total length classified	40 370		

[a] National Rivers Authority, London, Corporate Plan 1990/91, p. 32

more important aspects of how this is done and can be done better will now be considered. In the UK, unlike most other developed and developing countries, very little metering of domestic water supply use takes place, and water supply charges are levied not on water usage but on a notional annual value (rateable value) of each dwelling supplied. This particular arrangements is not conducive to economy in water use. There can be little doubt that from the environmental protection viewpoint, future charging schemes for domestic water supply must be based on metered water use as far as reasonably possible, and the metering arrangements must be highly efficient. Yet so many water mains in the UK and world-wide leak like sieves. A 25% leakage from mains is commonplace, and effort at reducing these losses is inadequate.

Both industrial and agricultural use of public water supply is usually metered, and the current performance of both industries in the UK in minimizing the waste of mains water is good. To achieve tight pollution control in industry and stock husbandry it is essential that the volume and quality of polluted wastewater produced is minimized at the factory or on the farm. Rainfall run-off from the premises must be collected and piped away separately from wastewater, although it may be necessary in some factory circumstances to allow the run-off from some inevitably polluted factory yards to be treated and disposed of along with the process wastewater stream. By ensuring that clean tap water is not allowed to run to waste or to add to the volume of wastewater produced reduces both water supply costs and wastewater disposal costs. By ensuring that relatively clean wastewater streams are not allowed to be contaminated by the stronger process liquors produced, not only are disposal costs reduced but very substantial re-use of water can be made, again reducing supply costs. It is usual in the developed countries that up to 85% of total water usage in factories should consist of re-used water. Much sewage effluent is also re-used direct by industry.[15] In many of the developing countries, *e.g.* in South America or China, this high level of economy in water use in industry is becoming increasingly necessary as profligate existing water use threatens to deny adequate water supply for new industry. While the foregoing methods of pursuing economy in water use in the home, factory, and farm are very important for the achievement of sustainable development of water resources, there is another major means of

[15] M. R. G. Taylor and J. M. Denner, *J. Inst. Water Eng. Sci.*, 1987, **40**.

screen, at the pollution-origin end of water management, the discharge of hazardous, or potentially hazardous, chemicals to the river system upstream of waterworks intakes. The first phase of the process was to establish, with the close co-operation of industry, exactly what chemicals were discharged from factories to rivers, directly or via public sewerage, and which of these discharges were of concern, assuming:

(a) that the substances as discharged were not biodegradable and remained unchanged in solution on passage through sewers, sewage works, along the river system, and through waterworks treatment;
(b) that if the estimated concentration of the substance in the finished water at a waterworks was less than $0.1\,\mu g\,l^{-1}$ the matter was of no further concern, unless it was known or suspected that the substance at that concentration was hazardous.

The second phase was to take the list of substances produced in Phase 1 and to establish by analysis of the discharge, the river water, and the public supply involved what concentrations of the Phase 1 listed chemicals could be found. Phase 3 involved seeking satisfactory explanations of differences between the estimated and actual concentrations of the chemicals, and deciding whether any remedial action was necessary to eliminate or reduce the concentration of any chemical passed to the river. The final part was to ensure, along with the management of the factory discharging the chemical, that the required remedial action was taken. In the very few cases where this position was reached the remedy applied was to use alternative, safer chemicals. Thereafter, the total position in the river basin was reviewed annually to confirm the continuing position of safety at premises already screened, and to make sure that the use of new or different chemicals was satisfactory. It should be noted that not one of all the industrial concerns in the Thames basin approached in this matter refused to co-operate on a confidential basis (there was no legal necessity to co-operate). Further that the whole process was complementary to, and in no way duplicative of, the monitoring of river water and water supply for the presence of hazardous chemicals carried out by the Water Research Centre as part of its Water Quality and Health research programme.

2.4.2 Case Study 4—Drought Management 1976. Following two very dry winters in 1974 and 1975, the worst drought for some 200 years occurred in 1976 over virtually the whole of the UK. In the early part of 1976, it became apparent that a serious drought and water shortage was likely and emergency preparations were made to minimize the impact of the threatening drought on domestic, industrial and agricultural affairs. There were only two approaches to be followed. One was to reduce demand for water by persuasion and by Drought Orders made by Government. The second approach was to bring into use such old and new sources of raw water, each of relatively small volume, as could be found and rendered fit for abstraction. In the Thames basin, all the usual steps were taken in following these two approaches, but it is the unusual actions taken that merit comment. On the administrative side of management, Drought Committees were formed representative of local authorities and customers in each of the six water supply Divisions. Each was chaired by a Member of the Thames Water Authority. The function of these committees was to decide the priorities for non-domestic use of water and to curtail non-priority uses, and also to monitor and enforce conformity with restrictions on domestic use of water. These Committees were of enormous value in leading local communities and their industry and commerce in an orderly distribution of a scarce commodity and the continuing protection of public health. Of the many novel situations and actions taken on the technical and scientific side of management the following were the most unexpected:

(a) In the late Spring it became apparent that the navigation on the freshwater river was losing some 10^7 l d^{-1} through lockage water in the Oxford area, and serious consideration was given to closing the navigation. However it was concluded that it was only the passage of cruisers along the river that was preventing almost complete stagnation of the river, and that if the navigation was closed water abstracted for supply could become untreatable. So the navigation was kept open, the frequency of lockage reduced, and the flow of the river downstream of Oxford was recycled upstream of the waterworks intakes by installation of temporary pumps. A similar situation existed at the tidal limit of the river at Teddington Lock and a similar recirculation of river flow was put into operation.

(b) In July, it was observed that the flow of the river below Abingdon had almost ceased, and continued in this state until the River Kennet conflowed at Reading. This was not noticeable generally, but it could be deduced by observation at weirs. What had happened was that the river passed over an extensive area of river gravels overlying chalk, and as groundwater levels fell, so the river flow went increasingly underground. Fortunately there were no public water supply abstractions along this length of river, and the many groundwater abstractions were actually protected by the lost river flows. The drought broke in late summer and so the embarrassing prospect of losing the River Thames was avoided.

3 WATER PURIFICATION FOR SUPPLY AND WASTEWATER TREATMENT

3.1 Water Purification for Supply

The primary requirement regarding the quality of public water supplies is the public health requirement that the water should be 'wholesome' for drinking. Wholesome in this context is generally interpreted to mean, as stated in dictionaries, 'promoting or conducive to health'. The most up-to-date legislation relating to water supply quality, the Water Act 1989, still requires water to be wholesome, but also requires that, so far as reasonably practicable, there is in general no sporadic or trend of deterioration in the quality of water which is supplied. Over the past 30 years or so, following the lead of the World Health

Organization, most of the countries of the world have adopted standards which prescribe the quality criteria with which water supplies should conform to be considered satisfactory for drinking. Some of these countries have simply adopted the WHO's recommendations.[17] Others have adopted most of the WHO's recommendations but have made variations to a few of the recommendations. The USA has adopted standards which are largely in line with the WHO criteria but are more broad ranging and detailed, and of course the European Community Directive on the Quality of Water Intended for Human Consumption came into effect in 1985. Given that a public water supply complies with requirements for drinking water, it can be taken as axiomatic that the water is suitable for agricultural purposes. However public water supplies are not sufficiently pure for use for many industrial purposes—for example in steam raising, in the chemical and pharmaceutical industries, and in the health care and electronics industries. Each industry is expected to take its own steps to produce, from the available public supply, the precise level of purity of water it requires for its particular purposes.

While the quality of public supply, like any other water, is described in physical, chemical, and biological terms, it is the microbiological quality which matters the most—simply because it is very easy for water to be contaminated by the micro-organisms causing human disease and major epidemics. The physical characteristics of public supply are measured in terms of its colour, turbidity, odour, taste, and pH value. The chemical characteristics cover a wide and complex field, embracing low concentrations of the acutely toxic inorganic chemicals, very low concentrations of the hazardous organic micro-pollutants such as polycyclic aromatic hydrocarbons, haloforms as produced in the chlorination of organic residuals in water, pesticides, mercury, and cadmium. Also specific inorganic chemicals such as fluoride, nitrate, nitrite, magnesium, and sulphate which at particular levels of concentration are or may be harmful to health are included. Other substances such as chloride, iron, manganese, and residuals of organic matter of vegetable or animal origin, which in one way or another could affect the acceptability of a water for drinking, are included. Table 14 sets out the normal chemical characteristics of public water supplies drawn from different sources, and Table 15 summarizes the more important chemical and microbiological characteristics of public water supply which feature in the quality standards of the EC directive on drinking water.

The importance of the microbiology of public supply concentrates on the organisms responsible for waterborne human disease. These may be classified as bacteria, viruses, and other organisms such as worms, flukes, and protozoa which are known to be associated with the waterborne spread of disease. Table 16 gives a very brief summary of the main aspects of this microbiology. In the day to day control of the microbiology of public water supply, since it is not practicable to examine waters frequently for the presence of all the organisms that cause disease, standard bacteriological examination of water in a relatively simple but effective way is carried out. This involves assessing the numbers of bacteria

[17] Guidelines for Drinking-Water Quality, Volume 1, Recommendations, WHO, Geneva, 1984.

Table 14 *Normal chemical characteristics of some public water supplies* (mg l^{-1} *except where otherwise stated)*

	Source of water			
	Chalk borehole (softened)	*Upland river*	*Lowland river*	*Lowland reservoir*
pH value	7.3	7.3	7.9	7.6
Colour (units)	0	2	5	4
NH_3 (N)	0	0.10	0.02	0.03
NO_3 (N)	3.0	1.2	4.0	1.7
Total solids	220	80	410	430
Total hardness ($CaCO_3$)	150	45	260	225
Non-CO_3 hardness	10	25	135	70
Alkalinity ($CaCO_3$)	110	20	130	160
Chloride	16	15	35	55
Iron	0.01	0	0.02	0
Manganese	<0.01	<0.01	<0.01	<0.01

Escherichia coli (*E.coli*) and of the coliform group present. The former is indicative of the faecal pollution of water while the presence of the other members of the group is indicative of pollution from animal sources, but not necessarily of faecal pollution. The EC directive and the WHO guidelines both require *E.coli* to be absent from public supply, whereas a little latitude, under prescribed circumstances, is permitted as regards the presence of coliform group organisms in water in distribution mains. Practice in the UK is regulated by Report 71 of the Department of Health and Social Security.[18] The macrobiology of public water supply is important in relation to the efficiency of water purification and to the aesthetic quality of the water. Concern here relates to the presence of algae and

Table 15 *Some aspects of water quality specified in the EC drinking water Directive*

Contaminant	*Maximum admissible concentration*
Acidity/alkalinity	pH range 5.5–9.5
Colour	20 units
Turbidity	4 units
Iron	200 $\mu g\ l^{-1}$
Manganese	50 $\mu g\ l^{-1}$
Aluminium	200 $\mu g\ l^{-1}$
Nitrate nitrogen	11.5 mg l^{-1}
Lead	50 $\mu g\ l^{-1}$
Pesticides (individual substances)	0.1 $\mu g\ l^{-1}$
Pesticides (total)	0.5 $\mu g\ l^{-1}$
E. coli	Not detectable in 100 ml of water
Coliform bacteria	95% of samples should be coliform free

[18] Department of Health and Social Security/Welsh Office/ DOE Report No. 71, 'The Bacteriological Examination of Water Supplies', HMSO, London, 1983.

Table 16 *List of the organisms mainly responsible for waterborne disease (temperate climate)*[a]

Type of organism	*Name*	*Source*	*Disease*
Bacterium	*Salmonella typhi*	Man	Typhoid fever
Bacterium	*Salmonella typhi*	Man	Paratyphoid fever
Bacterium	*Vibrio cholerae*	Man	Cholera
Bacterium	*Shigellas*	Man	Bacterial dysentry
Bacterium	*Salmonella* group	Man and animals	Gastro-enteritis (food poisoning)
Bacterium	*Leptospira interohaemorrhagiae*	Rats	Weil's disease
Protozoa	*Cryptosporidium parvum*	Man and animals	Diarrhoeal illness
Amoeba	*Entamoeba hystolica*	Man	Amoebic dysentry
Tapeworm	*Taenia saginata*	Man via cattle	Beef tapeworm

[a] No cases of waterborne infection with the viruses causing Poliomyelitis or Hepatitis A have yet been reported in the UK

invertebrate animals in the water, on which further information is given in Table 17.

The nature and extent of the treatment to be given to any particular raw water, to produce a public supply meeting specified quality standards at minimum cost, depends entirely on the nature and quality of the raw water. The minimum treatment, of disinfection using chlorine, would be appropriate for a deep groundwater source in a rural area, or for water from an upland impoundment not open to public access. At the opposite extreme, water drawn from the lowland reaches of a river, containing a high proportion of sewage or industrial wastewater discharged upstream, would need extensive and very carefully monitored treatment. The conventional techniques for the treatment of raw waters for public supply are set out in Table 18, and Figure 9 shows the layout of a typical waterworks treatment plant.

3.2 Wastewaters and Wastewater Treatment

The present system of wastewater disposal worldwide is based on the provision and operation of public sewerage reticulations and sewage treatment plants in urban areas, and the private provision of drains and wastewater treatment arrangements at industrial premises and stock rearing farms. The position in England and Wales is unique in that the urban sewerage systems (like the water supply systems) are now owned and managed by the new water services companies. Much of the industrial wastewater produced in urban areas is disposed of, with or without pre-treatment, into public sewers with the consent of

Table 17 *Algae and invertebrate animals of concern in public water supply*

Type of organism	*Common species or genera*	*Cause of presence in public supply*
Small, single-celled Chlorophyta (green algae)	*Chlamydomonas* *Arkistrodesmus* *Scenedesmus*	Passing through waterworks filters
Bacillariophyta (diatoms)	*Stephanodiscus*	Passing through waterworks filters
Cyanophyta (blue–green algae)	Various species	Passing through waterworks filters
Cyanophyta	Various species	Decay of algae causes taste and odour in water
Xanthophyta (yellow–green algae)	*Chryptomonas* *Rhodomonas*	Passing through waterworks filters
Xanthophyta	*Synura* *Peridinium*	Causes cucumber taste Causes fishy odour
Worms	*Nias*	Can pass filters and infest water mains
Rotifera	Various species	Passing through filters
Fly larvae	*Chironomid* sp.	Can pass filters and infest water mains
Crustacea	*Daphnia, Cyclops, Asellus*	Can pass filters and infest water mains

the water services plc's. The same general position applies in many overseas countries but in others, particularly developing ones where haste to increase industrial capacity inevitably outpaces the provision of modern urban sewerage, industrial wastewater disposal direct to rivers is the norm. Wastewaters produced on farms are usually too strong in the load of organic matter they carry for disposal to public sewerage systems. Instead, efforts are made to contain the farm wastes on the producing farm where the fertilizer value of the wastes can be realized. However it is a fact that farm waste disposal from intensive farming in England and Wales leads to much river pollution and presents a serious problem. There is no real solution to this problem in sight at present, other than a major switch being made to low intensity stock rearing.

Overall world-wide, most of the wastewater produced in urban areas is finally discharged to rivers or tidal waters. One of the major urban water pollution problems is that of disposal of storm sewage. This arises particularly in the older areas of developed countries where the sewerage systems are laid wholly or in part as 'combined sewers'—that is they receive both foul water (sewage and industrial wastewater) and rainfall run-off from roofs, and roads and other paved areas. In times of heavy rainfall these combined sewers become surcharged and to prevent serious and foul flooding the sewers have to be relieved of some of the overload of diluted sewage they carry. This is done via devices known as storm

Table 18 *Summary of treatment processes used in purification of public water supply*

Process	*Purpose*
Raw water storage (short term)	Sedimentation. Die-off of faecal organisms. Balancing of intake water quality. Raw water reserve
Raw water storage (long term)	Oxidation of organic matter. Partial removal of NO_3, HCO_3, PO_4, SiO_2, by algal uptake
Chemical precipitation using $Al_2(SO_4)_3$ or activated SiO_2, or Fe salts plus polyelectrolytes	Coagulation, flocculation, and settlement of turbidity and colour
Microstraining	Straining through very fine-mesh rotary screens
Rapid filtration	Rapid up-or down-flow filtration through sand
Slow sand-filtration	Filtration plus bio-oxidation by slow gravity downward flow
Chlorination and/or ozonation or UV	Disinfection
Softening by lime, lime-soda, or ion exchange	Removal of Ca and Mg hardness (no longer fashionable)
Activated carbon treatment by powder addition before filtration or passage through active carbon filters	Reduction in residual organic matter
Desalination by flash distillation or reverse osmosis	Production of freshwater from saltwater or super-purification of wastewater

sewage overflows which discharge to convenient watercourses. These are obnoxious ways of managing urban drainage and the only reasonable long term remedy is to steadily renew the old sewers on a separate system of foul sewers and surface water sewers. In the case of the old combined sewerage system in the City of London, a novel solution of the storm sewage disposal problem has been devised, as follows.

3.2.1 Case Study 5—Oxygenation of the Thames Tideway. As the major clean-up of the Thames Tideway in London proceeded in the mid-1970s, it became apparent that despite all the effort and expenditure being put in, sporadic pollution of the river arising from the surcharge of old combined sewers in summer storms would be heavy, and would be unacceptable. The problem was that fish were returning to the Tideway and it was possible that restoration of the salmon fishery would be feasible. However, when heavy summer storms overloaded the old London sewers after a long period of dry weather, the sewer overflows caused a sharp and deep oxygen sag in the Tideway. On several occasions this had resulted in the destruction of fish. The remedy of resewering much of the older areas of London, other than on a long timescale, was not and is still not economically feasible. The novel remedy selected was to fit out a barge with a pressure-swing plant to

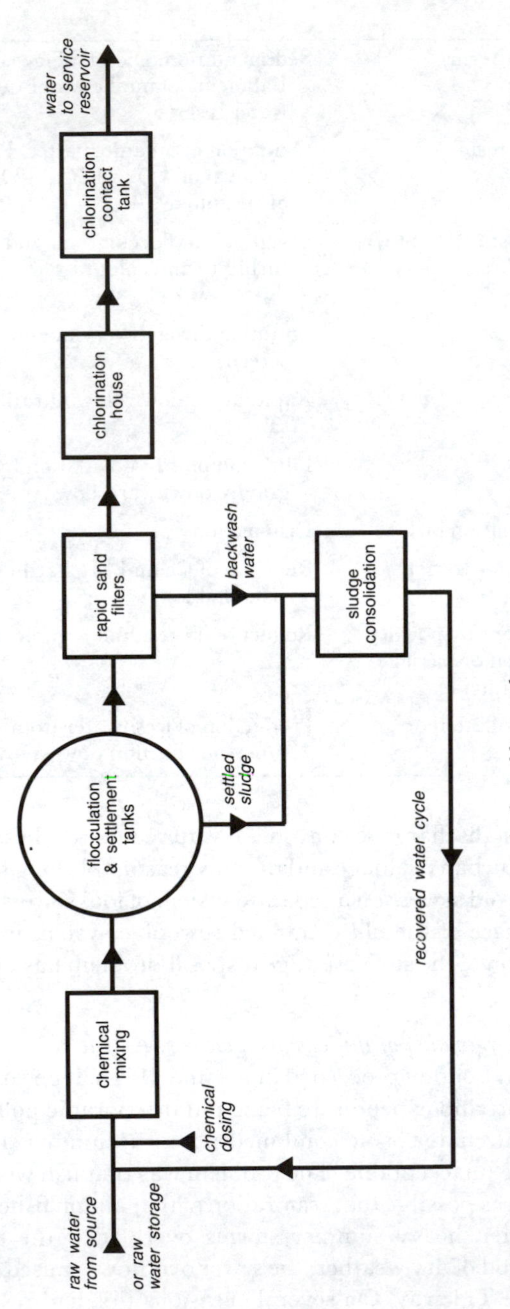

Figure 9 *Layout of a typical water treatment works for public supply*

produce 95% oxygen from the air, (10 tonnes day^{-1}) and to inject this into the river right over the point of maximum oxygen sag. This arrangement worked satisfactorily and the barge, affectionately named the 'Thames Bubbler' is now replaced by a much larger purpose-built vessel producing 30 tonnes of oxygen day^{-1}. This has prevented any serious destruction of fish, including salmon which have now been restored to the river.

3.3 Sewage Treatment

Modern sewage treatment employs three basic processes, namely:

(a) the removal of polluting matter from the sewage flow as solids, or slurries of solids in water (sludges);
(b) the removal of polluting matter from the sewage flow and from separated sludges by accelerated natural processes of biochemical breakdown;
(c) the separation of water from sludges to reduce the volume of sludge for disposal.

Table 19 sets out the conventional descriptions of these processes and the type and nature of the treatment given, and Figure 10 shows the layout of a typical sewage treatment works.

Primary settlement should reduce the suspended solids (dry) content of the sewage from about 300 mg l^{-1} to about 120 mg l^{-1}, and the BOD from about 250 mg l^{-1} to about 120 mg l^{-1}. Secondary treatment processes are based on placing the degradable carbonaceous and nitrogenous matter, carried in the settled sewage, into intimate contact with oxygen and a sufficient surface of biomass for an appropriate period of time. Where the reactor containing the process is a percolating filter the retention time is about 20 minutes, but the structure of the interior of the filter provides a large void space, a large reactive bio-surface,

Table 19 *Summary of main sewage treatment processes*

Process	*Type of treatment*	*Basic purpose*
Preliminary	Screening and grit removal	Removal of gross or abrasive solids
Primary settlement	Settlement in tanks	Removal of solids and grease
Secondary treatment	Activated sludge, percolating filter, other bio-reactors, plus settlement	Bio-oxidation of carbonaceous matter and ammonia, and removal of solids
Polishing treatment	Sand filtration, micro-straining, or lagooning	Removal of very fine solids
Tertiary treatment	Denitrification, chemical precipitation	Removal of N, P, and organic residuals
Sludge treatment	Digestion, thickening, dewatering, drying	CH_4 production. Preparation for disposal

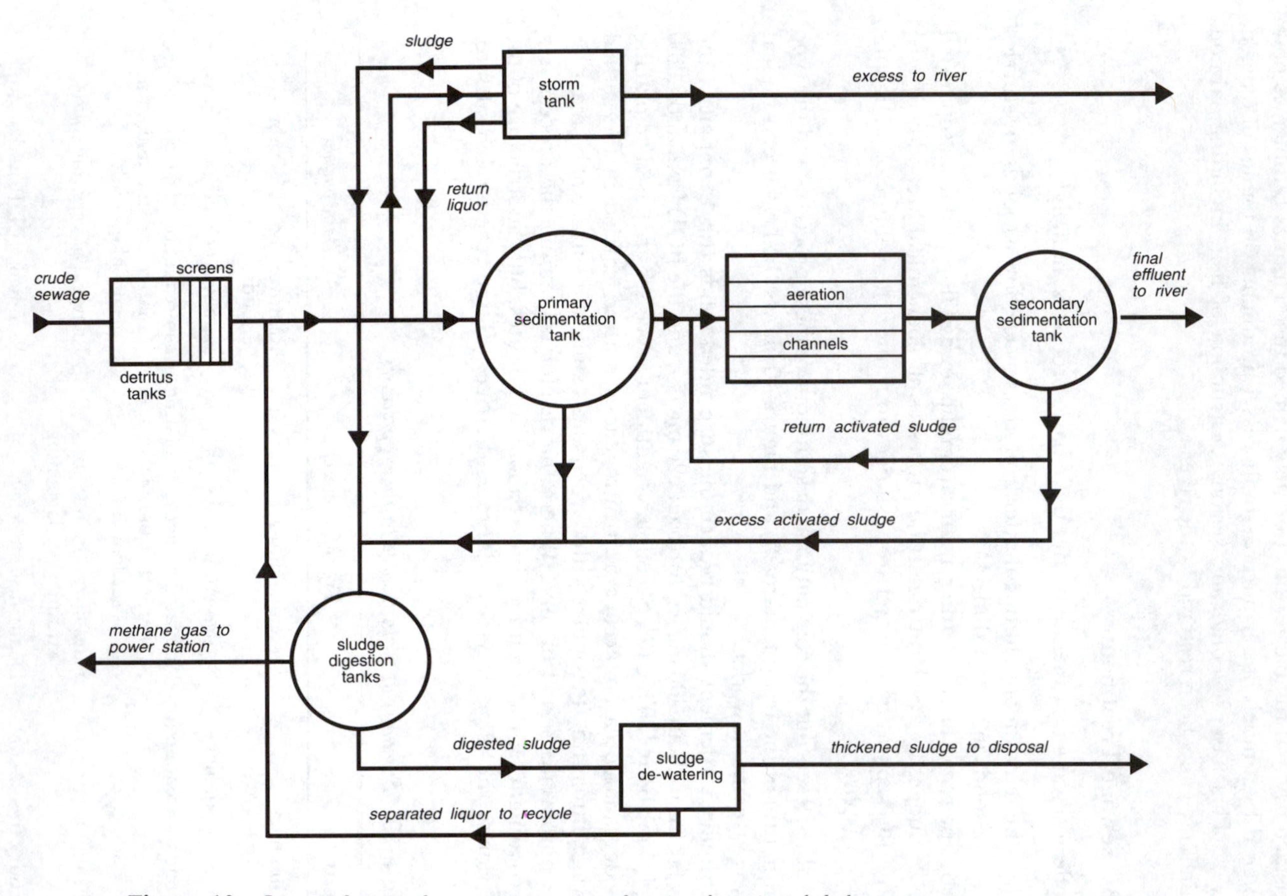

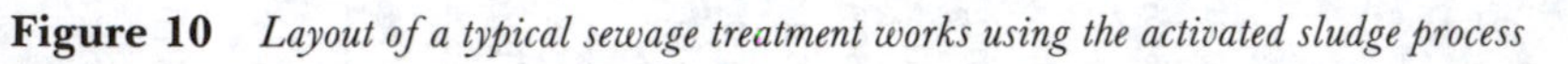

Figure 10 *Layout of a typical sewage treatment works using the activated sludge process*

natural ventilation through underdrains, a low volumetric loading of flow to capacity, and a large population of grazing fauna. Where the reactor is an activated sludge aeration tank (channel, square, or circular) the retention time is 4 to 6 hours. The air requirement, supplied as blown diffused or coarse bubble air, or mechanical surface aeration must be sufficient to maintain a residual of dissolved oxygen. A mixed liquor suspended solids concentration (activated sludge) of around 2000 to 4000 mg l^{-1} is the normal range. The latter figure applies where good nitrification (conversion of reduced nitrogen to nitrate) is required and is normally accompanied by the re-aeration of the activated sludge being returned to the flow of settled sewage at the inlet to the aeration tanks. The effluent from secondary treatment is passed to settlement tanks which are similar to, but about one-third of the capacity of, primary settlement tanks. Domestic sewage that has been given full primary and secondary treatment, including 6 hours aeration in a diffused air activated sludge plant should normally contain less than 15 mg l^{-1} suspended solids, about the same level of 5 day BOD and an ammonia nitrogen content of less than 10 mg l^{-1}.

Several major modifications of the activated sludge process have been developed, mainly for the smaller sized sewage works—such as the extended aeration system, the oxidation ditch, and the deep-shaft system. The development of a new type of aeration process for secondary treatment of sewage deserves note. This process involves bringing the settled sewage into intimate contact with pea-gravel sized, smooth, flattish-shaped stones and air in rectangular tanks. The entire contents of the tanks are maintained in fluidized suspension by air injection. The required reactive biomass forms on the surface of the stones and excess mature biomass breaks off into suspension. The effluent from the tanks is then given settlement in the usual way.

The factor that limits the rate at which carbonaceous oxidation can proceed is the rate of transfer of atmospheric oxygen into the activated sludge. This limitation can be raised substantially if oxygen is used instead of air because of the greater solubility of oxygen in water. The factor that limits the speed at which ammonia is oxidized is the rate of growth of the bacteria responsible for oxidizing ammonia to nitrate. These have a doubling time of about 1 day, whereas the bacteria responsible for the oxidation of carbonaceous matter have a doubling time of 20 minutes.

Domestic sewage sludge contains about 15% saponifiable fat, 40% fibre, 10% protein, 35% ash, 0.01% detergent, some toxic metals, *e.g.* cadmium at about 5 mg kg^{-1}, and residuals of pesticides from time to time. The sewage sludges produced in urban areas in industrial regions are an even worse mixture, the concentrations of toxic metals, pesticides, and biocides being particularly undesirable in the context of sludge disposal at sea. It is for this reason, rather than any other, that the sea disposal of sewage sludge is to cease within the EC by the turn of the century. The pretreatment of sludge is geared essentially to reducing the bulk of the sludge before disposal to land, sea, or incineration. Anaerobic digestion of sludge to yield methane for power generation reduces the bulk of the solids content of sludges but this does not give great advantage when the sludge contains only 3% solids. However, aeration of sludge after digestion followed by a quiescent period results in substantial separation of water, which if drawn off can give a 20% reduction in sludge volume. There is a variety of machinery available for the dewatering of sludge to yield a handleable sludge cake for use as a fertilizer, for landfill, or for incineration. All are based on chemical conditioning followed by centrifuging, vacuum filtration, or plate or filter belt pressing.

3.4 Treatment of Industrial Wastewaters

Industrial wastes can be divided into four categories, namely:

1. Wastewaters that have been changed in chemical quality during use in, or in connection with, manufacturing processes. This category consists of process waters and unclean cooling waters.
2. Contaminated run-off from roofs and yards at industrial premises.
3. Clean cooling waters.
4. Grossly contaminated wastewaters, waste chemicals, and liquid or semi-liquid sludges of small volume kept separate, or separated from, the main wastewater streams of manufacturing processes. These wastes are usually transported away for specialist disposal.

Clean cooling waters running to waste should as far as possible be recycled repeatedly and the bleed-off run to disposal with the main process stream. The remaining polluting waters in categories 1 and 2 are best disposed of, if practicable, to public sewers after such pre-treatment as may be necessary. Otherwise the wastewater will have to be given extensive treatment before disposal with the consent of the NRA in England and Wales, the EPA in the USA, the Agences Basin in France, and so on. Discharges to sewers in England and

Table 20 *Typical consent limits and pre-treatments given to industrial wastewaters before discharge to sewers*

Contaminant	*Consent limit* ($mg\ l^{-1}$)	*Pre-treatment (where necessary)*
Suspended solids	400–1000	Screening and settlement
BOD	500–1000	High rate bio-oxidation
Oil and grease	10	Passage through oil traps or separators
Cyanide	1–5	Chlorination or enzyme treatment for CN removal
Heavy metals	1–20	Alkaline precipitation and settlement
Acidity/alkalinity	pH 10 and 5	Neutralization
Solvents	Substantially absent	Recovery or activated carbon treatment
Strong dyestuff colour	Low colour	Bleaching with chlorine

Wales have to be consented to by the relevant water services plc, who will normally attach conditions to such consents and charge for the service given. Table 20 gives further information on consent conditions and methods of industrial wastewater treatment.

In considering the advantages and disadvantages of disposing of industrial wastes to sewers, there are several advantages and only one disadvantage. This is that occasionally mistakes in factories result in powerfully toxic materials being lost to sewers and greatly interfering with sewage treatment. This kind of mishap can be controlled by vigilance and co-operation between industry, the sewerage authority, and the river authority. It is upon this kind of co-operation that the safe management and improvement of the aquatic environment chiefly depends.

CHAPTER 4

The Oceans

S. J. DE MORA

1 INTRODUCTION

The world ocean is a complex solution. Seawater contains dissolved gases, and as a consequence is both well-oxygenated, although exceptional environments exist, and buffered at a pH of about 8. There are electrolytic salts, the ionic strength of seawater being approximately 0.7, and multitudinous organic compounds in solution. At the same time, there is a wide range of inorganic and organic particles in suspension. These comfortable distinctions become quite confused in seawater. Some molecules present in true solution are sufficiently large to be retained by a filter. Surface adsorption accumulates organic coatings and scavenges dissolved elements. Some elements, particularly those with biochemical functions, may be rapidly removed from solution. Concurrently, reactions involving geological time scales are proceeding slowly. Yet despite this apparent complexity, many aspects of the composition of seawater and chemical oceanography can now be explained with recourse to the fundamental principles of chemistry. This chapter serves to bridge the gap between those with environmental expertise and those with a traditional chemical background.

1.1 The Ocean as a Biogeochemical Environment

A traditional approach utilized in geochemistry, and now also in environmental chemistry, is to consider the system under investigation as a reservoir. For a given component, the reservoir has sources (inputs) and sinks (outputs). The system is said to be at equilibrium, or operating under steady-state conditions, when a mass balance between inputs and outputs is achieved. An imbalance could signify that an important source or sink has been ignored. Alternatively, the system may be perturbed, possibly anthropogenically mediated, and therefore changing towards a new equilibrium state.

Processes within the reservoir that affect the temporal and spatial distribution of a given component are transportation and transformations. The physics and biology within the system play a role. Clearly transport effects are dominated by the hydrodynamic regime. Although transformations could involve chemical (dissolution, redox reactions, speciation changes) or geological (sedimentation)

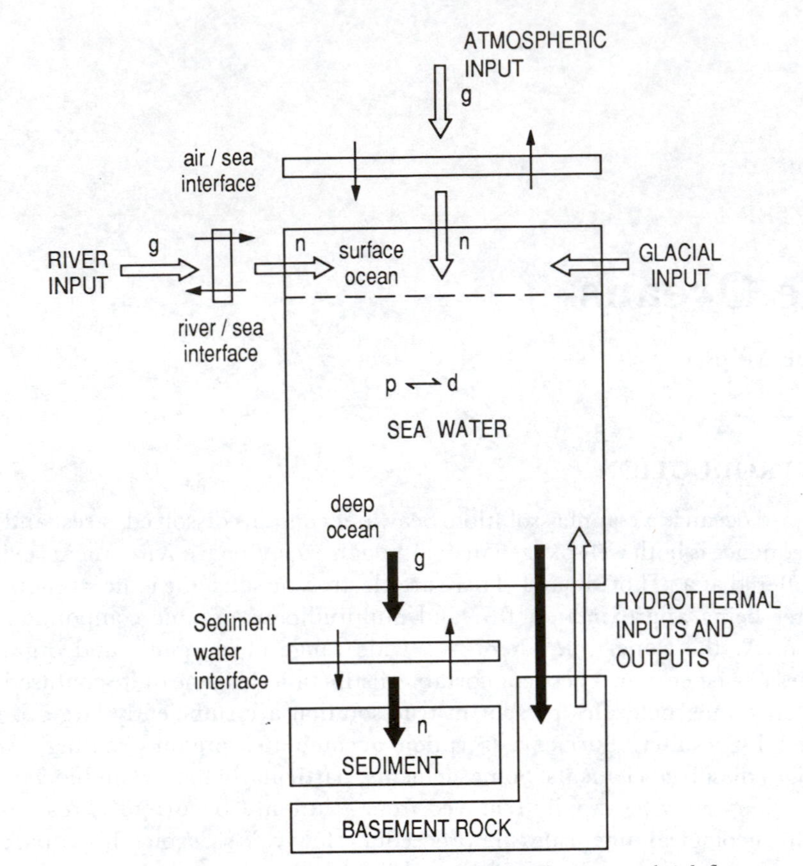

Figure 1 *A schematic representation of the ocean reservoir. The source and sink fluxes are designated as g and n, referring to gross and net fluxes, thereby indicating that interactions within the boundary regions can modify the mass transfer. Within seawater, the p ⇌ d term signifies that substances can undergo particulate–dissolved interactions. However, it must be appreciated that several transportation and transformation processes might be operative (from Chester, 1990)*[1]

processes, biological activity can control nutrient and trace metal distributions. Furthermore, the biota influence concentrations of O_2 and CO_2 which in turn determine the pH and p*e* (*i.e.* the redox potential see p. 108), respectively. For these reasons, some fundamental aspects of descriptive physical and biological oceanography are included in this chapter.

In terms of biogeochemical cycling, the ocean constitutes a large reservoir. The surface area is 361.11×10^6 km^2, nearly 71% of the earth's surface. The average depth is 3.7 km, but depths in the submarine trenches can exceed 10 km. The ocean contains about 97% of the water in the global hydrological cycle. A schematic representation of the ocean reservoir is presented in Figure 1.[1] The material within it can be operationally defined, usually on the basis of filtration, as dissolved or particulate. The ocean is divided into two layers, with distinct

[1] R. Chester, 'Marine Geochemistry', Chapman and Hall, London, 1990, pp. 698.

surface and deep waters. The boundary regions are also distinguished as the composition in these regions can be quite different to bulk seawater. Furthermore, interactions within these environments can alter the mass transfers across the boundary. The rationale for such features will be presented in subsequent sections.

Material supplied to the ocean originates from the atmosphere, rivers, glaciers, and hydrothermal waters. The relative importance of these pathways depends upon the component considered and geographic location. River run-off generally constitutes the most important source. Transported material may be either dissolved or particulate, but discharges are into surface waters and confined to coastal regions. Hydrothermal waters and vents are associated with seafloor spreading ridges. Seawater can circulate into the fissured rock matrix and come into contact with newly forming basalt. Compositional changes in the aqueous phase occur due to seawater–rock interactions and the release of material from the mantle into solution. Thus, hydrothermal activity releases dissolved components into the ocean at great depths. This is an important source of some elements, such as Li, Rb, and Mn. The atmosphere supplies particulate material globally to the surface of the ocean. This is the most prominent pathway for Pb to the world ocean. Aeolian transport is greatest in low latitudes and the Sahara Desert is known to act as an important source of dust. Conversely, glacial activity makes little impact on the world ocean. Glacier-derived material tends to be comprised of physically weathered rock residue, and so relatively insoluble, and also the input is largely confined to polar regions, with Antarctica responsible for approximately 90% of the material. Sedimentation acts as the major removal process. However, volatilization and subsequent evasion to the atmosphere can be important for elements such as Se and Hg that undergo bioalkylation.

Some definitions facilitate the interpretation of chemical phenomena in the ocean. Conservative behaviour signifies that the concentration of a constituent (or absolute magnitude of a property) varies only due to mixing processes. Components or parameters that behave in this manner can be used as conservative indices of mixing. Examples are salinity and potential temperature, the definitions for which are presented in subsequent sections. In contrast, non-conservative behaviour indicates that the concentration of a constituent may vary as a result of biological or chemical processes. Examples of parameters that behave non-conservatively are dissolved oxygen and pH. Residence time, τ, is defined as:

$$\tau = \frac{A}{(\mathrm{d}A/\mathrm{d}T)}$$

where A is the total amount of constituent A in the reservoir and $\mathrm{d}A/\mathrm{d}T$ can be either in rate of supply or the rate of removal of A. This represents the average life time of the component in the system and is, in effect, a reciprocal rate constant (see Chapter 6). Finally, the photic zone refers to the upper surface of the ocean in which photosynthesis can occur. This is typically taken to be the layer down to the depth at which sunlight radiation has declined to 1% of the magnitude at the surface.

1.2 Properties of Seawater

Water is a unique substance, with unusual attributes as a result of its structure. The molecule consists of a central oxygen atom with two attached hydrogen atoms forming a bond angle of about 105°. As oxygen is more electronegative than hydrogen, it attracts the shared electrons to a greater extent. Also, the oxygen atom has a pair of lone orbitals. The overall effect produces a molecule with a strong dipole moment, that is, having distinct negative (O) and positive (H) ends. While there are several important consequences, two will be considered here. Firstly, the positive H atoms of one molecule are attracted towards the negative O atoms in adjacent molecules giving rise to hydrogen bonding. This has important implications with respect to a number of physical properties, especially those relating to thermal characteristics. Secondly, the large dipole moment ensures that water is a very polar solvent.

Considering firstly the physical properties, water has much higher freezing and melting points than would be expected for a molecule of molecular weight 18. Water has high latent heats of evaporation and fusion. This means that considerable energy is required to stimulate phase changes, the energy being utilized in hydrogen bond rupturing. Moreover, it has a high specific heat and is a good conductor of heat. Consequently, heat transfer in water by advection and conduction gives rise to uniform temperatures. The density of pure water exhibits anomalous behaviour. In ice, O atoms have 4 H atoms orientated about them tetrahedrally. These units are packed together with a hexagonal symmetry. At the freezing point, 0 °C, ice is less dense than water. Heating breaks hydrogen bonds and molecules can achieve slightly closer packing which causes the density to increase. The maximum density occurs at 4 °C, as at higher temperatures thermal expansion compensates for this compression effect. As will be discussed later, seawater differs in this respect. Thus, fresh ice floats on water which in part explains how rivers and lakes can freeze over but remain liquid at depth. With respect to other properties, water has a high surface tension which is manifest in stable droplet formation and has a relatively low molecular viscosity and therefore is quite a mobile fluid.

Water is an excellent solvent. It is extremely polar and generally can dissolve a wider range of solutes and in greater amounts than any other substance. Water has a very high dielectric constant, a measure of the solvents' ability to keep apart oppositely charged ions. The solvation characteristics of individual ions influence their behaviour in solution, *i.e.* in terms of hydration, hydrolysis, and precipitation. Although water exhibits amphoteric behaviour, electrolytic dissociation is quite small. Furthermore, dissociation gives equal ion concentrations of both H_3O^+ and OH^- and so pure water is neutral. The amphoteric behaviour enhances dissolution of introduced particulate matter through surface hydrolysis reactions.

While the concept will be considered in detail below, the term salinity (S‰) is introduced here as a measure of the salt content of sea water, expressed in units g kg^{-1}. A typical value for oceanic waters is 35 g kg^{-1}. In an oceanographic context, the most important consequence of the addition of salt to water is the

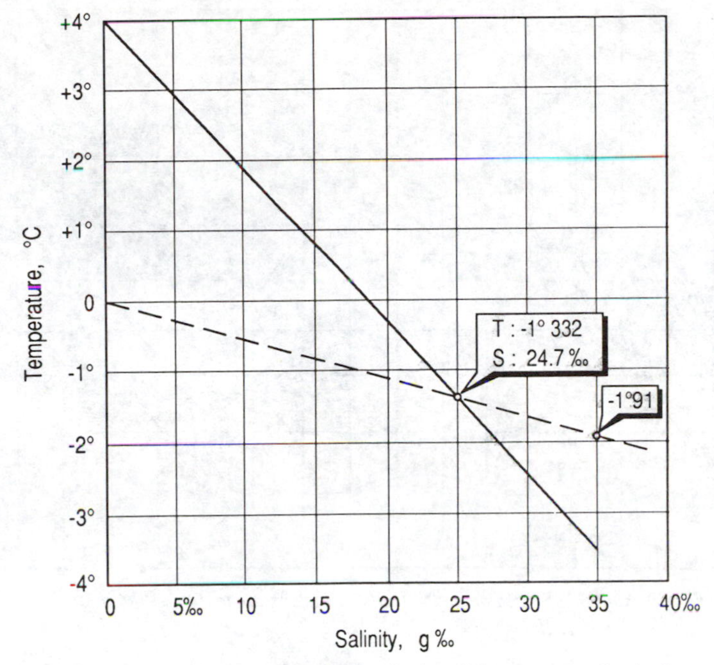

Figure 2 *The temperature of maximum density (—) and freezing point (– –) of seawater as a function of dissolved salt content (from Tchernia, 1980)*[2]

effect on density. However, many of the characteristics outlined above are also altered. The addition of electrolytes can cause a small increase in the surface tension. This effect is not generally observed in seawater due to the presence of surfactants, which decrease the surface tension and so facilitate foam formation. As illustrated in Figure 2,[2] the presence of salt does depress the temperature of maximum density and the freezing point of the solution relative to pure water. Thus, seawater with a typical salt content of 35 g kg^{-1} freezes at approximately −1.9 °C and the resulting ice is more dense than the solution. As a further consequence, the freezing process tends to produce fresh ice overlying a more concentrated brine solution. Salts can be precipitated at much lower temperatures, *i.e.* mirabilite ($Na_2SO_4.2H_2O$) at −8.2 °C and halite (NaCl) at −23 °C. Some brine inclusions and salt crystals can become incorporated into the ice.

The fundamental properties of seawater are temperature and salinity. Together with the pressure (*i.e.* depth dependent), these parameters control the density of the water. The density in turn determines the buoyancy of the water and pressure gradients. Small density differences integrated over oceanic scales cause considerable pressure gradients and result in currents.

Surface water temperatures are extremely variable, obviously influenced by location and season. The minimum temperatures found in polar latitudes are almost −2 °C. Equatorial waters can reach 30 °C. It would be expected that the

[2] P. Tchernia, 'Descriptive Regional Oceanography', Pergamon Press, Oxford, 1980, pp. 253.

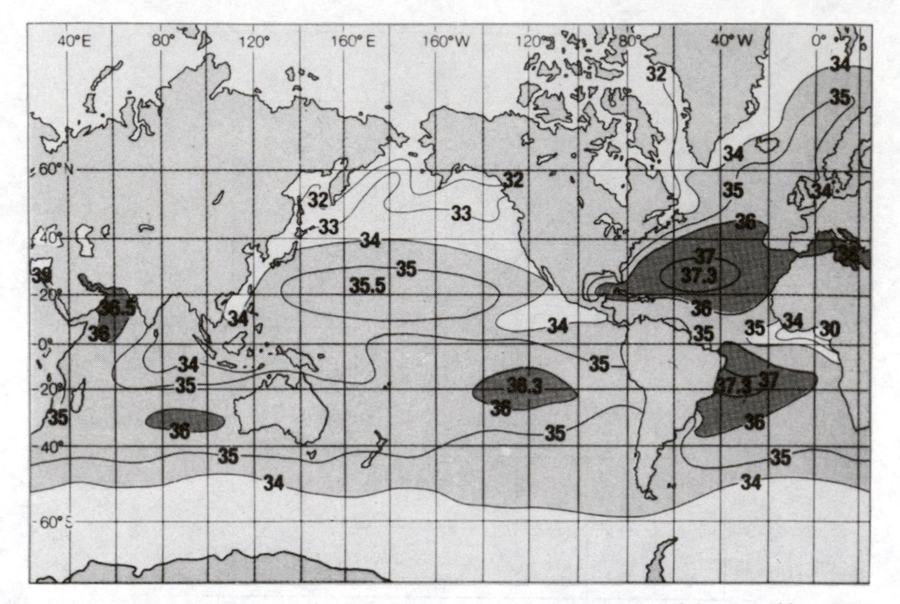

Figure 3 *The distribution of mean annual salinity in the surface waters of the ocean (from The Open University, 1989)*[3]

temperature decreases with depth; easily explained in that water density decreases with temperature and so, in the absence of other stabilizing influences, elevated temperatures at depth would produce a buoyant mass resulting in mixing. Temperature variations with depth are far from consistent. Regions in which mixing is prevalent, *i.e.* especially in the surface waters, produces a layer in which the temperature is relatively constant. The zone immediately beneath would normally exhibit a sharp change in temperature, known as the thermocline. The thermocline in the ocean extends down to about 1000 m within equatorial and temperate latitudes. It acts as an important boundary in the ocean, separating the surface and deep layers, and limiting mixing between these two reservoirs.

Below the thermocline, the temperature changes only little with depth. The temperature in seawater is non-conservative because adiabatic compression causes a slight increase in the *in situ* temperature measured at depth. For instance in the Mindanao Trench in the Pacific Ocean, the temperature at 8500 m and 10 000 m is 2.23 °C and 2.48 °C, respectively. The term potential temperature is defined to be the temperature that the water parcel would have if raised adiabatically to the ocean surface. For the examples above, the potential temperatures are 1.22 °C and 1.16 °C, respectively. Potential temperature is a conservative index.

Salinity in the surface waters in the open ocean range between 33 and 37‰ (Figure 3),[3] the main control being the balance between evaporation and

[3] The Open University, 'Seawater: its Composition, Properties and Behaviour', Open University and Pergamon Press, Oxford, 1989, pp. 165.

precipitation. The highest salinities occur in regional seas where the evaporation rate is extremely high, namely the Mediterranean Sea (38–39‰) and the Red Sea (40–41‰). Within the world ocean, the salinity is greatest in latitudes of about 20° where the evaporation exceeds precipitation. Lower salinities occur poleward as evaporation diminishes and near the equator where precipitation is very high. Local effects can be important, as evident in the vicinity of large riverine discharges which dilute the salinity. Salinity variations with depth are related to the origin of the deep waters and so will be considered in the section on oceanic circulation. A zone in which the salinity exhibits a marked gradient is known as a halocline.

Whereas the density of pure water is 1.000 g ml^{-1}, the density of sea water (S‰ = 35‰) is about 1.03 g ml^{-1}. The term 'sigma-tee', σ_t is used to denote the density (actually the specific gravity and hence a dimensionless number) of water at atmospheric pressure based on temperature and salinity *in situ*. Density increases, and so the buoyancy decreases, with an increase in σ_t. It is defined as:

$$\sigma_t = (\text{specific gravity}_{S‰,T} - 1) \times 1000$$

In a plot of temperature against salinity (a *T–S* diagram), constant σ_t appear as curved lines which denote waters of constant pressure and are known as isopycnals. A zone in which the pressure changes greatly is known as a pycnocline. Within the water column, a pycnocline therefore separates waters with distinctive temperature and salinity characteristics, generally indicative of different origins. A *T–S* diagram can also be used to estimate the properties resulting from the mixing of two water masses. As noted above, the temperature is not a conservative property, and therefore σ_t is also non-conservative. To circumvent the associated difficulties of interpretation, an analogous term known as the potential density, σ_θ, is defined on the basis of potential temperature (*i.e.* temperature if adjusted to a pressure of 1 atm.) instead of *in situ* temperature. The σ_θ is therefore a conservative index.

1.3 Salinity Concepts

Salinity is a measure of the salt content of seawater. Developments in analytical chemistry have led to an historical evolution of the salinity concept. Intrinsically it would seem to be a relatively straight forward task to measure. This is true for imprecise determinations which can be quickly performed using hand-held refractometers. The salinity affects seawater density, and thus, the impetus for high precision in salinity measurements came from physical oceanographers.

The first techniques utilized for the determination of salinity, involving the gravimetric analysis of salt left after evaporating seawater to dryness, were fraught with difficulties. Variable amounts of water of crystallization might be retained. Some salts, such as $MgCl_2$, can decompose leaving residues of uncertain composition. Other constituents, especially organic material, might be volatilized or oxidized. Overall, such methods led to considerable inconsistencies and inaccuracies.

The second set of procedures for salinity measurement made use of the observation from the *Challenger* expedition of 1872–76 that sea salt composition was apparently invariant. Hence, the total salt content could be calculated from any individual constituent, such as Cl^- which could be readily determined by titration with Ag^+. At the turn of the century, Knudsen defined salinity to be the weight in grammes of dissolved inorganic matter contained in 1 kg of seawater, after bromides and iodides were replaced by an equivalent amount of chloride and carbonate was converted to oxide. Clearly from the adopted definition, the method was not specific to Cl^- and so the term chlorinity was introduced. Chlorinity (Cl‰) is the chloride concentration in seawater, expressed as g kg^{-1}, as measured by Ag^+ titration (*i.e.* assuming Cl^- to be the only reactant). The relationship of interest was that between *S*‰ and Cl‰, given as:

$$S‰ = 1.805\text{Cl}‰ + 0.030$$

As a calibrant solution for the $AgNO_3$ titrant, Standard Seawater was prepared that had certified values for both chlorinity and salinity. This salinity–chlorinity relationship was derived on the basis of only 9 seawater samples that were somewhat atypical, and it has been redefined in recent years using a much larger set of samples representative of oceanic waters to become:

$$S‰ = 1.80655\text{Cl}‰$$

The third category of salinity methodologies was based on conductometry. These continue to be the most widely used methods, because electrical conductivity measurements can provide salinity values with a precision of ± 0.001‰. The conductivity of a solution is proportional to the salt content. High precision requires temperature control of samples and standards to within ± 0.001 °C. Standard Seawater, now also certified with respect to conductivity, provides the appropriate calibrant solution. Application of a non-specific technique like conductometry relies upon the assumption that the sea salt matrix is invariant, both spatially and temporally. Thus, the technique cannot be reliably employed in marine boundary environments where the sea water composition differs to the bulk characteristics. There are two types of procedures commonly used. Firstly, a Wheatstone Bridge circuit can be set up whereby the ratio of the resistance of unknown seawater to standard seawater balances the ratio of a fixed resistor to a variable resistor. The system uses alternating current to minimize electrode fouling. Alternatively, the conductivity can be measured by magnetic induction, in which case the sensor consists of a plastic tube containing sample seawater that links two transformers. An oscillator establishes a current in one transformer which induces current flow within the tube, the magnitude of which depends upon the salinity of the sample. This in turn induces a current in the second transformer which can then be measured. This design has been exploited for *in situ* conductivity measurements.

1.4 Oceanic Circulation

The distribution of components within the ocean is determined by both transportation and transformation processes. A brief outline of oceanic circulation is

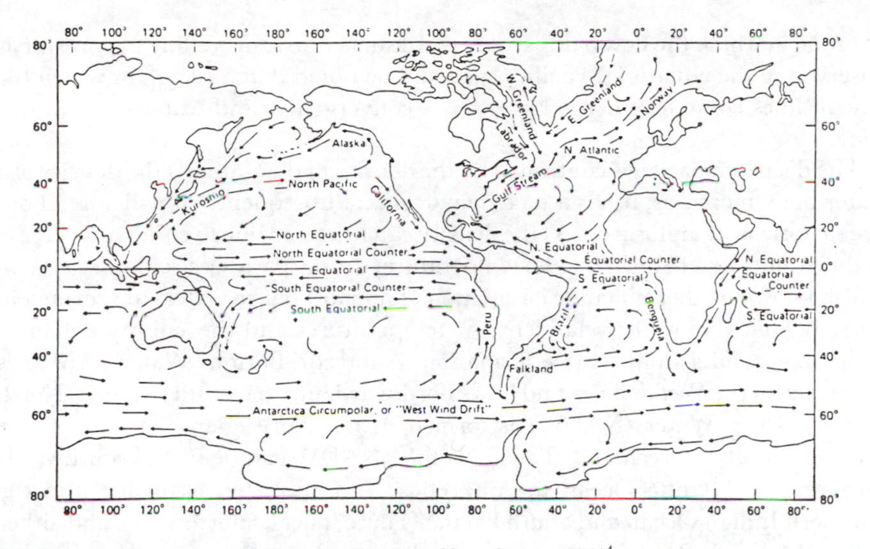

Figure 4 *The surface circulation in the ocean (from Stowe, 1979)*[4]

necessary to ascertain the relative influences. Two main flow systems must be considered. Surface circulation is established by the prevailing wind patterns but modified by Coriolis force and deep circulation is determined by gravitational forces. The Coriolis force is the acceleration due to the earth's rotation. It acts to deflect moving fluids (*i.e.* both air and water) to the right in the northern hemisphere and to the left in the southern hemisphere. The magnitude of the effect is a function of latitude, being nil at the equator and increasing polewards.

Surface oceanic circulation is depicted in Figure 4.[4] For the most part, the circulation patterns describe gyres constrained by the continental boundaries. The prevailing winds acting under the influence of Coriolis force result in clockwise and counter-clockwise flow in the northern and southern hemispheres, respectively. The flow fields are non-uniform, exhibiting faster currents along the western margins. These are manifest, for example, as the Gulf Stream, Kuroshio Current, and Brazil Current. Circulation within the Indian Ocean is exceptional in that there are distinct seasonal variations in accord with the monsoons. The absence of other continents within the immediate boundary region of Antarctica gives rise to a circumpolar current within the Southern Ocean.

The surface circulation is restricted to the upper layer influenced by the wind, typically about 100 m. However, underlying water can be transported up into this zone when horizontal advection is insufficient to maintain the superimposed flow fields. This process is called upwelling and is of considerable importance in that biochemical respiration of organic material at depth ensures that the ascending water is nutrient-rich. Upwelling occurs in the eastern oceanic boundaries where longshore winds result in the offshore transport of the surface water. Examples are found off Peru and West Africa. Similar processes cause upwelling off Arabia, but this is seasonal due to the monsoon effect. A divergence

[4] K. S. Stowe, 'Ocean Science', John Wiley and Sons, New York, 1979.

is a zone in which the flow fields separate. In such a case, upwelling may result as observed in the equatorial Pacific. It should be noted that a region in which the stream lines come together is known as a convergence, and water sinks in this zone.

The deep circulation is controlled by the density of the water. If the density of a water body increases, it has a tendency to sink. Subsequently it will spread out over a horizon of uniform σ_θ. As the density can be raised due to either an increase in the salinity or a decrease in the temperature, the deep water circulatory system is also known as thermohaline circulation. The most dense waters are formed in polar regions due to the relatively low temperatures and the salinity enhancement that results from sea ice formation. Antarctic Bottom Water (ABW) is generated in the Weddell Sea and flows northward into the South Atlantic. North Atlantic Deep Water (NADW) is formed in the Norwegian Sea and off the southern coast of Greenland. The flow of the NADW can be traced southwards through the Atlantic Ocean to Antarctica. It is diverted eastward into the Southern Indian Ocean and South Pacific. There it heads northwards and either enters the North Pacific or becomes mixed upward into the surface layer in the equatorial region. The transit time is on the order of 1000 years. As noted previously, the thermocline acts as an effective barrier against mixing of dissolved components in the ocean. Consequently, this deep water formation process in high latitudes is important because it facilitates the relatively rapid transport of material from the surface of the ocean down to great depths. The deep advection of atmospherically derived CO_2 is a pertinent example.

Intermediate waters within the water column can be formed by diverse processes. In the southern South Atlantic, the NADW overrides the more dense ABW. Antarctic Intermediate Water results from water sinking along the Antarctic Convergence (~50° S). Relatively warm, saline water exits the Mediterranean Sea at depth and can be identified as a distinctive layer within the North and South Atlantic.

2 SEAWATER COMPOSITION AND CHEMISTRY

2.1 Major Constituents

The major constituents in seawater are conventionally taken to be those elements present in typical oceanic water (35 g kg^{-1}) at concentrations greater than 1 mg kg^{-1}, excluding Si which is an important nutrient in the marine environment. The concentrations and main species of these elements are presented in Table 1.[5] One of the most significant observations from the *Challenger* expedition of 1872–76 was that these major components existed in constant relative amounts. As already explained, this feature was exploited for salinity determinations. Inter-element ratios are generally constant, and often expressed as a ratio to Cl‰ as shown in Table 1. This implies conservative behaviour, with concentrations

[5] D. Dyrssen and M. Wedborg, 'The Sea', ed. E. Goldberg, John Wiley and Sons, New York, 1974, p. 181.

Table 1 *Chemical species and concentrations of the major elements in seawater (based on Dyrssen and Wedborg, 1974)*[5]

Element	*Chemical species*	*Concentration for* S = 35‰ (mol dm^{-3})	(g kg^{-1})	*Ratio to chlorinity* (Cl = 19.374‰)
Na	Na^+	4.79×10^{-1}	10.77	5.56×10^{-1}
Mg	Mg^{2+}	5.44×10^{-2}	1.29	6.66×10^{-2}
Ca	Ca^{2+}	1.05×10^{-2}	0.4123	2.13×10^{-2}
K	K^+	1.05×10^{-2}	0.3991	2.06×10^{-2}
Sr	Sr^{2+}	9.51×10^{-5}	0.00814	4.20×10^{-4}
Cl	Cl^-	5.59×10^{-1}	19.353	9.99×10^{-1}
S	SO_4^{2-}, $NaSO_4^-$	2.89×10^{-2}	0.905	4.67×10^{-2}
C (inorganic)	HCO_3^-, CO_3^{2-}	2.35×10^{-3}	0.276	1.42×10^{-2}
Br	Br^-	8.62×10^{-4}	0.673	3.47×10^{-3}
B	$B(OH)_3$, $B(OH)_4^-$	4.21×10^{-4}	0.0445	2.30×10^{-3}
F	F^-, MgF^+	7.51×10^{-5}	0.00139	7.17×10^{-5}

depending solely upon mixing processes, and indeed salinity itself is a conservative index.

Because of this behaviour, individual seawater constituents can be utilized for source apportionment studies. For instance, an enrichment factor (EF) for a substance X is defined as:

$$EF_X = \frac{(X/\mathrm{Na}^+)_{\mathrm{sample}}}{(X/\mathrm{Na}^+)_{\mathrm{seawater}}}$$

An enrichment factor of 1 indicates that the substance exists in comparable relative amounts in the sample and in seawater, a good indication of a marine origin. If $EF_X > 1$, then it is enriched with respect to seawater. Conversely, depletion is signified when values $EF_X < 1$. Another example of this utility can be found in examining the geochemical cycle of sulfur. Concentrations of SO_4^{2-} and Na^+ in ice cores and marine aerosols exhibit a SO_4^{2-}:Na^+ greater than that observed in seawater. This excess can be readily calculated and is known as non-sea salt sulfate (NSSS). While there was much speculation as to the origin of this material, it is now recognized that NSSS is derived from dimethyl sulfide (DMS) of marine biogenic origin.

Not all the major constituents consistently exhibit conservative behaviour in the ocean. The most notable departures occur in deep waters where Ca^{2+} and HCO_3^- exhibit anomalously high concentrations due to the dissolution of calcite. The concept of relative constant composition does not apply in a number of atypical environments, generally associated with boundary regions. Inter-element ratios for major constituents can be quite different in estuaries and in the vicinity of hydrothermal vents. Obviously, these are not solutions of sea salt (with the implication that accuracy of salinity measurements by chemical and conductometric means are limited).

Table 2 *The residence time and speciation of some elements in the ocean (based on Brewer, 1975 and Bruland, 1983)*[6,7]

Element	*Principal species*	*Concentration* ($mol\ l^{-1}$)	*Residence time* (years)
Li	Li^+	2.6×10^{-5}	2.3×10^6
B	$B(OH)_3$, $B(OH)_4^-$	4.1×10^{-4}	1.4×10^7
F	F^-, MgF^+	6.8×10^{-5}	5.2×10^5
Na	Na^+	4.68×10^{-1}	6.8×10^7
Mg	Mg^{2+}	5.32×10^{-2}	1.2×10^7
Al	$Al(OH)_4^-$, $Al(OH)_3$	7.4×10^{-8}	1.0×10^2
Si	$Si(OH)_4$	7.1×10^{-5}	1.8×10^4
P	HPO_4^{2-}, PO_4^{3-}, $MgHPO_4$	2×10^{-6}	1.8×10^5
Cl	Cl^-	5.46×10^{-1}	1×10^8
K	K^+	1.02×10^{-2}	7×10^6
Ca	Ca^{2+}	1.02×10^{-2}	1×10^6
Sc	$Sc(OH)_3$	1.3×10^{-11}	4×10^4
Ti	$Ti(OH)_4$	2×10^{-8}	1.3×10^4
V	$H_2VO_4^-$, HVO_4^{2-}, $NaVO_4^-$	5×10^{-8}	8×10^4
Cr	CrO_4^{2-}, $NaCrO_4^-$	5.7×10^{-9}	6×10^3
Mn	Mn^{2+}, $MnCl^+$	3.6×10^{-9}	1×10^4
Fe	$Fe(OH)_3$	3.5×10^{-8}	2×10^2
Co	Co^{2+}, $CoCO_3$, $CoCl^+$	8×10^{-10}	3×10^4
Ni	Ni^{2+}, $NiCO_3$, $NiCl^+$	2.8×10^{-8}	9×10^4
Cu	$CuCO_3$, $CuOH^+$, Cu^{2+}	8×10^{-9}	2×10^4
Zn	$ZnOH^+$, Zn^{2+}, $ZnCO_3$	7.6×10^{-8}	2×10^4
Br	Br^-	8.4×10^{-4}	1×10^8
Sr	Sr^{2+}	9.1×10^{-5}	4×10^6
Ba	Ba^{2+}	1.5×10^{-7}	4×10^4
La	La^{3+}, $LaCO_3^+$, $LaCl^{2+}$	2×10^{-11}	6×10^2
Hg	$HgCl_4^{2-}$	1.5×10^{-10}	8×10^4
Pb	$PbCO_3$, $Pb(CO_3)_2^{2-}$, $PbCl^+$	2×10^{-10}	4×10^2
Th	$Th(OH)_4$	4×10^{-11}	2×10^2
U	$UO_2(CO_3)_2^{4-}$	1.4×10^{-8}	3×10^6

The residence times for some elements are presented in Table 2.[6,7] The major constituents generally have long residence times. The residence time is a crude measure of a constituent's reactivity in the reservoir. As noted by Raiswell *et al.* (1980),[8] the aqueous behaviour and rank ordering can be appreciated simply in terms of the ionic potential given by the ratio of electronic charge to ionic radius (Z/r). Elements with $Z/r < 3$ are strongly cationic. The positive charge density is relatively diffuse, but sufficient to attract and orientate an envelope of water molecules forming a hydrated cation. As the ionic potential increases, the force of attraction towards the water similarly rises to the extent

[6] P. Brewer, 'Chemical Oceanography' ed. J. P. Riley and G. Skirrow, Academic Press, London, 2nd Ed., 1975, Vol. 1., p. 415.
[7] K. W. Bruland, 'Chemical Oceanography' ed. J. P. Riley and R. Chester, Academic Press, London, 1983, Vol. 8., p. 157.
[8] R. W. Raiswell, P. Brimblecombe, D. L. Dent, and P. S. Liss, 'Environmental Chemistry', Edward Arnold, London, 1982, pp. 184.

that one oxygen–hydrogen bond in the molecule breaks. This causes the solution pH to fall and metal hydroxides to form. Neutral hydroxides tend to be relatively insoluble and so precipitate. However, in the more extreme case for which $Z/r >$ 12, the attraction toward the oxygen is so great that both bonds in the associated water molecules are broken. The reaction product is an oxyanion, usually quite soluble because of the associated anionic charge. Thus in seawater, those elements (Al, Fe) having a tendency to form insoluble hydroxides have short residence times. This is also true for elements that exist preferentially as neutral oxides (Mn, Ti). Hydrated cations (Na^+, Ca^{2+}) and strongly anionic species (Cl^-, Br^-, $UO_2(CO_3)_2^{4-}$) have long residence times. This treatment is, of course, somewhat of an over-simplification ignoring the rather significant role that biological organisms play in nutrient and trace element chemistry.

2.2 Dissolved Gases

2.2.1 Gas Solubility and Air–Sea Exchange Processes. The ocean contains a vast array of dissolved gases. Some of the gases such as Ar and chlorofluorocarbons behave conservatively and can be utilized as tracers for water mass movements and ventilation rates. Equilibrium processes at the air–sea interface generally lead to saturation, and then this concentration remains unchanged once the water sinks. Thus, the gas concentration is characteristic of the lost contact with the atmosphere. Deep waters usually contain no CFCs as such anthropogenic compounds have only a recent history of use. There are several important non-conservative gases, which exhibit wide variations in concentration due to biological activity. O_2 determines the redox potential in seawater and CO_2 buffers the ocean at pH 8. Ocean–atmosphere exchange processes for gases such as CO_2 and dimethyl sulfide may play an important role in climate change.

The solubility of gases in water is affected by both temperature and salinity. Empirical relationships can be found elsewhere (Weiss, 1970; Kester, 1975).[9,10] The trends are such that gas solubility increases with a decrease in temperature or an increase in salinity. The changes in solubility are non-linear and differ dramatically for various gases. Figure 5 depicts the solubility of several gases as a function of temperature.[11]

At the ocean–atmosphere interface, exchange of gases occurs to achieve an equilibrium between the two systems and as a result, gases become saturated. However, super-saturation can be achieved by several mechanisms. Firstly, bubbles that form from white cap activity can be entrained and dissolved at depth. Secondly as evident from Figure 5, if two water masses that have been equilibrated at different temperatures are mixed, then the resulting water body would be super-saturated. Thirdly, gases that are produced *in situ* by biological

[9] R. F. Weiss, *Deep-Sea Res.*, 1970, **17**, 721.
[10] D. Kester, 'Chemical Oceanography' ed. J. P. Riley and G. Skirrow, Academic Press, London, 2nd Ed., 1975, Vol. 1., p. 497.
[11] W. S. Broecker and T. H. Peng, 'Tracers in the Sea', Lamont-Doherty Geological Observatory, Palisades, 1982, pp. 690.

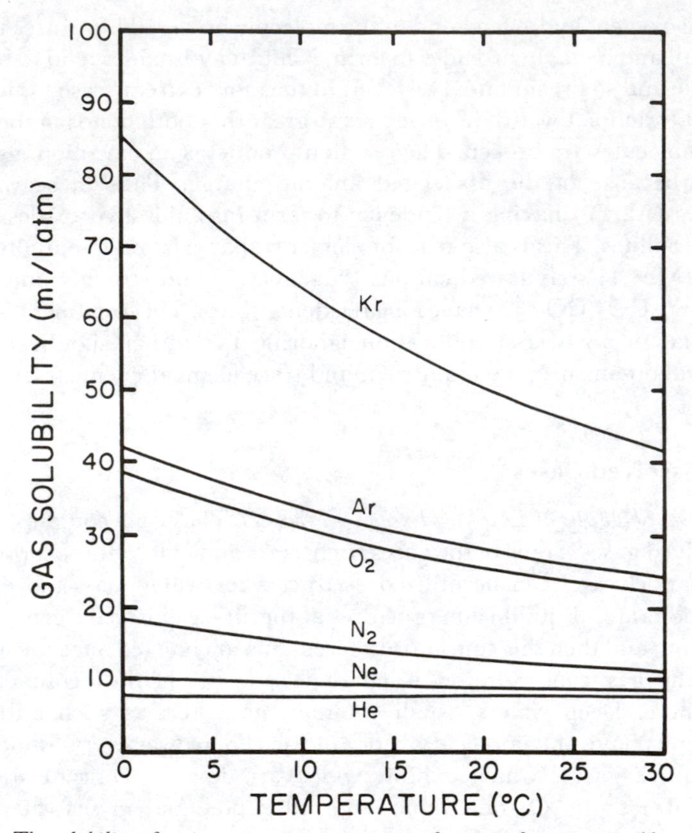

Figure 5 *The solubility of various gases in seawater as a function of temperature (from Broecker and Peng, 1982)*[11]

activity may become super-saturated, particularly when evasion to the atmosphere is not favoured.

The gas solubility for a water body in equilibrium with the overlying air mass can be expressed in several ways. It is convenient to consider Henry's Law which states:

$$H = c_a c_w^{-1}$$

where c_a and c_w refer to the concentration of a gas in air and water, respectively. As discussed by Liss (1983),[12] air–sea exchange occurs when a concentration gradient exists (*i.e.* $\Delta C = c_a H^{-1} - c_w$) and the magnitude of the consequential flux, F, is given as:

$$F = K\,\Delta C$$

where the proportionality constant, K has dimensions of velocity and so is generally referred to as the transfer velocity (see also Chapter 6).

[12] P. S. Liss, in 'Air–Sea Exchange of Gases and Particles', ed. P. S. Liss and W. G. N. Slinn, Reidel, Dordrecht, 1983, p. 241.

Table 3 *The net global fluxes of some trace gases across the air/sea interface (from Chester, 1990)*[1]

Gas	*Global air–sea direction*[a]	*Flux magnitude*[b]
CH_4	+	10^{12}–10^{13}
Man-made CO_2	−	6×10^{15}
N_2O	+	6×10^{12}
CCl_4	−	10^{10}
CCl_4	=	~0
CCl_3F	−	5×10^9
CCl_3F	=	~0
CH_3I	+	3–13 $\times 10^{11}$
CO	+	$100 \pm 90 \times 10^{12}$
H_2	+	$4 \pm 2 \times 10^{12}$
Hg	+	$\sim 2 \times 10^9$

[a] + indicates sea → air flux direction, − indicates air → sea flux direction, = indicates no net flux
[b] Units in g (of the compound, where applicable) year^{-1}

Air–sea exchange processes are thus dependent upon the concentration gradient and the transfer velocity. The transfer velocity is not a constant, but rather depends upon several physical parameters such as temperature, wind speed, and wave state. The exchange can also be attenuated by the presence of a surface film or slick. Alternatively, the exchange can be facilitated by bubble formation. The concentration gradient determines the direction of the flux, into or out of the ocean. Net global fluxes for some gases are presented in Table 3. The atmosphere serves as the source of material for conservative gases, especially those of anthropogenic origin, but several gases produced *in situ* by biological activity evade from the ocean.

2.2.2 *Oxygen.* Oxygen is a non-conservative gas and a typical oceanic profile is shown in Figure 6.[13] The concentration varies throughout the water column, its distribution being greatly influenced by biological activity. The generalized chemical equation for carbon fixation is often given as:

$$nCO_2 + nH_2O \rightleftharpoons (CH_2O)_n + nO_2$$

During photosynthesis this reaction proceeds to the right, thereby producing organic material, as designated by $(CH_2O)_n$, and O_2. The surface waters become equilibrated with respect to atmospheric O_2. However, it is possible for these waters to become super-saturated with O_2 during periods of intense photosynthetic activity. Respiration occurs as the above reaction proceeds to the left and O_2 is consumed. Photosynthesis is obviously restricted to the upper ocean (in the photic zone) and generally exceeds respiration. However, the relative importance of the two processes changes with depth. The oxygen compensation depth is the

[13] K. W. Bruland, *Earth Planet. Sci. Lett.*, 1980, **47**, 176.

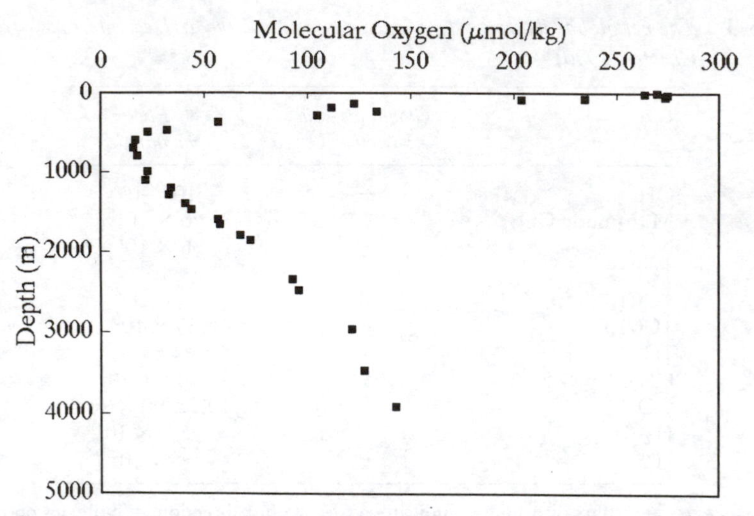

Figure 6 *A profile of molecular oxygen in the North Pacific Ocean (data from Bruland, 1980)*[13]

horizon in the water column at which the rate of O_2 production by photosynthesis equals the rate of respiratory O_2 oxidation.

Below the photic zone, O_2 is utilized in chemical and biochemical oxidation reactions. As evident in Figure 6, the concentration diminishes with depth to develop an oxygen minimum zone. Thereafter, the O_2 concentration in deeper waters begins to increase because these waters originated from polar regions. They were cold and in equilibrium with atmospheric gases at the time of sinking, but subsequently lost little of the dissolved O_2 because the flux of organic material to deep waters is relatively small.

The dissolved O_2 content of seawater has a significant control on the redox potential, often designated in environmental chemistry by pe. This is defined with reference to electron activity in an analogous fashion to pH and thus:

$$\mathrm{p}e = -\log\{e^-\}$$

The relationship between pe and the more familiar electrode potential E or E_H is:

$$\mathrm{p}e = \frac{F}{2.303RT}E$$

and for the standard state:

$$\mathrm{p}e^{\ominus} = \frac{F}{2.303RT}E^{\ominus}$$

Whereas a high value of pe indicates oxidizing conditions, a low value signifies reducing conditions. Oxygen plays a role via the reaction:

$$O_2 + 4H^+ + 4e^- \rightleftharpoons 2H_2O$$

At 20 °C, $K = 10^{8.1}$, and so water of pH = 8.1 in equilibrium with atmospheric O_2 ($pO_2 = 0.21$ atm) has pe = 12.5. This conforms to surface conditions but the *pe* decreases as the O_2 content diminishes with depth. The oxygen minimum is particularly well developed beneath the highly productive surface waters of the eastern tropical Pacific Ocean. There is a large flux of organic material to depth and subsequent oxidation. The O_2 becomes sufficiently depleted that the resulting low redox conditions causes NO_3^- to be reduced to NO_2^-. In some special environments characterized by poor flushing and hence stagnant water, the oxygen may be completely utilized producing anoxic conditions. Such regions represent atypical marine environments where reducing conditions prevail.

O_2 can be used as a tracer to help identify the origin of water masses. The warm, saline intrusion in to the Atlantic Ocean from the Mediterranean Sea is relatively O_2 deficient. Alternatively, the waters descending from polar regions have elevated O_2 concentrations.

2.2.3 Carbon Dioxide and Alkalinity. Marine chemists sometimes adopt activity conventions quite different to those traditionally used in chemistry. It is useful to preface a discussion on the carbon dioxide–calcium carbonate system in the oceans with a brief outline of pH scales. Although originally introduced in terms of ion concentration, today the definition of pH is based on hydrogen ion activity and is:

$$\text{pH} = -\log a_{\text{H}}$$

where a_{H} refers to the relative hydrogen ion activity (*i.e.* dimensionless as is pH). Defined using concentration scales, the pH can be:

$$\text{pH} = -\log\,(c_{\text{H}}\gamma_{\text{H}}/c^\circ)$$

or

$$\text{pH} = -\log\,(m_{\text{H}}\gamma_{\text{H}}/m^\circ)$$

where c_{H} and m_{H} represent molar and molal concentrations, c° and m° are the respective standard state conditions (1 mol l^{-1} and 1 mol kg^{-1}), and γ_{H} is the appropriate activity coefficient. Obviously γ_{H} differs in these two expressions as $c^\circ \neq m^\circ$. However, different activity scales may also be used. In the *infinite dilution activity scale*, $\gamma_{\text{H}} \rightarrow 1$ as the concentration of hydrogen ions and all other ions approach 0. For analyses, pH meters are calibrated using dilute buffers prepared in pure water. Alternatively, in the *constant ionic medium activity scale*, $\gamma_{\text{H}} \rightarrow 1$ as the concentration of hydrogen ions approaches 0 while all other components are maintained at some constant level. Calibrant buffers are prepared in solutions of constant ionic composition, and in marine chemistry this is often a solution of synthetic seawater. While these two methodologies are equally justifiable from a thermodynamic point of view, it is important to appreciate that pH scales so-defined are quite different. As a further consequence, the absolute values for dissociation constants also differ. Therefore, this apparently esoteric distinction between activity scales is fundamental to a quantitative appreciation of marine acid–base chemistry.

The biogeochemical cycle of inorganic carbon in the ocean is extremely complicated. It involves the transfer of gaseous carbon dioxide from the atmosphere into solution. Not only is this a reactive gas that readily undergoes hydration in the ocean, but also it is fixed as organic material by marine phytoplankton. Several marine organisms utilize calcium carbonate to form shells. Surface waters are super-saturated with respect to aragonite and calcite, forms of $CaCO_3$, but precipitation is limited to coastal lagoons such as found in the Bahamas. Inorganic carbon removed from surface waters by biological processes can be regenerated at depth. This may result through either the respiratory oxidation of organic material or the dissolution of shells in the under-saturated waters found at depth. Nonetheless, calcitic oozes of biogenic origin constitute a major component in marine sediments. Finally, the inorganic carbon equilibrium is responsible for buffering seawater at a pH near 8 on time scales of centuries to millennia.

There are several equilibria to be considered. Firstly, CO_2 is exchanged across the air–sea interface:

$$CO_{2g} \rightleftharpoons CO_{2aq}$$

The equilibrium process obeys Henry's Law, but the dissolved CO_2 reacts rapidly with water to become hydrated as:

$$CO_2 + H_2O \rightleftharpoons H_2CO_3$$

Relative to the exchange process, the hydration reaction forming carbonic acid occurs quite quickly. This means that the concentration of dissolved CO_2 is extremely low. The two processes can be considered together as:

$$CO_{2g} + H_2O \rightleftharpoons H_2CO_3$$

The equilibrium constant is then:

$$K_{CO_2} = \frac{\{H_2CO_3\}}{p_{CO_2}\{H_2O\}}$$

where p_{CO_2} is the partial pressure of CO_2 in the marine troposphere. Carbonic acid undergoes dissociation:

$$H_2CO_3 + H_2O \rightleftharpoons H_3O^+ + HCO_3^-$$

$$HCO_3^- + H_2O \rightleftharpoons H_3O^+ + CO_3^{2-}$$

for which the first and second dissociation constants (using $\{H^+\}$ rather than $\{H_3O^+\}$) are:

$$K_1 = \frac{\{H^+\}\{HCO_3^-\}}{\{H_2CO_3\}}$$

$$K_2 = \frac{\{H^+\}\{CO_3^{2-}\}}{\{HCO_3^-\}}$$

The hydrogen ion activity can be established with a pH meter. However as discussed above, this measurement must be operationally defined. On the other

Table 4 *Equilibrium constants for the carbonate system (adapted from Stumm and Morgan, 1981)*[14]

T (°C)	pK_{CO_2}*	pK'_1	pK'_2
0	1.19	6.15	9.40
5	1.27	6.11	9.34
10	1.34	6.08	9.28
15	1.41	6.05	9.23
20	1.47	6.02	9.17
25	1.53	6.00	9.10
30	1.58	5.98	9.02

*Equilibrium constants given here are based on the constant ionic medium scale with a reference state of seawater having a chlorinity of 19‰ (see Section 5.3.1 for further details)

hand, the individual ion activities of bicarbonate and carbonate ions cannot be measured. Instead, ion concentrations are determined, as outlined below, by titration. Accordingly, the equilibrium constants are redefined in terms of concentrations. These are then known as apparent rather than true equilibrium constants and distinguished using a prime notation. It must be appreciated that apparent equilibrium constants are not invariant, but rather are affected by temperature, pressure, salinity, and, as outlined previously, the pH scale adopted. The apparent dissociation constants are:

$$K'_1 = \frac{\{H^+\}[HCO_3^-]}{[H_2CO_3]}$$

$$K'_2 = \frac{\{H^+\}[CO_3^{2-}]}{[HCO_3^-]}$$

It should be noted that whereas ion activities are denoted by curly brackets { }, concentrations are designated by square brackets [ ]. Analogous to the pH, pK conventionally refers to $-\log K$. Numerical values for the constants pK_{CO_2}, pK'_1, and pK'_2 based on a constant ionic medium scale (*i.e.* seawater with chlorinity = 19‰) are given in Table 4.[14] This provides sufficient information to calculate the speciation of carbonic acid in seawater at a given temperature as a function of pH. This is shown for carbonic acid at 15 °C in seawater equilibrated with atmospheric carbon dioxide ($\sim 3.5 \times 10^{-4}$ atm) in Figure 7. While there are several confounding features, the pH of seawater can be considered to be buffered by the bicarbonate:carbonate pair. The pH is generally about 8, but is sensitive to the concentration ratio $[HCO_3^-]:[CO_3^{2-}]$ as evident from rearranging the expression for K'_2 to become:

[14] W. Stumm and J. J. Morgan, 'Aquatic Chemistry', John Wiley and Sons, New York, 2nd Ed., 1981, pp. 780.

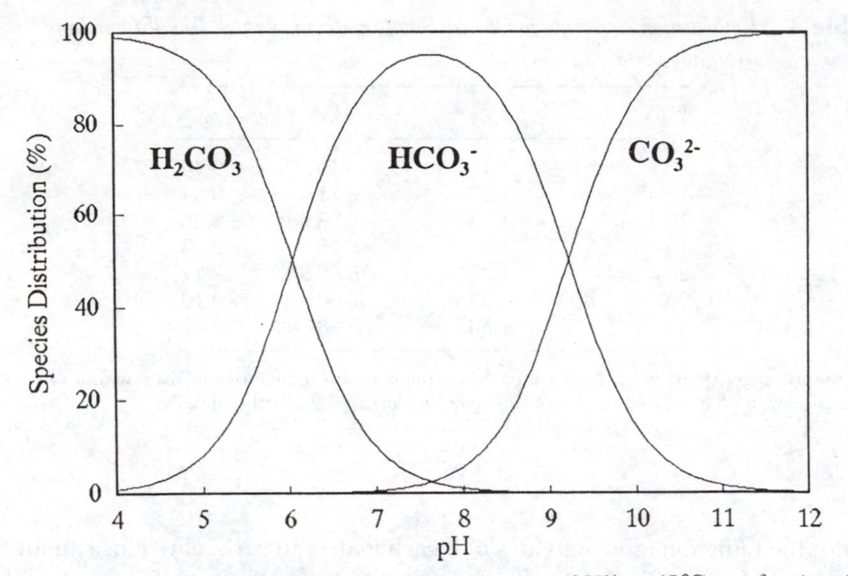

Figure 7 *The distribution of carbonic acid species in seawater of 35‰ at 15 °C as a function of pH*

$$\{H^+\} = K_2' \frac{[HCO_3^-]}{[CO_3^{2-}]}$$

To understand the response of the oceanic CO_2 system to *in situ* biological activity or enhanced CO_2 concentrations in the atmosphere, it is necessary to consider in more detail the factors influencing the inorganic carbon cycle. Two useful parameters can be introduced. Firstly, the total concentration of inorganic carbon, ΣCO_2, in seawater is:

$$\Sigma CO_2 = [CO_2] + [H_2CO_3] + [HCO_3^-] + [CO_3^{2-}]$$

The first term is negligible and as evident in Figure 7, the major species at pH 8 are HCO_3^- and CO_3^{2-}.

Alkalinity is defined as a measure of the proton deficit in solution, and not to be confused with basicity. Alkalinity is operationally defined by titration with a strong acid to the carbonic acid end point. This is known as the titration alkalinity (TA). Seawater contains weak acids other than bicarbonate and carbonate and so TA is given as:

$$TA = [HCO_3^-] + 2[CO_3^{2-}] + [B(OH)_4^-] + [OH^-] - [H^+]$$

The influence of $[OH^-]$ and $[H^+]$ on the TA are small and can generally be ignored. The borate contributes about 3% of the TA, and if not determined independently, can be estimated from the apparent boric acid dissociation constants and the salinity, relying upon the relative constancy of composition of sea salt. This would give the carbonate alkalinity (CA):

$$CA = [HCO_3^-] + 2[CO_3^{2-}]$$

Considering the dissociation constants above, this can be alternatively expressed as:

$$CA = \frac{K_1'[H_2CO_3]}{\{H^+\}} + \frac{2K_1'K_2'[H_2CO_3]}{\{H^+\}^2}$$

or

$$CA = \frac{K_1'K_{CO_2}p_{CO_2}}{\{H^+\}} + \frac{2K_1'K_2'K_{CO_2}p_{CO_2}}{\{H^+\}^2}$$

This equation can be rearranged to give the following quadratic expression that can be solved for the pH:

$$CA\{H^+\}^2 - K_1'K_{CO_2}p_{CO_2}\{H^+\} - 2K_1'K_2'K_{CO_2}p_{CO_2} = 0$$

Using constants from Table 4 at 15 °C and assuming a typical seawater alkalinity of 2·30 meq l^{-1}, the pH of seawater in equilibrium with atmospheric $CO_2 = 3.5 \times 10^{-4}$ atm is calculated to be 8.20.

Consider now the effect of altering the pCO_2 in the water. The alkalinity should not change in response to variations in CO_2 alone because the hydration and dissociation reactions give rise to equivalent amounts of H^+ and anions. CO_2 can be lost by evasion to the atmosphere (a process usually confined to equatorial regions) or by photosynthesis. This causes the ΣCO_2 to diminish and the pH to rise, an effect that can be quite dramatic in tidal rock pools in which pH may then rise to 9. Conversely, an increase in pCO_2, either by invasion from the atmosphere or release following respiration, prompts an increase in ΣCO_2 and a fall in pH. Thus, the depth profiles of pH would mimic that of O_2 but the ΣCO_2 would exhibit a maximum at the oxygen minimum.

There are further confounding influences, in particular concerning $CaCO_3$. $CaCO_3$ in the form of aragonite or calcite is used by many organisms to form calcareous shells (tests). The shells sink and dissolve when the organism dies. The solubility is governed by:

$$Ca^{2+} + CO_3^{2-} \rightleftharpoons CaCO_{3s}$$

Surface waters are super-saturated with respect to $CaCO_3$, but precipitation rarely occurs, possibly due to an inhibitory effect by Mg^{2+} forming ion pairs with CO_3^{2-}. The solubility of $CaCO_3$ increases with depth, due to both a pressure effect and the decrease in pH following respiratory release of CO_2, with the result that the shells dissolve. This behaviour not only increases the alkalinity but accounts for the non-conservative nature of Ca^{2+} and inorganic carbon in deep waters. The depth at which appreciable dissolution begins is known as the lysocline. At a greater depth, designated as the carbonate compensation depth (CCD), no calcareous material is preserved. The depths of the lysocline and CCD are influenced by the flux of organic material and shells, and tend to be deeper under high productivity zones.

The CO_2 and $CaCO_3$ systems are coupled in that the pH buffering in the ocean is due to the reaction:

$$CO_2 + H_2O + CaCO_3 \rightleftharpoons Ca^{2+} + 2HCO_3^-$$

In addition to the effects noted previously, an input of CO_2 promotes the dissolution of $CaCO_3$. The reaction does not proceed to the right without constraint, but rather meets a resistance given by the Revelle factor, R:

$$R = \frac{dp_{CO_2}/p_{CO_2}}{d\Sigma CO_2/\Sigma CO_2}$$

This value is approximately 10, indicating that the ocean is relatively well buffered against changes in ΣCO_2 in response to variations in atmospheric pCO_2. Although the ocean does respond to an increase in the atmospheric burden of CO_2, the time scales involved are quite considerable. The surface layer can become equilibrated on the order of decades, but as the thermocline inhibits exchange into deep waters, the equilibration of the ocean as a whole with the atmosphere proceeds on the order of centuries. The ventilation of deep water by descending water masses in polar latitudes only partly accelerates the overall process.

2.2.4 Dimethyl Sulfide and Climatic Implications. The Gaia hypothesis states that the biosphere regulates the global environment for self-interest (Lovelock, 1979).[15] This presupposes that controls, perhaps poorly understood or unknown, serve to maintain the present *status quo*. Charlson *et al.* (1987)[16] have made use of this hypothesis to suggest that biogenic production of dimethyl sulfide (DMS) and the consequent formation of atmospheric cloud condensation nuclei (CCN, *i.e.* small particles onto which water can condense) acts as a feedback mechanism to counteract the global warming that results from elevated greenhouse gas concentrations in the atmosphere. The cycle is illustrated in Figure 8. Global warming, with concurrent warming of the ocean surface, leads to enhanced phytoplankton productivity. This promotes the production and evasion to the atmosphere of DMS. The DMS undergoes oxidation to form CCN which promote cloud formation and increase the planetary albedo (*i.e.* reflectivity with respect to sunlight) thereby causing a cooling effect. From a biogeochemical perspective, the two key features are the controls on the biogenic production of DMS and the formation of CCN following aerial oxidation of DMS. These will be considered in more detail below. With respect to the physics, the most important aspects of the proposed climate control mechanism are that the enhancement of the albedo is due to an increase in the number and type of CCN, and that this CCN production occurs in the marine boundary layer. The albedo of calm sea water is very low (~2%) in comparison to vegetated regions (10–25%), deserts (~35%), and snow covered surfaces (~90%).

That biological processes within the oceans act as a major source of reduced sulfur gases is well established (Andreae, 1986).[17] Of particular importance is

[15] J. Lovelock, 'Gaia. A New Look at Life on Earth', Oxford University Press, Oxford, 1979.
[16] R. J. Charlson, J. E. Lovelock, M. O. Andreae, and S. G. Warren, *Nature (London)*, 1987, **326**, 655.
[17] M. O. Andreae, 'The Role of Air–Sea Exchange in Geochemical Cycling', ed. P. Buat-Ménard, Reidel, Dordrecht, 1986, p. 331.

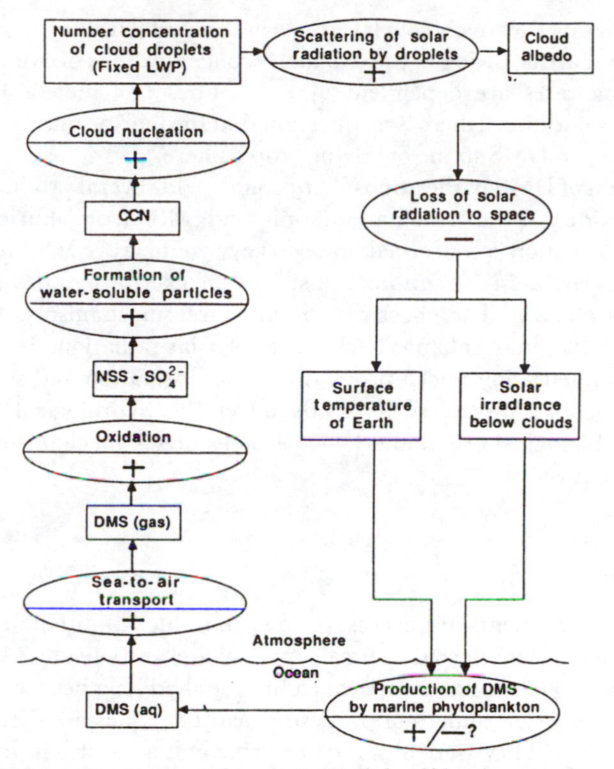

Figure 8 *The possible climatic influence of dimethyl sulfide of marine biogenic origin (from Charlson* et al., *1987)*[16]

the generation of DMS. Surface concentrations, approximately in the range of 0.7 to 17.8 nmol l^{-1}, exhibit wide temporal and geographic variations. Oceanic distributions indicate that DMS is produced within the photic zone, that is consistent with a phytoplankton source, but DMS concentrations are poorly correlated with normal indicators of primary productivity. While *Phaeocystis* and *Coccolithoporidae* have been identified as important DMS producers, there is still uncertainty as to the full potential for biological DMS formation. With respect to climate modification, questions remain as to the biological response to global warming. For the model of Charlson *et al.* (1987)[16] to hold, organisms might either increase DMS formation or biological succession could change in such a way as to favour DMS producers. Thus, marine biogenic source strengths and the controlling factors remain important unresolved issues in sulfur biogeochemistry.

DMS concentrations in the remote marine troposphere vary in the range 0.03 to 32 nmol m^{-3}. Not surprisingly and as with the seawater concentrations, considerable temporal and geographic variations occur. Furthermore, atmospheric DMS concentrations, exhibit diurnal variations, with night-time maxima and afternoon minima, consistent with a photochemical sink. Oxidation involves HO free radicals during the day, while a reaction with NO_3 radicals may be important at night. Relatively low levels are associated with air masses derived from continental areas, owing to the enhanced concentrations of oxidants. While oceanic venting rates are dependent upon a number of meteorological and oceanographic conditions, there is no question that the marine photic zone acts as the major source of DMS to the overlying troposphere.

The oxidation of DMS in the atmosphere could yield several products, namely dimethyl sulfoxide (DMSO), methanesulfonic acid (MSA), or sulfuric acid. In so far as aerosol formation is concerned, the two key products are MSA and H_2SO_4. Atmospheric particles in the sub-micron size range exert a significant influence on the earth's climate. The effect can be via three mechanisms. Firstly, the particles themselves may enhance backscatter of solar radiation. Secondly, they act as cloud condensation nuclei promoting cloud formation and so increasing the earth's albedo. Thirdly, such clouds affect the hydrological cycle. The evidence for a biological origin and hence a biofeedback mechanism for global warming is circumstantial.

2.3 Nutrients

Although several elements are necessary to sustain life, traditionally in oceanography, 'nutrients' has referred to nitrate, phosphate, and silicate. The rationale for this classification was that analytical techniques had long been available that allowed the precise determination of these constituents despite their relatively low concentrations. They were observed to behave in a consistent manner, but quite differently to the major constituents in seawater.

The distributions of these three nutrients are determined by biological activity. Nitrate and phosphate become incorporated into the soft parts of organisms. As evident in the modified carbon fixation equation (Redfield, 1958)[18] given below:

$$106CO_2 + 16NO_3^- + HPO_4^{2-} + 122H_2O + 18H^+ \rightleftharpoons C_{106}H_{263}O_{110}N_{16}P + 138O_2$$

the uptake of these nutrients into tissues occurs in constant relative amounts. The ratio (*i.e.* Redfield ratio) for C:N:P is 106:16:1. Silicate is utilized by some organisms, particularly diatoms (phytoplankton) and radiolaria (zooplankton), to form siliceous skeletons. Such skeletons consist of an amorphous, hydrated silicate, $SiO_2 \cdot nH_2O$, often called opaline silica.

A depth profile of nitrate, phosphate, and silicate in the North Pacific Ocean is presented in Figure 9. Nutrients behave much like ΣCO_2 and are removed in the surface layer, especially in the photic zone. Thus, concentrations can become quite low, and indeed sufficiently low to limit further photosynthetic carbon fixation. Following the death of the organisms, they sink. The highest concentrations are found where respiration and bacterial decomposition of the falling organic material are greatest, that is at the oxygen minimum. As a consequence of

[18] A. C. Redfield, *Am. J. Sci.*, 1958, **46**, 205.

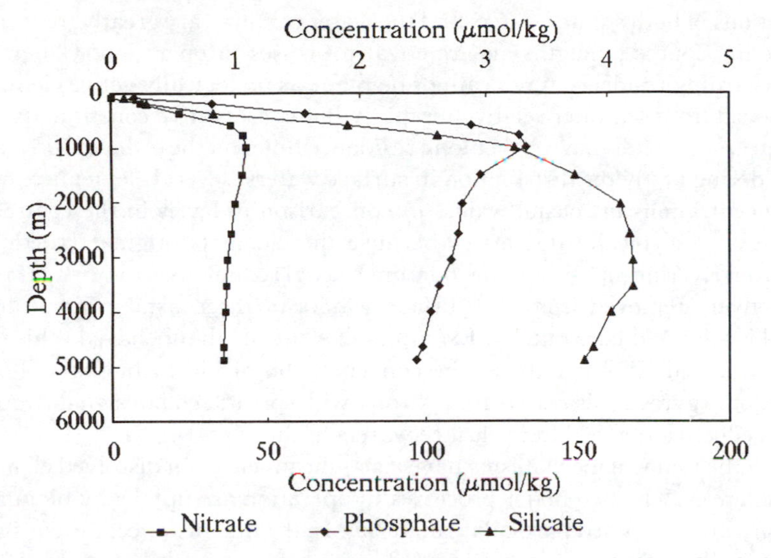

Figure 9 *The depth distribution of nitrate, phosphate, and silicate in the North Pacific Ocean (data from Bruland 1980)*[13]

these processes, the nutrients, including silica, are regenerated. Thus, the concentrations in deep waters are much greater than those observed in the surface waters, and account for the fertilizing effect of upwelling. It should be noted that the siliceous remains behave differently than the calcareous shells discussed previously. The oceans are everywhere under-saturated with respect to silica. Its solubility exhibits no pronounced variation with depth and there is no horizon analogous to the CCD (see Section 2.2.3). Silica is preserved to any great extent only in deep-sea sediments associated with the highly productive upwelling zones in the ocean.

2.4 Trace Elements

Trace elements in seawater are taken to be those that are present in quantities less than 1 mg l^{-1}, excluding the nutrient constituents. The distribution and behaviour of minor elements have been reviewed in the light of data which conforms to an oceanographically consistent manner (Bruland, 1983; Chester, 1990).[7,1] Analytical difficulties are readily comprehensible when it is appreciated that the concentration for some of these elements can be extremely low, *i.e.* a few pg l^{-1} for platinum group metals (Goldberg *et al.*, 1988).[19] Some trace elements, such as Cs^+, behave conservatively and therefore absolute concentrations depend upon salinity. More often, elements are non-conservative and their

[19] E. Goldberg, M. Koide, J. S. Yang, and K. K. Bertine, 'Metal Speciation: Theory, Analysis and Applications', ed. J. R. Kramer and H. E. Allen, Lewis Publishers, Chelsea, 1988, p. 201.

distributions in both surface waters and the water column vary greatly, reflecting the differing source strengths and removal processes in operation. Generalizations regarding residence times cannot be made as biologically active elements are removed from seawater relatively rapidly but conservative constituents and platinum group metals have rather long residence times on the order of 10^5 years.

Considering firstly the distribution in surface waters, several elements exhibit high concentrations in coastal waters in comparison to levels in the centres of oceanic gyres. Typically this arises because the elements originate predominantly from riverine inputs or diffusion from coastal sediments. However, as they are effectively removed from the surface waters in the coastal regions, little material is advected horizontally. Examples of elements that behave in this way are Cd, Cu, and Ni. In contrast, the concentration of Pb, including ^{210}Pb, is greater in the gyres. This results from strong widespread aeolian signal coupled with less effective removal from surface waters in the gyres.

Clearly the removal mechanisms have a significant effect on dissolved elemental abundances. The two major processes in operation are uptake by biota and scavenging by suspended particulate material. In the first instance, the constituent mimics the behaviour of nutrients. This is evident in the metal : nutrient correlations for Cd:P and Zn:Si (Figure 10).

No consistent pattern for depth profiles exist. Conservative elements trend with salinity variations, provided they have no significant submarine sources. Non-conservative elements may exhibit peak concentrations at different depths in oxygenated waters as:

(1) surface enrichment;
(2) maxima at O_2 minimum;
(3) mid-depth maximum not associated with O_2 minimum;
(4) bottom enrichment.

The criteria for an element such as Pb to exhibit a maximum concentration in surface waters are that the only significant input must be at the surface (aeolian supply) and it must be effectively removed from the water column. Constituents such as As, Ba, Cd, Ni, and Zn exhibit nutrient type behaviour. Those elements (*e.g.* Cd) associated with the soft parts of the organism are strongly correlated with phosphate and are regenerated at the O_2 minimum. Elements (*e.g.* Zn) associated with the skeletal material may exhibit a smooth increasing trend with depth. The third case pertains to elements, notably Mn, that have a substantial input from hydrothermal waters. These are released into the ocean from spreading ridges. Ocean topography is such once these waters are advected away from such regions to the abyssal plains, they are then found at some intermediate depth. Bottom enrichment is observed for elements (*e.g.* Mn) that are remobilized from marine sediments. The behaviour of Al combines features outlined above, resulting in a mid-depth minimum concentration. Surface enrichment evident in mid (41 °N) but not high (~60 °N) latitudes in the North Atlantic results from the solubilization of aeolian material. Removal occurs via scavenging and incorporation into siliceous skeletal material. Subsequent regeneration by shell dissolution increases deep water Al levels.

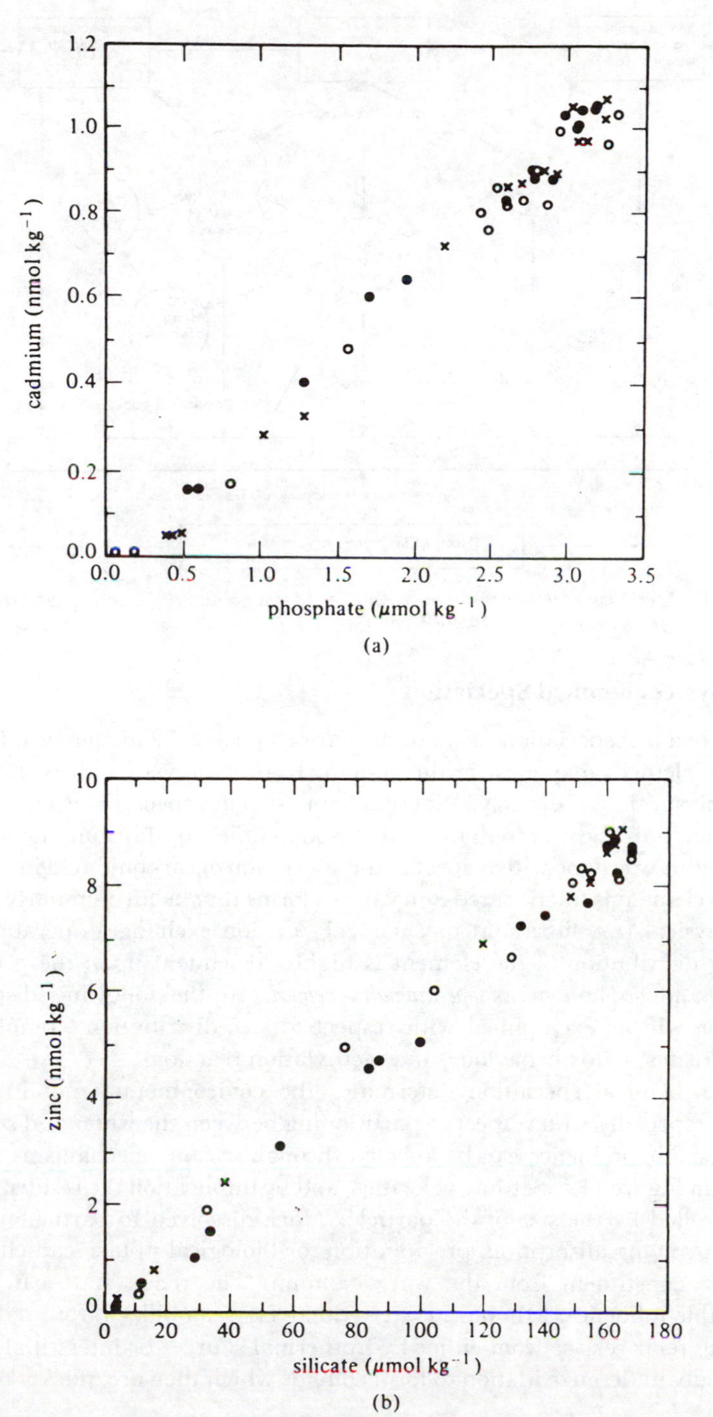

Figure 10 *Metal : nutrient correlations in the North Pacific for Cd:P and Zn:Si (from Bruland, 1980)*[13]

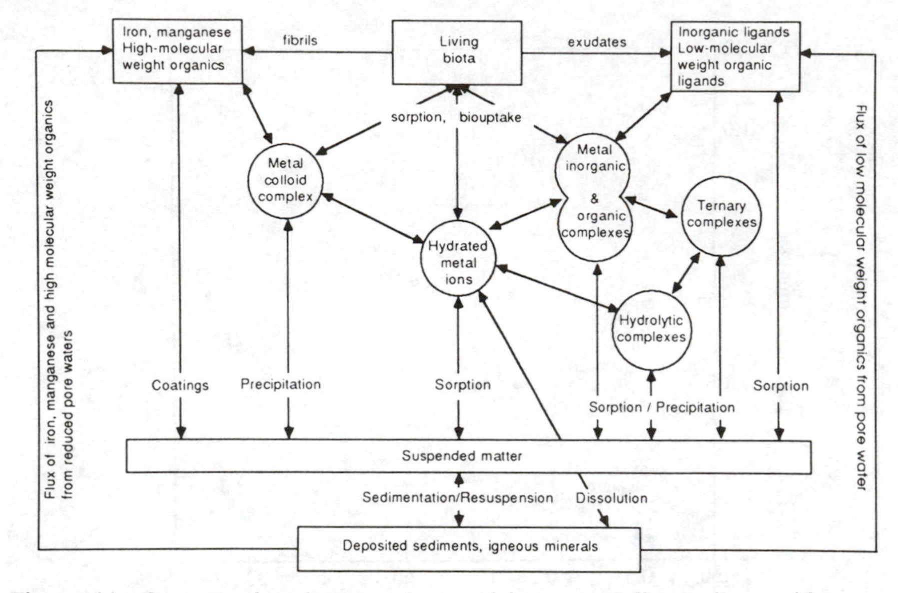

Figure 11 *Speciation of metal ions in seawater and the main controlling mechanisms (from Öhman and Sjöberg, 1988)*[20]

2.5 Physicochemical Speciation

Physicochemical speciation refers to the various physical and chemical forms in which an element may exist in the system. In oceanic waters, it is difficult to determine speciation directly. Whereas some specific species can be analysed, others can only be inferred from thermodynamic equilibrium models. For instance, this was done with respect to the speciation of carbonic acid in Figure 7. Often an element is fractionated into various forms that behave similarly under a given physical (*e.g.* filtration) or chemical (*e.g.* ion exchange) operation. The resulting distribution of the element is highly dependent upon the procedure utilized, and so known as *operationally defined*. In the following discussion, speciation will be exemplified with respect to size distribution, complexation characteristics, redox behaviour, and methylation reactions.

Physicochemical speciation determines the environmental mobility of an element, especially with respect to partitioning between the water and sediment reservoirs. The influence can be affected through various mechanisms as summarized in Figure 11.[20] Settling velocities, and by implication the residence time, are controlled by the size of the particle. Thus, dissolved to particulate interactions involving adsorption, precipitation, or biological uptake can effectively remove a constituent from the water column. The redox state can have a comparable influence. Mn and Fe are reductively remobilized from sediments. Following their release from either hydrothermal sources or interstitial waters, they rapidly undergo oxidation to form colloids which then are quickly removed

[20] L. Öhman and S. Sjöberg, 'Metal Speciation: Theory, Analysis and Applications', ed. J. R. Kramer and H. E. Allen, Lewis Publishers, Chelsea, 1988, p. 1.

from the water column. Speciation also determines the bioavailability of an element for marine organisms. It is generally accepted that the uptake of trace elements is limited to free ions and some types of lipid-soluble organic complexes. This is important in that high concentrations of some elements are toxic but, of course, some substances are essential for life.

An element can exist in natural waters in a range of forms that exhibit a size distribution as indicated in Table 5.[21] Entities in true solution include ions, ion pairs, complexes, and a wide range of organic molecules which can span several size categories. At the smallest extreme are ions, which exist in solution with a co-ordinated sphere of water molecules as discussed previously. Na^+, K^+, and Cl^- exist predominantly as free hydrated ions. Collisions of oppositely charged ions occur due to electrostatic attraction, and can produce an ion pair. An ion pair is the transient coupling of a cation and anion, in which each retain its co-ordinated water envelope. While impossible to measure directly, concentrations can be calculated with knowledge of the ion activities and stability constant. For the formation of the ion pair $NaSO_4^-$ via:

$$Na^+ + SO_4^{2-} \rightleftharpoons NaSO_4^-$$

the stability constant is defined as:

$$K = \frac{\{NaSO_4^-\}}{\{Na^+\}\{SO_4^{2-}\}}$$

Ion pair formation is important for Ca^{2+}, Mg^{2+}, SO_4^{2-}, and HCO_3^-.

Table 5 *The size distribution of trace metal species in natural waters (from de Mora and Harrison, 1984)*[21]

Size range	*Metal species*	*Examples*	*Phase state*
<1 nm	Free metal ions	Mn^{2+}, Cd^{2+}	Soluble
1–10 nm	Inorganic ion pairs, inorganic complexes, low molecular weight organic complexes	$NiCl^+$, $HgCl_4^{2-}$, Zn-fulvates	Soluble
10–100 nm	High molecular weight organic complexes	Pb-humates	Colloidal
100–1000 nm	Metal species adsorbed onto inorganic colloids, metals associated with detritus	Co–MnO_2, Pb–$Fe(OH)_3$	Particulate
>1000 nm	Metals adsorbed into living cells, metals adsorbed onto, or incorporated into mineral solids and precipitates	Cu-clays, $PbCO_{3s}$	Particulate

[21] S. J. de Mora and R. M. Harrison, 'Hazard Assessment of Chemicals, Current Developments', ed. J. Saxena, Academic Press, London, Vol. 3, 1984, p. 1.

If the attraction is sufficiently great, a dehydration reaction can occur leading to covalent bonding. A complex comprises a central metal ion sharing a pair of electrons donated by another constituent, termed a ligand, acting as a Lewis base. The metal ion and ligand share a single water envelope. Ligands can be neutral (*e.g.* H_2O) or anionic species (*e.g.* Cl^-, HCO_3^-). A metal ion can co-ordinate with one or more ligands, which need not be the same chemical entity. Alternatively, the cation can share more than one electron pair with a given ligand thereby forming a ring structure. This type of complex is known as a chelate. They have enhanced stability, largely due to the entropy effect of releasing large numbers of molecules from the water envelopes.

Complex formation is an equilibrium process, and ignoring charges for the general case of a metal, M and ligand, L complex formation occurs as:

$$M + L = ML$$

for which the formation constant, K_1 is given by:

$$K_1 = \frac{\{ML\}}{\{M\}\{L\}}$$

A second ligand may then be coordinated as:

$$ML + L = ML_2$$

$$K_2 = \frac{\{ML_2\}}{\{ML\}\{L\}}$$

The equilibrium constant for ML_2 can be expressed solely in terms of the activities of the M and L:

$$\beta_2 = \frac{\{ML_2\}}{\{M\}\{L\}^2}$$

where β_2 is the product of K_1K_2 and is known as the stability constant. The case can be extended to include n ligands as:

$$\beta_n = \frac{\{ML_n\}}{\{M\}\{L\}^n}$$

The speciation of constituents in solution can be calculated if the individual ion activities and stability constants are known. This information is relatively well known with respect to the major constituents in seawater, but not for trace elements. There are some important confounding variables that create considerable difficulties in speciation modelling. Firstly, it is assumed that equilibrium is achieved, that is, neither biological interference nor kinetic effects prevent this state. Secondly, seawater contains appreciable amounts (at least in relation to the trace metals) of organic matter. However, the composition of the organic matrix, the number of available binding sites, and the appropriate stability constants are

[22] K. Hunter, *Geochim. Cosmochim. Acta*, 1983, **47**, 467.

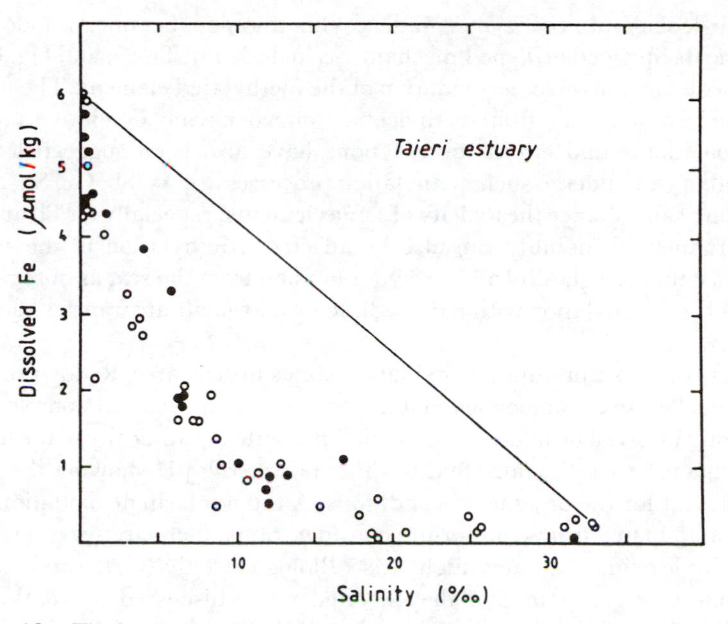

Figure 12 *The distribution of dissolved Fe* versus *S‰ in the Taieri Estuary, New Zealand;* ● *21 October 1980,* ○ *4 December 1980 (from Hunter 1983)*[22]

poorly known. Nonetheless, speciation models can include estimates of these parameters. Organic material can form chelates with relatively high stability constants and dramatically decrease the free ion activity of both necessary and toxic trace elements. Organisms may make use of such chemistry, producing compounds either to sequester metals in limited supply or to detoxify contaminants. Thirdly, surface adsorption onto colloids or suspended particles may remove them from solution. As with organic matter, an exact understanding of the complexation characteristics of the suspended particles is not available, but approximations can also be incorporated into speciation models.

Elements may be present in a variety of phases other than in true solution. Colloidal formation is particularly important for elements such as Fe and Mn which form amorphous oxyhydroxides with very great complexation characteristics. Adsorption processes cannot be ignored in biogeochemical cycling. Particles tend to have a much shorter residence time in the water column than do dissolved constituents. Scavenging of trace components by falling particles accelerates deposition to the sediment sink. Interactions between phases lead to non-conservative behaviour, as evident for dissolved Fe in Figure 12.[22]

Several elements in seawater may undergo alkylation via either chemical or biological mechanisms (Craig, 1986).[23] Type I mechanisms involve methyl radical or carbonium ion transfer and no formal change in the oxidation state of the acceptor element. The incoming methyl group may be derived for example

[23] P. Craig, 'Organometallic Compounds in the Environment', ed. P. S. Craig, Longman, Harlow, 1986, p. 1.

from methylcobalamin coenzyme, *S*-adenosylmethionine, betaine, or iodomethane. Elements involved in Type I mechanisms include Pb, Tl, Se, and Hg. Other reaction sequences involve the oxidation of the methylated element. The methyl source can be a carbanion from methylcobalamin coenzyme. Oxidative addition from iodomethane and enzymatic reactions have also been suggested. Some elements that can undergo such methylation processes are As, Sb, Ge, Sn, and S. Methylation can enhance the toxicity of some elements, especially for Pb and Hg. The environmental mobility can also be affected. Methylation in the surface waters can enhance volatility and so favour evasion from the sea, as observed for S, Se, and Hg. Methylation within the sediments may facilitate transfer back into overlying waters.

Elements may exhibit multiple oxidation states in seawater. Redox processes can be modelled in an analogous manner to the ion pairing and complexation outlined previously. The information is often presented graphically in the form of a predominance area diagram, that is a plot of p*e* *versus* pH showing the major species present for the designated conditions. Although a single oxidation state might be anticipated from equilibrium considerations, there are several ways in which multiple oxidation states might arise. Biological activity can produce non-equilibrium species, as evident in the alkylated metals discussed above. Whereas Mn^{IV} and Cu^{II} might be expected, photochemical processes in the surface waters can lead to the formation of significant amounts of Mn^{II} and Cu^{I}.

Goldberg *et al.* (1988)[19] has presented impressive information (given that seawater concentrations are as low as 1.5 pg l^{-1} for Ir and 2 pg l^{-1} for Ru) on the speciation, including redox state, of platinum group metals as a means of interpreting distributions in seawater and marine sediments. Pt and Pd are stabilized in seawater as tetrachloro-divalent anions. Their relative abundance of 5 Pt:1 Pd agreeing with a factor of five difference in β_4. Pt is enriched in ferromanganese nodules following oxidation to a stable (IV) state, behaviour not observed for Pd. Rh exists predominantly in the heptavalent state, but accumulates in reducing sediments as lower valence sulfides. Au and Ag are present predominantly in solution as monovalent forms. Ag, but not Au, accumulates in anoxic coastal sediments.

3 SUSPENDED PARTICLES AND MARINE SEDIMENTS

3.1 Description of Sediments and Sedimentary Components

The sediments represent the major sink for material in the oceans. The main pathway to the sediments is the deposition of suspended particles. Such particles may be only in transit through the ocean from a continental origin or formed *in situ*. Sinking particles can scavenge material from solution. Accordingly, this section introduces the components found in marine sediments, but emphasizes processes that occur within the water column that lead to the formation and alteration of the deposited material.

Marine sediments cover the ocean floor to a thickness averaging 500 m. The deposition rates vary with topography. The rate may be several mm per year in

Table 6 *The four categories of marine sedimentary components with examples of mineral phases (from Harrison* et al., *1991)*[24]

Classification	*Mineral example*	*Chemical formula*
Lithogenous	Quartz	SiO_2
	Microcline	$KAlSi_3O_8$
	Kaolinite	$Al_4Si_4O_{10}(OH)_8$
	Montmorillonite	$Al_4Si_8O_{20}(OH)_4 \cdot nH_2O$
	Illite	$K_2Al_4(Si, Al)_8O_{20}(OH)_4$
	Chlorite	$(Mg, Fe^{2+})_{10}Al_2(Si, Al)_8O_{20}(OH, F)_{16}$
Hydrogenous	Fe–Mn minerals	$FeO(OH)–MnO_2$
	Carbonate fluoroapatite	$Ca_5(PO_4)_{3-x}(CO_3)_xF_{1+x}$
	Barite	$BaSO_4$
	Pyrite	FeS_2
	Aragonite	$CaCO_3$
	Dolomite	$CaMg(CO_3)_2$
Biogenous	Calcite	$CaCO_3$
	Aragonite	$CaCO_3$
	Opaline silica	$SiO_2 \cdot nH_2O$
	Apatite	$Ca_5(F, Cl)(PO_4)_3$
	Barite	$BaSO_4$
	Organic matter	
Cosmogenous	Cosmic spherules	
	Meteoric dusts	

nearshore shelf regions, but is only 0.2 to 7.5 mm per 1000 years on the abyssal plains. Oceanic crustal material is formed along spreading ridges and moves outwards eventually to be lost in subduction zones, the major trenches in the ocean. As a consequence of this continual movement, the sediments on the sea floor are no older than Jurassic, about 166 million years.

The formation of marine sediments depends upon chemical, biological, geological, and physical influences. There are four distinct processes that can be readily identified. Firstly, the source of the material obviously is important. This is generally the basis on which sediment components are classified and will be considered below in more detail. Secondly, the material and its distribution on the ocean floor are influenced by its transportational history, both to and within the ocean. Thirdly, there is the deposition process which must include particle formation and alteration in the water column. Finally, the sediments may be altered after deposition, a process known as diagenesis. Of particular importance are reactions leading to changes in the redox state of the sediments.

The components in marine sediments are classified according to origin. Examples are given in Table 6.[24] Lithogenous (or terrigenous) material comes from the continents as a result of weathering processes. The relative contribution of lithogenous material to the sediments will depend upon proximity to the

[24] R. M. Harrison, S. J. de Mora, S. Rapsomanikis, and W. R. Johnston, 'Introductory Chemistry for the Environmental Sciences', Cambridge University Press, Cambridge, 1991, 354 pp.

continent and the source strength of material derived elsewhere. The most important components in the lithogenous fraction are quartz and the clay minerals (kaolinite, illite, montmorillonite, and chlorite). The distribution of the clay minerals varies considerably. Illite and montmorillonite tend to be ubiquitous in nature, although the latter has a secondary origin associated with submarine volcanic activity. Kaolinite typifies intense weathering observed in tropical and desert conditions. Therefore, it is relatively enriched in equatorial regions. On the other hand, chlorite is indicative of the high latitude regimes where little chemical weathering occurs. The lithogenous components tend to be inert in the water column and represent detrital deposition. Nonetheless, the particle surfaces can be important sites for adsorption of organic material and trace elements.

Hydrogenous components, also known as chemogenous or halmeic material, are those formed within the water column. This may comprise primary material formed directly from seawater upon exceeding a given solubility product, termed authigenic precipitation. The best known example of authigenic material are ferromanganese nodules found throughout the oceans. Alternatively, secondary material may be formed as components of continental or volcanic origin become altered by low temperature reactions in seawater, a mechanism known as halmyrolysis. Halmyrolysis reactions occur in the estuarine environment, being essentially an extension of chemical weathering of lithogenous components. Such processes continue at the sediment–water interface. Accordingly, there are considerable overlaps between the terms weathering, halmyrolysis, and diagenesis. Owing to the importance surface chemistry has on the final composition, authigenic precipitation and halmyrolysis are considered further in Section 3.2.

Biogenous (or biotic) material is produced by the fixation of mineral phases by marine organisms. The most important phases are calcite and opaline silica, although aragonite and magnesian calcite are also deposited. As indicated in Table 7,[25] several plants and animals are involved, but the planktonic organisms are the most important with respect to the world ocean. The source strength depends upon the species composition and productivity of the overlying oceanic waters. For instance, siliceous oozes are found in polar latitudes (diatoms) and along the equator (radiolaria). The relative contribution of biogenous material to the sediments depends upon its dilution by material from other sources and the extent to which the material is dissolved in seawater. As noted previously, both calcareous and siliceous skeletons are subject to considerable dissolution in the water column and at the sediment–water interface.

There are two sources that give rise to minor components in the marine sediments. Cosmogenous material is that derived from an extra-terrestrial source. Such material tends to be small (*i.e.* <0.5 mm) black micrometeorites or cosmic spherules. The composition is either magnetite or a silicate matrix including magnetite. They are ubiquitous but scarce, with relative contributions to the sediments decreasing with an increase in sedimentation rate. Finally, there

[25] E. K. Berner and R. A. Berner, 'The Global Water Cycle', Prentice-Hall, New Jersey, 1987, 397 pp.

Table 7 *Quantitatively important plants and animals that secrete calcite, aragonite, Mg-calcite, and opaline silica (from Berner and Berner, 1987)*[25]

Mineral	*Plants*	*Animals*
Calcite	Coccolithophorids[a]	Foraminifera[a] Molluscs Bryozoans
Aragonite	Green algae	Molluscs Corals Pteropods[a] Bryozoans
Mg-calcite	Coralline (red) algae	Benthic foraminifera Echinoderms Serpulids (tubes)
Opaline silica	Diatoms[a]	Radiolaria[a] Sponges

[a] Planktonic organisms

are anthropogenic components, notably heavy metals and Sn, which can have a significant influence on sediments in coastal environments.

As noted in Section 1.1, the principal modes of transport of particulate material to the ocean is by rivers or via the atmosphere. Within the oceans, distribution is further affected by the oceanic currents. There are two additional mechanisms, namely turbidity currents and ice-rafting, which can transport substances, including relatively coarse material, to the deep-sea. Ice-rafting is confined to polar latitudes, 40 °N and 55 °S at the present time.

3.2 Surface Chemistry of Particles

3.2.1 Surface Charge. Particles in seawater tend to exhibit a negative surface charge. There are several mechanisms by which this might arise. Firstly, the negative charge can result from crystal defects (*i.e.* vacant cation positions) or cation substitution. Clay minerals are layered structures of octahedral AlO_6 and tetrahedral SiO_4. Either substitution of Mg^{II} and Fe^{II} for the Al^{III} in octahedral sites or replacement of Si^{IV} in tetrahedral location by Al^{III} can cause a net negative charge. Secondly, a surface charge can result from the differential dissolution of an electrolytic salt such as barite ($BaSO_4$). A charge will develop whenever the rate of dissolution of cations and anions differs. Thirdly, organic material can be negatively charged due to the dissociation of acidic functional groups.

Adsorption processes can also lead to the development of a negatively charged particle surface. One example is the specific adsorption of anionic organic compounds onto the surfaces of particles. Another mechanism relates to the acid–base behaviour of oxides in suspension. Metal oxides (most commonly Fe, Mn)

Table 8 *The point of zero charge (PZC) for some mineral phases (from Stumm and Morgan, 1981)*[14]

Mineral	pH_{PZC}
α-Al_2O_3	9.1
α-$Al(OH)_3$	5.0
γ-AlOOH	8.2
CuO	9.5
Fe_3O_4	6.5
α-FeOOH	7.8
γ-Fe_2O_3	6.7
'$Fe(OH)_3$' (amorph)	8.5
MgO	12.4
δ-MnO_2	2.8
β-MnO_2	7.2
SiO_2	2.0
$ZrSiO_4$	5
Feldspars	2–2.4
Kaolinite	4.6
Montmorillonite	2.5
Albite	2.0
Chrysotile	>12

and clay minerals have frayed edges resulting from broken metal–oxygen bonds. The surfaces can be hydrolysed and exhibit amphoteric behaviour:

$$\text{—X—O}_s^- + H_{aq}^+ \rightleftharpoons \text{—X—OH}_s$$

$$\text{—X—OH}_s + H_{aq}^+ \rightleftharpoons \text{—X—OH}_{2s}^+$$

The hydroxide surface exhibits a different charge depending upon the pH. Cations other than H^+ can act as the potential determining ion. The point of zero charge (PZC) is the negative log of the activity at which the surface exhibits no net surface charge. At the PZC:

$$[\text{—X—O}_s^-] = [\text{—X—OH}_{2s}^+]$$

The PZC for some mineral solids found in natural waters are shown in Table 8. Clearly, the extent to which such surfaces can adsorb metal cations will be dependent upon the pH of the solution. At the pH typical of seawater, most of the surfaces indicated in Table 8 would be negatively charged and would readily adsorb metal cations.

3.2.2 Adsorption Processes. Physical or non-specific adsorption involves relatively weak attractive forces, such as electrostatic attraction and van der Waals forces. Adsorbed species retain their co-ordinated sphere of water, and hence cannot approach the surface closer than the radius of the hydrated ion. Adsorption is favoured by ions having a high charge density, *i.e.* trivalent ions in preference to univalent ones. Additionally, an entropy effect promotes the physical adsorption of polymeric species, such as Al and Fe oxides, because a large number of monomeric species are displaced.

Specific adsorption, or chemisorption, involves greater forces of attraction than physical adsorption. As hydrogen bonding or π-orbital interactions are utilized, the adsorbed species lose their hydrated spheres and can approach the surface as close as the ionic radius. Whereas multilayer adsorption is possible in physical adsorption, chemisorption is necessarily limited to monolayer coverage.

As outlined previously, hydrated oxide surfaces have sites that are either negatively charged or readily deprotonated. The oxygen atoms tend to be available for bond formation. This is favoured with respect to transition metals. Several mechanisms are possible. An incoming metal ion, M^{z+}, may eliminate an H^+ ion as:

$$\text{—X—O—H} + M^{z+} \rightleftharpoons \text{—X—O—}M^{(z-1)+} + H^+$$

Alternatively, two or more H^+ ions may be displaced, thereby forming a chelate as:

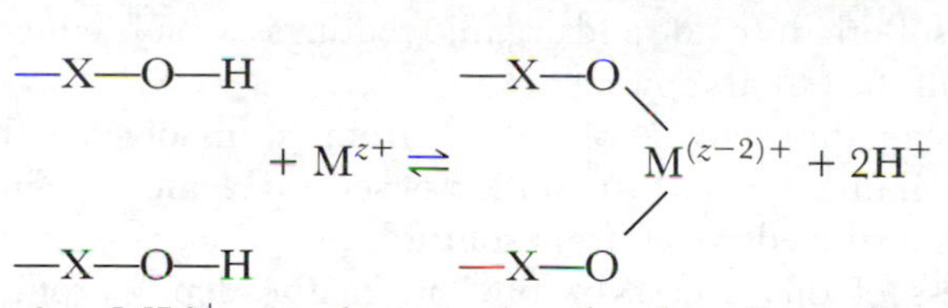

A metal complex, ML_n^{z+}, may be complexed rather than a free ion by displacement of one or more H^+ ions in a manner analogous to the reactions above. Also, the metal complex might eliminate a hydroxide group giving rise to a metal–metal bond as:

$$\text{—X—O—H} + ML_n^{z+} \rightleftharpoons \text{—X—}ML_n^{(z-1)+} + OH^-$$

It should be noted that all of these reactions are equilibria for which an appropriate equilibrium constant can be defined and measured. This data can be incorporated into the speciation models discussed in Section 2.5.

3.2.3 Ion Exchange Reactions. Both mineral particles and particulate organic material can take up cations from solution and release an equivalent amount of another cation into solution. This process is termed cation exchange and the cation exchange capacity (CEC) for a given phase is a measure of the number of exchange sites present per 100 g of material. This is operationally defined by the uptake of ammonium ions from 1 mol l^{-1} ammonium acetate at pH 7. The specific surface area and CEC are given in Table 9 for several sorption active materials.[26]

There are several factors that influence the affinity of cations towards a given surface. Firstly, the surface coverage will increase as a function of the cation concentration. Secondly, the affinity for the exchange site is enhanced as the oxidation state increases. Finally, the higher the charge density of the hydrated

[26] U. Förstner and G. T. W. Wittman, 'Metal Pollution in the Aquatic Environment', Springer-Verlag, Berlin, 2nd Edn., 1981, pp. 486.

cation, the greater its affinity for the exchange site. In order of increasing charge density, the Group I and II cations are :

$$Ba < Sr < Ca < Mg < Cs < Rb < K < Na < Li$$

3.2.4 Role of Surface Chemistry in Biogeochemical Cycling. Reactions at the aqueous–particle interface have several consequences for material in the marine environment, from the estuarine zone to the deep-sea sediment–water boundary. Within the estuary, suspended particles experience a dramatic change in the composition and concentration of dissolved salts. A number of halmyrolysates can be formed. Clay minerals undergo cation exchange as Mg^{2+} and Na^{+} replace Ca^{2+} and K^{+}. Alternatively, montmorillonite may take up K^{+} becoming transformed to illite. Hydrogenous components, in particular Mn and Fe oxides, may be precipitated onto the surfaces of suspended particles. The particulate material generally accumulates organic coatings within estuaries, which together with an increase in the ionic strength of the surrounding solution, leads to the formation of stable colloids. Both the oxide and organic coatings can subsequently scavenge other elements in the estuary.

Within the ocean, the exchange of material from the dissolved to the suspended particulate state influences the distribution of several elements. This scavenging process removes dissolved metals from solution and accelerates their deposition. The effectiveness of this process is obvious in the depth profiles of metals, especially those of the surface enrichment type. Furthermore, the removal can be expressed in terms of a deep water scavenging residence time as indicated in Table 10.

The scavenging mechanism can be particularly effective in the sediment boundary region. The resuspension of fine sediments generates a very large surface area for adsorption and ion exchange processes. Within the immediate vicinity of hydrothermal springs, reduced species of Mn and Fe are released which subsequently are oxidized to produce colloidal oxyhydroxides, having a

Table 9 *The specific surface area and cation-exchange capacities of several sorption active materials (from Förstner and Wittman 1981)*[26]

Material	*Specific surface area* (m^2/g)	*Cation-exchange capacity* (meq/100 g)
Calcite (<2 μm)	12.5	—
Kaolinite	10–50	3–15
Illite	30–80	10–40
Chlorite	—	20–50
Montmorillonite	50–150	80–120
Freshly precipitated $Fe(OH)_3$	300	10–25
Amorphous silicic acid	—	11–34
Humic acids from soils	1900	170–590

Table 10 *The deep water scavenging residence times of some trace elements in the oceans (from Chester, 1990)*[1]

Element	*Scavenging residence time* (years)	*Element*	*Scavenging residence time* (years)
Sn	10	Mn	51–65
Th	22–33	Al	50–150
Fe	40–77	Sc	230
Co	40	Cu	385–650
Po	27–40	Be	3700
Ce	50	Ni	15850
Pa	31–67	Cd	177800
Pb	47–54	Particles	0.365

large surface area and very great sorptive characteristics. Finally, ferromanganese nodules form at the sediment–water interface and are considerably enriched in a number of trace metals via surface reactions.

Ferromanganese nodules result from the authigenic precipitation of Fe and Mn oxides on the sea floor. Two morphological types are recognized, depending upon the growth mechanism. Firstly, spherical encrustations formed atop oxic deep-sea sediments grow slowly, accumulating material from seawater. These seawater nodules exhibit a relatively low Mn:Fe ratio and are enriched especially with respect to Co, Fe, and Pb. Secondly, discoid shaped nodules develop in nearshore environments deriving material via diffusion from the underlying anoxic sediments. Such diagenetic nodules grow faster than deep sea varieties and metals tend to be in lower oxidation states. They have a high Mn:Fe ratio and enhanced content of Cu, Mn, Ni, and Zn. Ferromanganese nodules have concentric light and dark bands in cross-section, related to Fe and Mn oxides, respectively. Patterns of trace element enrichment in the nodules are determined by mineralogy, the controlling mechanisms being related to cation substitution in the crystal structure. Mn phases preferentially accumulate Cu, Ni, Mo, and Zn. Alternatively, Co, Pb, Sn, Ti, and V are enriched in Fe phases.

3.3 Diagenesis

Diagenesis refers to the collection of processes that alters the sediments following deposition. These mechanisms may be physical (compaction), chemical (cementation, mineral segregation, ion exchange reactions), or biological (respiration). The latter are of particular importance as the bacteria control pH and p*e* in the interstitial waters, master variables that affect a wide range of equilibria. They influence the composition of the interstitial water, which in turn can exert a feedback effect on the overlying seawater. Also, they can ultimately control the mineralogical phases that are lost to the sediment sink.

Organic material accumulates with other sedimentary components at the time of deposition. High biological activity in surface waters and rapid sedimentation ensures that most nearshore and continental margin sediments contain significant amounts of organic matter. Biochemical oxidation of this material exhausts the available O_2 creating anoxic conditions. The oxic/anoxic boundary occurs at the horizon where the respiratory consumption of O_2 balances the downward diffusion. Upon depleting the O_2, other constituents are used as oxidants leading to the stepwise depletion of NO_3^-, NO_2^-, and SO_4^{2-}. Thereafter, organic matter itself may be utilized with the concurrent production of CH_4 (see also Chapter 6).

This series of reactions causes progressively greater reducing conditions, with consequent influences on the chemistry of several elements. Metals are reduced and so are present in lower oxidation states. In particular, Mn undergoes reductive dissolution from MnO_{2s} to Mn_{aq}^{2+}. The divalent state being much more soluble, Mn is effectively remobilized under anoxic conditions and can be released back into overlying seawater. As seen in the previous section, this is one pathway to ferromanganese nodule formation. This can also be true for other elements that had been deposited following incorporation into the Fe and Mn oxide phases. On the other hand, some elements can be preserved very effectively in anoxic sediments. Interstitial waters in marine sediments, in contrast to freshwater deposits, have high initial concentrations of SO_4^{2-}. Bacterial sulfate reduction proceeds via the reaction:

$$2CH_2O + SO_4^{2-} \rightarrow H_2S + 2HCO_3^-$$

Thus, sulfide levels in interstitial waters increase. A number of elements form insoluble sulfides which under these anoxic conditions are precipitated and retained within the sediments. A notable example is the accumulation of pyrite, FeS_2, but also Ag, Cu, Pb, and Zn are enriched in anoxic sediments in comparison with oxic ones.

4 PHYSICAL AND CHEMICAL PROCESSES IN ESTUARIES

Rivers transport material in several phases: dissolved, suspended particulate, and bedload. Physical and chemical processes within an estuary influence the transportation and transformation of this material, thereby affecting the net supply of material to the oceans. Several definitions for an estuary exist, but from a chemical perspective an estuary is the mixing zone between river water and seawater characterized by sharp gradients in the ionic strength and chemical composition. Geographic distinctions can be made between drowned river valleys, fjords, and bar built estuaries. Alternatively, they can be classified in terms of the hydrodynamic regime as:

(1) salt wedge;
(2) highly stratified;
(3) partially stratified;
(4) vertically well mixed.

The aqueous inputs into the system are river flow and the tidal prism. The series above is ranked according to the diminishing importance of the riverine flow and the increasing marine contribution. Thus, a salt wedge estuary represents the extreme case dominated by river flow, in which there occurs very little mixing. A fresh, buoyant layer flows outward over denser, saline waters. In contrast, the vertically well mixed estuary is one dominated by the tidal prism. The inflowing river water mixes thoroughly with and dilutes the seawater, but the effective dilution diminishes with distance along the mixing zone.

The position of the mixing zone in the estuary exhibits considerable temporal variations. There can be a strong seasonal effect, largely due to non-uniform river discharge. High winter rainfall leads to a winter time discharge maximum. However, winter precipitation as snow creates a storage reservoir, such that the river flow maximum occurs following snow melt in spring or even early summer if the catchment area is of high elevation as for the Fraser River in Western Canada. On shorter time scales, the mixing zone is influenced by tidal cycles. Thus, the penetration of seawater into the estuary depends upon the spring–neap tidal cycle and the diurnal nature of the tides. Together these influences determine the geographic extent of sediments experiencing a variable salinity regime. Variations in the river discharge affects the mass loading of the discharge, both in terms of suspended sediment and bedload material. The hydrodynamic regime in the estuary influences the deposition of the riverine sedimentary material and the mixing of dissolved material.

The estuary is a mixing zone for river and seawaters. Their characteristics differ considerably. River water is slightly acidic and of low ionic strength with a salt matrix predominantly of $Ca(HCO_3)_2$ ($\sim$120 mg l^{-1}). In contrast, seawater has a higher pH ($\sim$8), higher ionic strength ($\sim$0.7), and consists primarily of NaCl ($\sim$35 g l^{-1}). As a consequence, the salt matrix within the estuary is dominated by the sea salt end member throughout the mixing zone except for a small proportion at the dilute extreme. Salinity can be used as a conservative index, although conductivity is better, not being subject to systematic conversion errors in the initial mixing region.

In a plot of concentration *versus* some conservative index (*i.e.* S‰, Cl‰, or conductivity), the theoretical dilution curve would comprise a straight line between the river and seawater end members. A dissolved constituent that exhibits such a distribution is said to behave conservatively in the estuary. Whereas a negative slope shows that the riverine end member is progressively diluted during mixing with seawater, a positive slope indicates that the seawater end member has the greater concentration. Conservative behaviour is exhibited, for example, by Na^+, K^+, and SO_4^{2-}. Non-conservative behaviour can result from an additional supply of material (*i.e.* causing positive deviations from the theoretical dilution curve). Elements that may show a maximum concentration at some intermediate salinity are Mn and Ba. Alternatively, the removal of dissolved material during mixing (*i.e.* negative deviation from the theoretical dilution curve) can be caused by biological activity or by dissolved to particulate transformations. Biological activity can cause non-conservative behaviour for nutrient elements. Dissolved constituents typically transformed to the particulate phase include Al, Fe, and Mn in some estuaries (see Figure 12). The pH distribution is usually characterized by a pH minimum in the initial mixing zone, resulting from the non-linear salinity dependence of the first and second dissociation constants of H_2CO_3.

There are considerable physicochemical speciation alterations that a component can undergo in the estuary. With respect to dissolved constituents, the composition and concentration of available ligands changes. As a result, the importance of chlorocomplexes for metals such as Cd, Hg, and Zn increases as the salinity increases. Conversely, the competitive influence of seawater derived Ca and Mg for organic material decreases the relative importance of humic complexation for Mn and Zn.

Estuaries are particularly well known for dissolved–particulate interactions. Phase changes come about via several mechanisms. Firstly, dissolution–precipitation processes may occur. This is especially important for the authigenic precipitation of Fe and Mn oxyhydroxides. Secondly, components may experience adsorption–desorption reactions. Desorption can occur in the initial mixing zone, partly in response to the pH minimum. Adsorption, particularly in association with the Fe and Mn phases, can accumulate material within the suspended sediments. Thirdly, flocculation and aggregation processes can remove material from solution. This occurs as particulate material with negatively charged surfaces adsorb cations in the estuary. The surface charge diminishes and as the ionic strength increases, the particles experience less electrostatic repulsion. Eventually the situation arises whereby particle collisions lead to aggregation due to weak bonding. This process can be facilitated if particle surfaces are coated with organic material (then known as flocculation rather than aggregation). The three types of processes outlined above often happen simultaneously in the estuarine environment. Such non-conservative behaviour is typified by dissolved Fe, as shown in Figure 12. The transformation of dissolved into particulate phases can then be followed by deposition to the estuarine sediments. Thus, the flux of material to the ocean can be considerably modified, particularly as such sediments may be transported landward rather than seaward.

Biological activity in the estuarine environment can also influence the speciation of constituents. A complex regeneration cycle determines distributions and modifies the riverine flux, especially with respect to nutrients. Upon encountering brackish water, freshwater organisms may die with consequent cell rupture and the release of cell content into solution. However, some biological debris (from both freshwater and estuarine origin) will be deposited onto the floor of the estuary. Of this material, some is regenerated by benthic organisms and the remainder becomes incorporated into the estuarine sediments. Estuaries are themselves regions of high biological productivity as rivers, especially polluted ones, have elevated nutrient concentrations. Organisms may absorb the riverine derived dissolved constituents, including nutrients. They can then be flushed to sea or settle within the estuarine environment, undergoing subsequent nutrient regeneration as outlined previously.

5 MARINE POLLUTION

Pollution comprises the perturbation of the natural state of the environment by anthropogenic activity. In the marine environment, the man-induced disturbances take many forms. Due to source strengths and pathways, the greatest effects tend to be in coastal regions. Such waters and sediments bear the brunt of industrial and sewage discharges, and are subject to dredging and spoil dumping. Agricultural run-off may contain pesticide residues and elevated nutrients, the latter of which may over-stimulate biological activity producing anoxic conditions. However, the deep sea has not escaped contamination. Although this has usually been in the form of crude oil, petroleum products, and plastic pollutants, the aeolian transport of heavy metals has enhanced natural elemental cycles, particularly with respect to lead. Two case studies are discussed below to illustrate some aspects of marine pollution.

5.1 Oil Slicks

Massive releases of oil have been caused by the grounding of tankers (*i.e. Torrey Canyon*, Southwest England, 1967; *Argo Merchant*, Nantucket Shoals, USA, 1976; *Amoco Cadiz*, Northwest France, 1978; *Exxon Valdiz*, Alaska, 1990) or by the accidental discharge from offshore oil platforms (*i.e. Chevron MP-41C*, Mississippi Delta, 1970; *Ixtox I*, Gulf of Mexico, 1979). The resulting oil slicks are essentially surface phenomena, with greatest impact where they impinge on coastal ecosystems. The slicks are affected by several transportation and transformation processes (Murray, 1982).[27] With respect to transportation, the principal agent for the movement of slicks is the wind, but length scales are important. Whereas small (*i.e.* relative to the slick size) weather systems, such as thunderstorms, tend to disperse the slick, cyclonic systems can move the slick essentially intact. Advection of a slick is also affected by waves and currents. Oil slicks spread as a buoyant lens under the influence of gravitational forces, but generally separate into distinctive thick and thin regions. Such pancake formation is due to the fractionation of the components within the oil mixture. Diffusion can also act to transport the oil.

Several transformation mechanisms are operative. Evaporative loss of the more volatile components is a significant loss process, especially for light crude oil. Sedimentation can play a role in coastal waters. This occurs when rough seas bring dispersed oil droplets into contact with suspended particulate material, and the density of the resulting aggregate exceeds the specific density of seawater. Colloidal suspensions can comprise either water-in-oil or oil-in-water emulsions. These behave distinctly differently. Water-in-oil emulsification creates a thick, stable colloid which can persist at the surface for months. The volume of the slick increases and it aggregates into large lumps known as 'mousse', thereby acting to retard weathering. Conversely, oil-in-water emulsions comprise small droplets of oil in seawater. This aids dispersion and increases the surface area of the slick which can accelerate weathering processes, including microbial degradation.

[27] S. Murray, 'Pollutant Transfer and Transport in the Sea', ed. G. Kullenberg, CRC Press, Boca Raton, 1982, Vol 2., p. 169.

5.2 Tributyltin

Tributyltin (TBT) is perhaps the most toxic constituent that has been knowingly introduced into the marine environment (Goldberg, 1986).[28] TBT has been utilized as the active ingredient in marine anti-fouling paint formulations. Its potent toxicity and environmental persistence has resulted in a wide range of deleterious biological effects on non-target organisms. TBT is lethal to some shellfish at concentrations as low as 0.02 μg TBT-Sn l^{-1}. Lower concentrations result in sub-lethal effects, such as poor growth rates and reduced recruitment leading to the decline of shellfisheries. The most obvious manifestations of TBT contamination have been shell deformation in Pacific oysters (*Crassosterea gigas*), and the development of imposex (*i.e.* the formation of male sex organs in females) in marine gastropods. This has caused dramatic population decline of gastropods at locations throughout the world. Laboratory experiments and field observations of deformed oysters and imposex indicate that adverse biological effects occur at concentrations below those detectable. Thus, a 'no effect' concentration has yet to be demonstrated. It is worth noting that TBT has been observed to accumulate in salmon, but it has not been shown to pose a public health risk.

TBT exists in solution as a large univalent cation and forms a neutral complex with Cl^-. It is extremely surface active and so is readily adsorbed onto suspended particulate material. Such adsorption and deposition to the sediments limits its lifetime in the water column. Degradation, via photochemical reactions or microbially mediated pathways, obeys first order kinetics. Stepwise debutylation produces di- and mono-butyltin which are much less toxic in the marine environment than is TBT. As degradation rates in the water column are on the order of days to weeks, they are slow relative to sedimentation. TBT accumulates in the sediments where degradation rates are much slower, with the half-life being on the order of years (Stewart and de Mora 1990).[29]

The economic consequences of the shellfisheries decline led to a rapid political response globally. The first publication suggesting TBT to be the causative agent appeared in only 1982 (Alzieu *et al.*, 1982),[30] but already TBT has been banned from use in several countries. Other nations have imposed partial restrictions, its use being permitted only on vessels >25 m in length or on those with aluminium hulls and outdrives. This has certainly had the effect to decrease the TBT flux to the marine environment. However, TBT continues to exist within sediments. Furthermore, concentrations are highest in those areas, such as marinas and harbours, that are most likely to undergo dredging. The intrinsic toxicity of TBT, its persistency in the sediments and its periodic remobilization by anthropogenic activity are likely to retard the long-term recovery of the marine ecosystem.

[28] E. Goldberg, *Environment*, 1986, **28**, 17.
[29] C. Stewart and S. J. de Mora, *Environ. Technol.*, 1990, **11**, 565.
[30] C. Alzieu, M. Heral, Y. Thibaud, M. J. Dardignac, and M. Feuillet, *Rev. Trav. Inst. Marit.*, 1982, **45**, 100.

CHAPTER 5

Land Contamination and Reclamation

B. J. ALLOWAY

1 INTRODUCTION

Land, or more precisely its key component the soil, differs in many ways from the other parts of terrestial ecosystems, such as water and the atmosphere, in that it acts as a sink, a filter, and a bioreactor for contaminants. For the purposes of defining the contaminative uses of land, the UK Department of the Environment (DOE) defines land as 'any ground, soil, or earth, houses or other buildings, the airspace above it, but excluding all mines and minerals beneath the land'.[1]

It is the convention to use the term 'contaminated land' whether or not apparent harmful effects have been caused by the contaminants. This is in contrast to the widespread use of the term 'polluted' when substances of anthropogenic origin have caused obvious harm and 'contaminated' when they have not. Contaminated land is defined by the DOE as 'land which represents an actual or potential hazard to health or the environment as a result of current or previous use'. On the other hand, 'Derelict land' is defined by the DOE as 'land so damaged by industrial or other development that it is incapable of beneficial use without treatment'.[1]

The composition of the 'soil' on contaminated land needs to be determined in order that the concentrations of contaminants comply with the critical ('trigger') concentrations stipulated for various end uses based on estimates of risk. The risks to materials and services include: sulfate attack of concrete, degradation of plastics and rubber by hydrocarbons, and fires or explosions from methane or hydrocarbons. The risks to humans include respiratory disorders and other illnesses from the inhalation of toxic fumes or dusts, direct ingestion of toxic compounds in contaminated drinking water or particles of soil, skin disorders from contact with chemicals, and consumption of food plants that have accumulated significant amounts of harmful substances.

In addition to contamination, anomalously high concentrations of cations and anions potentially harmful to either construction materials or crops, livestock,

[1] House of Commons Environment Committee, 1st Report: 'Contaminated Land', Volume 1, HMSO, London, 1990.

and humans, may be present in soils as a result of the weathering of some types of rocks. Examples include: high concentrations of SO_4^{2-} (which attacks concrete) from weathering sulfides in clay and shales, and heavy metals, such as Cd, Pb, and Zn, from weathering marine black shales or unexploited ore minerals. Since these potentially harmful substances are not anthropogenic in origin, they should not be classed as pollutants or contaminants.

The phytotoxic effect of contaminants is of obvious economic importance where crops are being grown commercially but the greatest impact on human health is likely to occur where relatively bioavailable but potentially hazardous metals, such as Cd, are present in contaminated soil at sub-phytotoxic concentrations. Food crops on these soils may accumulate significant amounts of the metals without the contamination problem being recognized and this could have a chronic effect on the health of regular consumers of these crops.

It is important to recognize that although acute cases of land contamination attract a lot of attention and pose major risk, almost all the world's land surface is contaminated to some extent. The transport and deposition of atmospheric pollutants affects even very remote areas, although the extent of contamination is generally worst in the more industrialized northern hemisphere. Contaminants can be carried hundreds or thousands of kilometres by moving air masses. The Southern parts of Norway and Sweden receive a wide range of metals and sulfate from coal combustion in the UK and other Northern European countries. Likewise, Canada receives acidic atmospheric pollutants from industrial emissions in the Northern States of the USA.

Particles of soil blown thousands of kilometres from China caused a 'brown snow' event in the Canadian Arctic in April 1988. Around 4000 tonnes of soil was deposited over an area of 20 000 km^2 contributing appreciable quantities of organic micropollutants (PAHs and PCBs) in a very remote area.[2] Radionuclides were dispersed over much of the earth's surface by the atmospheric testing of nuclear weapons in the 1950s and 1960s whilst large areas were affected by the accident at the Chernobyl nuclear reactor in the Ukraine in 1986 and an earlier accident at Windscale in England in 1957.

Nriagu[3] estimated that the annual global emissions of heavy metals which reached the soil during the 1980s were (in tonnes): Hg 8300, Cd 22 000, Pb 796 000, and Zn 1 372 000. However, marked differences occur in different parts of the world. For example Pb enrichment in the atmosphere can be more than a hundred times greater in rural England than in rural parts of Africa.

Apart from atmospheric deposition, the soil receives contaminants through direct placement. This can occur from the agricultural use of fertilizers and pesticides, the addition of large amounts of livestock manures and sewage sludges to agricultural land, and other types of waste disposal. However, direct deposition tends to give rise to localized contamination, for example a burnt-out dumped car on wasteland severely contaminates a few square metres, whereas the fallout from a metal smelter will have affected tens or hundreds of square kilometres, but with smaller quantities of contaminants.

[2] H. E. Welch, D. C. G. Muir, and B. M. Lemoine. *Environ. Sci. Technol.*, 1991, 280.
[3] J. Nriagu, *Environ. Pollut.*, 1988 **50**, 139.

Although almost all the soil on the earth's land surface is contaminated to some extent, it is the more acute cases of contamination which pose the most risk to living organisms and construction materials. It is therefore important to determine the extent of contamination, the contaminants involved, the risks they pose, and the most suitable way of either cleaning up the soil or managing it appropriately. This chapter will address these points briefly but the reader will need to refer to other more specialized books and papers for greater detail on the subject.

2 SOIL: ITS FORMATION, CONSTITUENTS, AND PROPERTIES

Soil is the complex biogeochemical material which forms at the interface between the earth's crust and the atmosphere and differs markedly in physical, chemical, and biological properties from the underlying weathering rock from which it has developed. Soil comprises a matrix of mineral particles and organic material, often bound together as aggregates, and populated by a wide range of micro-organisms, soil animals, and plant roots. The spaces between particles and aggregates form a system of pores which are filled with soil solution and gases. Since soil acts as a sink for contaminants and also a filter for water infiltrating to the water table, its physical, chemical, and biological properties are highly relevant to considerations of contaminated land and its reclamation.

2.1 Soil Formation

The soil profile (a vertical section from the surface to the underlying weathered rock) is the unit of study in pedology. Natural pedogenic processes bring about the differentiation of the soil material into distinct horizontal layers, called horizons, and the characteristics of these horizons forms the basis of soil morphological classification. A typical soil profile with a summary of the distribution of contaminants is shown in Figure 1.

The type of soil which forms at any site is determined by the climate, the nature of the parent rock material on which it forms, the landscape (especially whether it is a free-draining or water collecting site), the vegetation, and the time period over which the soil has been forming. These factors control the intensity and type of pedogenic processes operating at any site, and these processes determine the nature of the soil profile which develops. The major pedogenic processes include: leaching, elution, podzolization, calcification, ferralitization, laterization, salination, solodization, gleying, and peat formation.[4] With the exception of the last two, most pedogenic processes involve the movement of material (usually the products of weathering) either down or up the soil profile, depending on the overall balance of precipitation and evaporation. Movement is predominantly downwards in humid regions and upward where evaporation exceeds precipitation but the extent will depend on the climate, the permeability of the soil and the site. Gleying is the creation of reducing conditions in soils, due to

[4] E. M. Bridges, 'World Soils', 2nd Edn., Cambridge University Press, Cambridge, 1978.

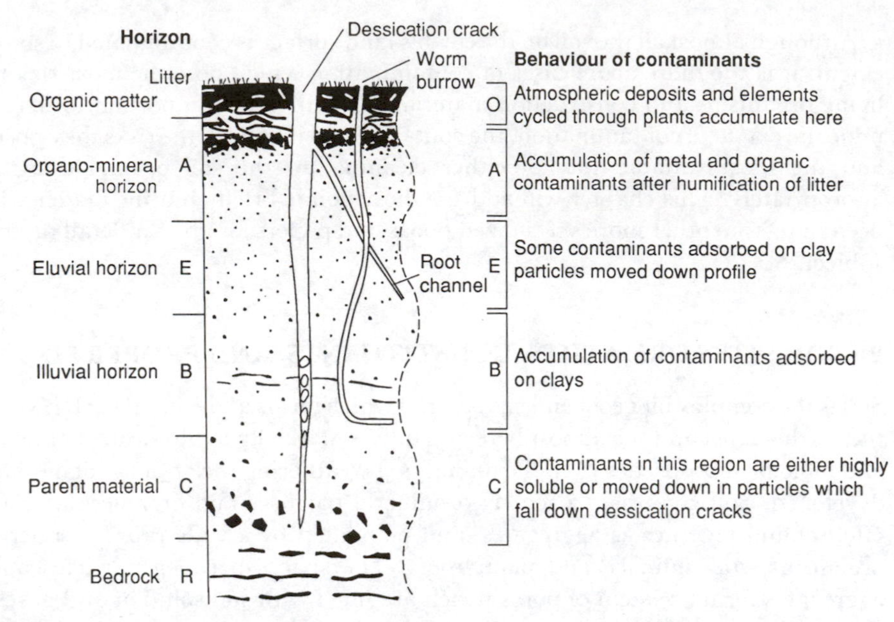

Figure 1 *Diagram of a soil profile showing horizon nomenclature and the general distribution of contaminants*

intermittent or permanent waterlogging, and leads to the reduction and dissolution of hydrous oxides of Fe and Mn.

The pedogenic processes occurring within a soil will have a major influence on the behaviour of contaminants. For example, soluble or polar contaminants are more likely to be leached down the profiles of highly permeable soils and reach the water table than they are in impermeable clay soils or soils with relatively high organic matter contents in their uppermost horizon. Soil organic matter, especially colloidal humus, has a strong capacity to adsorb a wide range of substances. Soils in arid environments tend to have low contents of organic matter in their topsoils ($<1\%$) and are therefore less likely to adsorb organic non-polar pollutants, such as PAHs and chlorinated hydrocarbons, than soils in humid temperate regions which have higher organic matter contents (1–10%). However, the relatively high free calcium carbonate content of arid soils increases their ability to adsorb many heavy metals.

2.2 Soil Constituents

2.2.1 Soil Minerals. The soil clay fraction is <2 μm in diameter and is formed mainly of clay minerals with some small particles of other minerals. Clay minerals are secondary minerals which have formed from the weathering products of primary minerals. Clay minerals are sheet silicates formed from two basic components: a silica sheet formed of Si—O tetrahedra, and a gibbsite sheet comprising Al—OH octahedra. Isomorphous substitution of either Al^{3+} for Si^{4+}, or Mg^{2+} for Al^{3+} in the structural sheets of some clay minerals gives rise to excess negative charges on their surface. This gives most clay minerals except kaolinite their cation adsorptive capacity which is independent of the soil pH. The four most commonly occurring types of clay minerals are:

Kaolinite —a highly stable 1:1 (silica:gibbsite) mineral, which is non-swelling with a relatively small surface area (5–100 m^2 g^{-1}) and a low adsorptive capacity (cation exchange capacity [CEC] 2–20 c mol_c kg^{-1}).

Smectites —2:1 (silica:gibbsite) minerals which swell on wetting and shrink on drying, they have a large surface area (700–800 m^2 g^{-1}) due to the access of soil solution to all lamellae surfaces which, together with isomorphous substitution, contributes to their relatively high adsorptive capacity (CEC 80–120 c mol_c kg^{-1}).*

Illite —a 2:1 clay like the smectites with a lower adsorptive and swelling/shrinking capacity and properties intermediate between kaolinite and smectites (surface area 100–200 m^2 g^{-1} and CEC of 10–40 c mol_c kg^{-1}).

Vermiculite—a 2:1 mineral with a very high degree of isomorphous substitution of Mg^{2+} for Al^{3+} in the gibbsite sheet giving it a high CEC (100–150 c mol_c kg^{-1}) with an intermediate surface area (300–500 m^2 g^{-1}).

The other ubiquitous secondary minerals in soils are the hydrous oxides of Fe and Al which are the ultimate residual weathering products of soils; the anhydrous forms of the oxides are highly stable under the oxidizing conditions found in freely draining soil environments. The amount of these oxides present in a soil depends on the mineralogy of the parent rock, the degree of weathering, and the oxidation/reduction (redox) conditions. The ferromagnesian minerals such as olivine, augite, and biotite mica found in basic igneous rocks are the richest primary source of Fe. Poorly drained and waterlogged soils have reducing conditions which cause the dissolution of hydrous oxides. The main forms of Fe oxide are ferrihydrite ($Fe_2O_3 \cdot 2FeOOH \cdot nH_2O$), the freshly deposited form, and goethite (α-FeOOH). The Al oxides include amorphous $Al(OH)_3$ which slowly crystallizes to gibbsite (γ-$Al(OH)_3$). These hydrous oxides tend to occur as precipitates within the clay-sized fraction of soil minerals (<2 μm). The hydrous oxides have pH-dependent surface charges; in general they are positively charged under acid conditions and negatively charged under alkaline conditions.

Many other minerals may be found in soils, the most ubiquitous being resistant quartz grains which comprise much of the sand-size fraction of soils and are relatively unreactive. Other minerals tend to occur in fragments of weathering rock, either as clasts of parent material or erratics in fluvioglacial deposits.

2.2.2 Soil Organic Matter. The presence of organic matter and living organisms distinguishes a soil from regolith (decomposed rock). All soils, in the pedological

*1 c mol_c kg^{-1} = 10^{-2} mol cation exchange capacity per kilogram of soil.

sense of the word, contain organic matter but the amount and type may vary considerably. The organic matter present in soils can be classified as either humic, or non humic material. Humic compounds are highly polymerized, colloidal, products of microbial decomposition of plant material, especially lignins. Non-humic material includes undecomposed or partially decomposed fragments of plant tissues and soil organisms. These latter include a wide range of species of bacteria, fungi, protozoa, actinomycetes, and algae, and many species of mesofauna, such as earthworms, which fill an important ecological niche in the comminution of plant litter and its incorporation into the soil profile.

Although soil organic matter only tends to form a small percentage of the mass of the soil, it has a very great influence on soil chemical and physical properties especially with regard to the behaviour of contaminants. Humus has a high but pH-dependent CEC (<200 c mol_c kg^{-1}) and a strong adsorptive capacity for non-polar organic molecules. Soil micro-organisms can adapt to being able to degrade many different types of persistent contaminant molecules, such as chlorinated hydrocarbons and PAHs by the secretion of extracellular enzymes. However, the C—Cl bond does not occur in natural compounds and hence the chlorinated organic molecules are more difficult to degrade. Nevertheless, several species of micro-organism have been found to be able to degrade chlorinated compounds but there is normally a time lag while they adapt to the new molecules. During this period there is localized selection and multiplication of strains which have a mutation enabling them to break the bonds of the contaminant molecule and to tolerate the toxicity of either the contaminant or its intermediate decomposition products.[5]

2.3 Soil Properties

Space does not permit a detailed discussion of soil properties here and so only those properties which have a significant effect on the behaviour of soil contaminants will be discussed. Readers are recommended to consult other specialized text books, such as White,[6] Brady,[7] and Singer and Munns[8] for more details of soil science.

2.3.1 Soil Permeability. The voids between soil particles and aggregates form a continuous system of pores within the soil profile. Pores with a diameter greater than 30 μm tend to drain under gravity and are normally filled with air in dry weather. Pores smaller than 30 μm tend to retain water against gravity and much of this may be available to plant roots. Soluble contaminants infiltrating the soil profile will be subject to both internal drainage within the profile, diffusion between regions of different solute concentration, and, in most cases, adsorption onto the organo-mineral colloidal complex. In addition to the inter-particle

[5] S. Ross, 'Soil Processes', Routledge, London, 1989.
[6] R. E. White, 'Introduction to the Principles and Practice of Soil Science', 2nd Edn., Blackwells, Oxford, 1987.
[7] N. Brady, 'The Nature and Properties of Soils', 10th Edn., Macmillan, New York, 1990.
[8] M. J. Singer and D. N. Munns, 'Soils: An Introduction', Macmillan, New York, 1987.

pores, worm burrows, root channels, desiccation cracks, or excavations will lead to more rapid movement of contaminants down the profile, either in solution, or adsorbed on soil particles.

The permeability of soil can vary widely as a result of the nature of the soil constituents and, or, soil management; for example, clay soils have a relatively low permeability and sandy soils tend to be highly permeable. Compaction of soils by machinery, cultivations, or excavations when the soil is too wet, can reduce its permeability and may induce temporary reducing conditions in the upper part of the profile especially in soils with a high clay content.

2.3.2 Soil Chemical Properties. Soil pH. The pH of the soil is the most important physico-chemical parameter affecting plant growth and the behaviour of contaminants in soils. The soil pH is a measurement of the concentration of H^+ ions in the soil solution present in the pores of a soil which is in equilibrium with the negatively charged surfaces of the soil particles. Unless stated otherwise soil pH values are usually measured in distilled water but the use of dilute electrolytes (*e.g.* $CaCl_2$) more closely reflects the field situation.

Soil pHs are normally within the range 4–8.5 although the extreme range found over the world is pH 2–10.5. In general, soils in humid regions tend to have pHs between 5 and 7 and those in arid regions between 7 and 9. In temperate regions, such as the UK, the optimum pH for arable soils is 6.5 and 6.0 for grassland. Soil pHs can be raised by liming with $CaCO_3$.

In general, bacteria do not tolerate very acid conditions and so soils in which the microbial degradation of organic pollutants is desirable should be maintained at a pH of between 6 and 8. The mobility and bioavailability of most divalent metals are greatest under acid conditions and therefore liming is a way of reducing their bioavailability (except for the MoO_4^{2-} anion which is most available at high pH).

Redox conditions. The balance of oxidation–reduction conditions in soils mainly affects the species of elements such as C, N, O, S, Fe, and Mn although Ag, As, Cr, Cu, Hg, and Pb are also affected. The redox conditions in a soil are also a reflection of the oxygen supply for plant roots and soil micro-organisms. Redox equilibria are controlled by the aqueous free electron activity and can be expressed either as p*e* (negative log of electron activity) or E_H (millivolt difference in potential between a Pt electrode and H electrode). The conversion factor for the two expressions is: E_H (mV) = 5.2 p*e*.[9] Micro-organisms catalyse redox reactions in soils and respiration by plant roots, soil fauna, and micro-organisms consumes a relatively large amount of oxygen. In situations of water logging, or exclusion of air by over-compaction, micro-organisms with anaerobic respiration predominate causing a change in the products of decomposition of organic matter (volatile fatty acids and ethylene, *etc.*) and in the speciation of susceptible metals.

The combined effects of redox conditions and pH on the behaviour of hydrous oxides in soils can be summarized by stating that either low pH or negative redox values result in the dissolution of hydrous oxides of Fe and Mn and, conversely

[9] W. L. Lindsay, 'Chemical Equilibria in Soils', John Wiley, Chichester, 1979.

small increases in either pH or redox values can lead to the precipitation of ferrihydrite or Mn hydrous oxides. Sulfate anions are reduced to sulfide below p*e* −2.0 and this can lead to the formation of insoluble sulfides of a wide range of metals. These sulfides tend to be insoluble and act as a temporary sink of the metal until redox conditions change and the sulfides are oxidized. The oxidation of sulfides, such as iron pyrites (FeS_2) results in a marked increase in the acidity of the soil.

3 SOURCES OF LAND CONTAMINANTS

Soils receive contaminants from a wide range of sources, including:

(1) Atmospheric fallout from:
- —fossil fuel combustion (oxides and acid radicals of S and N);
- —Pb, PAHs, *etc.* from automobile exhausts (see also Chapter 1);
- —metal smelting operations (As, Cd, Cu, Cr, Ni, Pb, Sb, Tl, and Zn);
- —chemical industries (organic micropollutants, Hg);
- —waste disposal by incineration (TCDDs, TCDFs);
- —radioisotopes from reactor accidents (*e.g.* Windscale, UK, 1957 and Chernobyl, USSR, 1986) and atmospheric testing of nuclear weapons;
- —large fires (*e.g.* soot, PAHs, *etc.* from burning oil wells).

(2) Agricultural chemicals:
- —herbicides (*e.g.* 2,4-D, 2,4,5-T containing TCDD, B, and As compounds);
- —insecticides (chlorinated hydrocarbons, *e.g.* DDT, BHC);
- —fungicides (Cu, Zn, Hg, and organic molecules);
- —acaricides (*e.g.* 'Tar Oil');
- —fertilizers (*e.g.* Cd and U impurities in phosphates).

(3) Waste Disposal (intentional/unintentional input to soil):
- —farm manures (As and Cu in pig and poultry manures);
- —sewage sludges (rich in heavy metals and organic pollutants—PAHs and PCBs, *etc.*);
- —composts from domestic wastes (metals, *etc.*);
- —mine wastes (coal mines—SO_4, *etc.*, metalliferous mines Cd, Cu, Pb, Zn, Ba, U, *etc.*);
- —seepage of leachate from landfills;
- —ash from fossil fuel combustion, incinerators, bonfires, and accidental fires;
- —burial of diseased livestock on farmland.

(4) Incidental Accumulation of contaminants:
- —corrosion of metal in contact with soil (*e.g.* Zn from galvanized metal, Cu and Pb from roofing, scrapyards, *etc.*);
- —wood preservatives from fencing (PCP, creosote, As, and Cu);
- —leakage from underground storage tanks (petrol, chlorinated solvents);
- —warfare (organic pollutants from fuels, smoke, and fires, metals from munitions and vehicles);

—sports and leisure activities (Pb from gun shot and fishing weights, Pb, Cd, Ni, and Hg from discarded batteries, hydrocarbons from spilt petrol and lubricating oil).

(5) Derelict industrial sites—wide range of contaminants from production, waste disposal, and building demolition, *e.g.*:
—Gas works—phenols, tars, cyanides, As, Cd;
—Electrical industries—Cu, Pb, Zn, PCBs, solvents;
—Tanneries—Cr;
—Scrapyards—metals, PCBs, hydrocarbons.

A more comprehensive list of the most common contaminating uses of land includes: waste disposal sites, gas works, oil and petroleum refineries, and petrol stations, electricity generating stations, iron and steel works, non-ferrous metals processing, metal products fabrication and metal finishing, chemical works, glass-making and ceramics, textile plants, leather tanning works, timber and timber products treatment works, manufacture of integrated circuits and semi-conductors, food processing, sewage works, asbestos works, docks and railway land, paper and printing works, heavy engineering installations, installations processing radioactive materials, and burial of diseased farm livestock.[1]

4 CHARACTERISTICS OF SOME MAJOR GROUPS OF LAND CONTAMINANTS

4.1 Heavy Metals

Metals, such as Ag, Cd, Cr, Cu, Hg, Mn, Mo, Ni, Pb, Sb, Tl, U, V, and Zn tend to be strongly adsorbed by soil constituents, especially organic matter, 2:1 clays and hydrous oxides of Fe and Mn. Their mobility and bioavailability depends on the soil organic matter content, pH, and redox conditions. With the exception of Mo, metals are generally more mobile (bioavailable) under acid and reducing conditions and/or where the soil organic matter content is low (<2%). In addition to the relatively obvious sources of heavy metals, another major non-point source of Pb is game bird and clay pigeon shooting which can result in the dispersal of >1000 t Pb into the terrestrial environment annually in most technologically advanced countries.

The range of heavy metal concentrations found in sewage sludges ($\mu g\ g^{-1}$ in dry matter): Ag (<960), As (<30), Cd (<3410), Cr (<40 600), Cu (50–8000), Hg (<55), Mn (60–3900), Mo (<40), Ni (<5300), Pb (29–3600), Sb (<34), Se <10), Sn (40–700), V (20–400), Zn (91–49 000).[10]

4.2 Organic Contaminants

There are more than 20 000 organic contaminants known already and this number will increase as analytical methods are further refined and more studies

[10] B. J. Alloway, 'Heavy Metals in Soils', ed. B. J. Alloway, Blackie, Glasgow, 1990.

made of materials containing wide ranges of organic pollutants, such as industrial wastes, sewage sludges, and landfill leachates.

Pesticides can be soil contaminants as a result of persistence after use on crops, run-off from treated land, accidental spillages, or pesticide manufacture. Although there are over ten thousand commercial pesticide formulations of around 450 compounds in use, they can be classified under relatively few groups of molecules. These compounds are used because of their inherent toxicity towards specific pests but there is a risk of the toxic properties affecting soil organisms, other beneficial plants and animals, and humans. This can occur by uptake through plants, through the food chain, or in contaminated drinking water which the pesticide, or its toxic decomposition product, reached either by leaching or in run-off.

Three main groups of insecticides are currently used in agriculture: organochlorines, organophosphates, and carbamates. The organochlorines have been widely used for up to 50 years and are the most persistent of all groups of pesticides. The persistence decreases in the order: DDT > dieldrin > lindane (BHC) > heptachlor > aldrin with half-lives of eleven years for DDT down to four years for aldrin.[5] Organophosphates are highly toxic to humans and other mammals but are less persistent in the soil than organochlorines (six month half-lives for parathion, diazinon, and demeton). Carbamates are used to control a wide range of pests including molluscs, fungi, and insects but have a similar persistence to organophosphates. Aldicarb (or 'Temik') is a highly toxic carbamate that is used both as an insecticide and nematicide on potato and sugar beet crops. It is readily oxidized in the soil but its oxidation products are also highly toxic and readily leached and can cause ecological and human health problems.[11] Some examples of the molecular structures of pesticides and other organic contaminants are shown in Figure 2.

There are six major groups of compounds used as herbicides: phenoxyacetic acids, toluidines, triazines, phenylureas, bipyridyls, and glycines. The most important phenoxyacetic acids are 2,4-D and 2,4,5-T which have a persistence of up to eight months but are of particular environmental significance because they can be contaminated with dioxins (TCDDs). 'Agent Orange' the defoliant used by the US forces in Vietnam comprised a mixture of 2,4-D and 2,4,5-T and caused widespread contamination of soils by dioxin. The toluidines and triazines are fairly strongly adsorbed and have a persistence of up to twelve months although atrazine contamination of groundwaters is a serious problem in many intensive arable farming areas. The phenylureas tend to be fairly soluble and are rapidly leached. In contrast, the bipyridyls such as paraquat and diquat are cationic and strongly adsorbed on soil colloids (K_d values of $<4.2 \times 10^4$ on montmorillonite) and are therefore very persistent. Fungicides comprise a more diverse group of compounds including inorganics, such as copper and mercury compounds, and a wide range of organic compounds.

Apart from pesticides, which are synthesized intentionally, a wide range of chlorinated compounds including dioxins (TCDDs) and dibenzodifurans

[11] F. A. M. de Haan, 'Scientific Basis for Soil Protection in Europe', ed. H. Barth and P. L'Hermite, Elsevier, Amsterdam, 1987, p. 211.

2,4,5-T

O-CH$_2$-CO$_2$H

Cl Cl Cl

2,4,5-trichlorophenoxyacetic acid

ATRAZINE

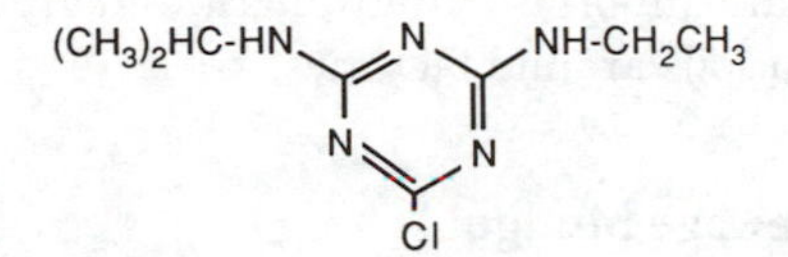

2-chloro-4-ethylamino-6-isopropylamino-1,3,5-triazine

PARAQUAT(dichloride)

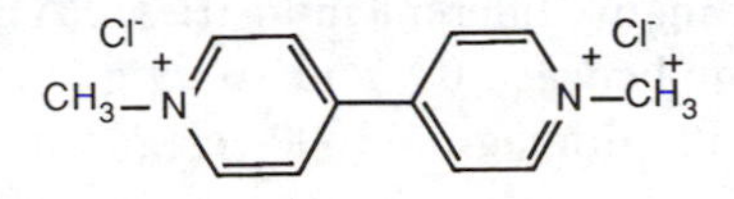

1,1'-dimethyl-4,4'-bipyridylium dichloride

DDT

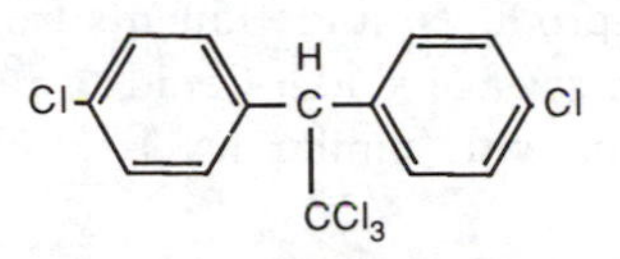

1,1-di(4-chlorophenyl)-trichloroethane

PARATHION

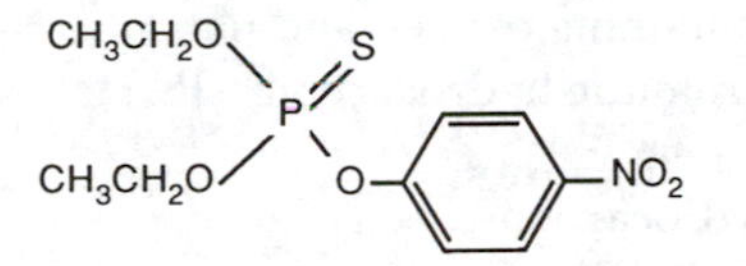

O,O -diethyl-O-(4-nitrophenyl)-phosphorothioate

TCDD

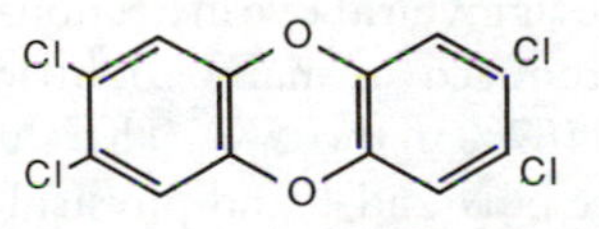

2,3,7,8-tetrachlorodibenzodioxin

TCDF

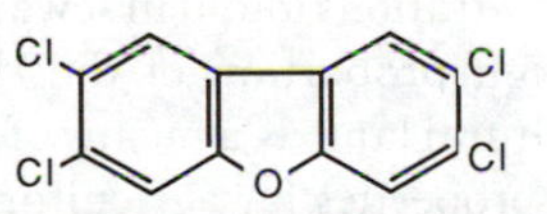

2,3,7,8-tetrachlorodibenzofuran

Figure 2 *Examples of the molecular structures of pesticides and organic contaminants*

(TCDFs) have been widely dispersed in the environment as a result of their accidental synthesis at relatively high temperatures, their highly stable structure, and their slow rate of degradation. Typical situations in which they are formed include: synthetic reactions where the temperature conditions become too hot (*e.g.* production of 2,4,5-T), the incineration of rubbish containing PVC and other sources of chlorides and aromatic compounds. Polychlorinated biphenyls (PCBs) are very stable and were manufactured for use in electrical transformers and capacitors and as plasticizers. They are found in soils around industrial and domestic waste tips and electronic component factories. Incineration temperatures need to be >1200 °C in order to ensure the destruction of stable molecules like PCBs. Solvents used for degreasing electrical components such as trichloroethane are important atmospheric and groundwater contaminants and may also occur in soils around factories.

4.3 Sewage Sludge

Sewage sludges contain a wide range of environmental contaminants owing to the diverse sources of effluents discharged into sewers. This includes human excretion products, household chemicals, automobile fuels, lubricants and cleaning compounds, stormwater run-off from highways containing PAHs and other fuel combustion products, and effluents from many different industries. PAH concentrations in sewage sludges tend to range between 0.5 and 10 $\mu g\ g^{-1}$ in sludge dry matter with similar levels of PCBs although >1000 $\mu g\ g^{-1}$ are sometimes found.[12]

The organic contaminants frequently found in sewage sludges include:

- Halogenated aromatics (PCBs—polychlorinated biphenyls, PCTs—polychlorinated terphenyls, PCNs—polychlorinated naphthalenes, and polychlorobenzenes);
- Aromatic amines and nitrosamines;
- Halogenated aromatics containing oxygen and phenols;
- Polyaromatic and heteroaromatic hydrocarbons (PAHs);
- Halogenated aliphatics;
- Aliphatic and aromatic hydrocarbons;
- Phthalate esters;
- Pesticides.

Of all the organic contaminants listed here PAHs and PCBs are currently considered to constitute the greatest hazard to human health.[12] The range of heavy metal concentrations found in sewage sludges is given above in Section 4.1.

A relatively high proportion of the sludge produced in many countries is applied to agricultural land as a means of disposal (67% in the UK). This sludge has many useful properties for agriculture (source of N and P and physical soil conditioner) but its use is limited by its concentrations of persistent contaminants. Sewage sludges are frequently used in the landscaping of derelict land where they act as a growth medium for plants grown for amenity purposes rather than food.

5 THE ADSORPTION OF CONTAMINANTS IN SOILS

5.1 Adsorption Reactions

Soils act as a sink for contaminants due to several adsorption processes which bind contaminants with varying strengths to the surfaces of the colloidal

[12] D. Sauerbeck, 'Scientific Basis for Soil Protection in Europe' ed. H. Barth and P. L'Hermite, Elsevier, Amsterdam, 1987, p. 181.

constituents of soils. This adsorption therefore delays or prevents the leaching of the contaminant down the soil profile to the water table, reduces its bioavailability to plants, and can affect the rate of decomposition of organic contaminants.

The concentration of solutes in the soil solution is determined by the interaction of adsorption and desorption processes. When contaminants are deposited on the soil surface they either react with the colloids in the soil aggregates at the surface or are washed into the soil profile in rain, irrigation water, or snow melt. Soluble contaminants will infiltrate the topsoil and enter the system of pores, whereas insoluble and hydrophobic organic molecules will bind to sites on the soil surface and become incorporated into the topsoil by movement of soil particles during cultivation, or excavations.

Several different adsorption reactions can occur between the surfaces of organic and mineral colloids and the contaminants. The extent to which the reactions occur will be determined by the composition of the soil (especially the clay mineral, hydrous oxide, and organic matter contents) and the soil pH, and the nature of the contaminants. The more strongly adsorbed contaminants are less likely to be leached down the soil profile and will tend to have relatively low bioavailabilities. Ionic contaminants such as metals, inorganic anions, and certain organic molecules, such as the bipyridyl herbicides Paraquat and Diquat are adsorbed onto surface charges on soil colloids. Non-ionic organic molecules, which includes most of the organic micropollutants and pesticides are adsorbed onto humic polymers by both chemical and physical adsorption mechanisms.

5.2 Adsorption of Cations and Anions in Soils

Ion exchange (or non-specific adsorption) refers to the exchange between the counter-ions balancing the surface charge on the soil colloids and the ions in the soil solution. In the case of cations, it is the negative charges on soil colloids which are responsible for cation exchange. The extent to which adsorbing soil constituents can act as cation exchangers is expressed as the cation exchange capacity (CEC) measured in c mol_c kg^{-1}. Some examples of the typical CEC values for soil colloidal constituents are:[5,10]

Soil organic matter	150–300 (c mol_c kg^{-1})
Kaolinite (clay)	2–5 (c mol_c kg^{-1})
Illite (clay)	15–40 (c mol_c kg^{-1})
Montmorillonite (clay)	80–100 (c mol_c kg^{-1})
Vermiculite (clay)	150 (c mol_c kg^{-1})
Hydrous oxides (Fe, Al, Mn)	4 (c mol_c kg^{-1})

Soil organic matter has a higher CEC at pH 7 than other soil colloids and therefore plays a very important part in all adsorption reactions in most soils even though it is normally present in much smaller amounts (1–10%) than clays (<80%). However, sandy soils with low contents of both organic matter and clay tend to have low adsorptive capacities and are a greater danger for contaminants infiltrating through the soil profile to the water table.

The negative charges on the surfaces of soil colloids are of two types:

(a) permanent charges resulting from the isomorphous substitution of a clay mineral constituent by an ion with a lower valency;
(b) the pH dependent charges on oxides of Fe, Al, Mn, and Si and organic colloids which are positive at pHs below their isoelectric points and negative above their isoelectric points.

Hydrous iron and aluminium oxides have relatively high isoelectric points (>pH 8) and so tend to be positively charged under most conditions whereas clay and organic colloids are predominantly negatively charged under alkaline conditions. With most colloids, increasing the soil pH, at least up to neutrality, tends to increase their CEC. Humic polymers in the soil organic matter fraction become negatively charged due to the dissociation of protons from carboxyl and phenolic groups.

The concept of cation exchange implies that ions will be exchanged between the soil solution and the zone affected by the charged colloid surfaces (double diffuse layer). The relative replacing power of any ion on the cation exchange complex will depend on its valency, its diameter in hydrated form and the type and concentration of other ions present in the soil solution. With the exception of H^+, which behaves like a trivalent ion, the higher the valency, the greater the degree of adsorption. Ions with a large hydrated radius have a lower replacing power than ions with smaller radii. For example, K^+ and Na^+ have the same valency but K^+ will replace Na^+, owing to the greater hydrated size of the Na^+ ion. The commonly quoted relative order of replaceability on the cation exchange complex of metal cations is:[8]

$$\begin{aligned} Li^+ = Na^+ > K^+ = NH_4^+ > Rb^+ > Cs^+ > Mg^{2+} > Ca^{2+} > Sr^{2+} \\ = Ba^{2+} > La^{3+} = H^+(Al^{3+}) > Th^{4+} \end{aligned}$$

Anion adsorption occurs when anions are attracted to positive charges on soil colloids. As stated above, hydrous oxides of Fe and Al are usually positively charged below pH 8 and so tend to be the main sites for anion exchange in soils. In general, most soils tend to have far smaller anion exchange capacities than cation exchange capacities. Some anions, such as nitrates and chlorides are not adsorbed to any marked extent but others, such as orthophosphates tend to be strongly adsorbed.

Some organic pesticides, such as the phenoxyalkanoic acid herbicides, such as 2,4-D, 2,4,5-T, and MCPA, exist as anions at normal soil pHs and are adsorbed to a limited extent by hydrous oxides and by H-bonding to humic polymers. However, this sorption is less marked than that which occurs with the cationic bipyridyl herbicides which are inactivated on contact with soil colloids.[5,8]

Specific adsorption is a stronger form of adsorption, involving several heavy metal cations and most anions with surface ligands to form partly covalent bonds with lattice ligands on adsorbents, especially hydrous oxides of Fe, Mn, and Al. This adsorption is strongly pH specific and the metals and anions which are most able to form hydroxy complexes are adsorbed to the greatest extent. The order for the increasing strength of specific adsorption of selected heavy metals is:[13]

$$Cd > Ni > Co > Zn \gg Cu > Pb > Hg$$

5.3 Adsorption and Decomposition of Organic Contaminants

The adsorption of non-ionic and non-polar pesticides and other organic contaminants occurs mostly on soil humic material. Since the highest content of soil organic matter in soils occurs in the surface horizons of soils, there is a tendency for most organic contaminants to be concentrated in the topsoil. Migration of organic contaminants down the profiles of soils occurs to the greatest extent in highly permeable sandy or gravelly soils with low organic matter contents. High concentrations of water soluble soil organic matter can cause enhanced mobility and leaching of organic contaminants in soils due to the binding of the contaminant to the soluble ligand. Soils will vary in their contents of water soluble soil organic matter, but applications of sewage sludge, animal manure, and compost results in increased concentrations.[14]

In the case of pesticide contaminants, most are relatively insoluble and do not move down the soil profile but exceptions to this are the organophosphate insecticides, phenoxyacetic acid herbicides, and bipyridylium herbicides which can be leached. In general, most pesticide and other organic contaminants that reach water courses from soils have been washed into the water adsorbed on soil particles in run-off and not leached down through the profile[5,8] (see Figure 2 for structures of some pesticide and other organic contaminant molecules).

In some cases where organic wastes containing organic contaminants, such as sewage sludges and some composts, are applied to soils, they act as both a source and a sink of contaminants (*e.g.* PAHs, PCBs, and heavy metals). A range of distribution coefficients are found for the same contaminant molecules in different soils and this is considered to be largely due to variations in the soil organic matter content. For example, k^{oc} (distribution coefficient for organic carbon) values ranging from 540–1730 have been reported for Lindane and 240–1290 for naphthalene.[15] Positive correlations are frequently found between the concentrations of pesticide residues and the soil organic matter content.

Adsorption of organic contaminants depends on their surface charge and their aqueous solubility, both of which are affected by the soil pH. Adsorption of non-polar organic contaminants onto soil organic matter will not occur in the presence of oils. Microbially synthesized surfactants can help to accelerate the rate of degradation of hydrocarbon oils in contaminated soils. For many organic contaminants, adsorption onto soil colloids and the presence of water are important factors promoting decomposition by micro-organisms. Ross,[8] lists the

[13] G. W. Brummer, 'The Importance of Chemical Speciation in Environmental Processes', Springer Verlag, Berlin, 1986.

[14] I. Kogel-Knabner, P. Knabner, and H. Deschauer, 'Contaminated Soil '90', ed. F. Arendt, M. Hinsenveld, and W. J. van den Brink, Kluwer Academic, Dordrecht, 1990, p.323.

[15] H. Kishi and Y. Hasimoto, 'Contaminated Soil '90', ed. F. Arendt, M. Hinsenveld, and W. J. van den Brink, Kluwer Academic, Dordrecht, 1990, p.331.

types of degradation of organic molecules as: (a) Non-biological degradation which includes: hydrolysis, oxidation, reduction, and photodecomposition; and (b) Microbial decomposition, often involving specially adapted micro-organisms. This type of decomposition normally follows a first order (exponential) type of reaction after an initial lag period while the micro-organisms become adapted to the substrate. In most cases, non-biological degradation processes, such as photodecomposition and volatilization can occur at the same time as microbially catalysed reactions.

The range of factors affecting the degradation of organic contaminants by micro-organisms include: soil pH, temperature, supply of oxygen and nutrients, the structure of the contaminant molecules, their toxicity and that of their intermediate decomposition products, the water solubility of the contaminant, and its adsorption to the soil matrix (and therefore the organic matter content of the soil).[16] Temperature effects tend to result in adsorption decreasing with increased temperature. Most pesticide adsorption reactions are exothermic. Volatilization losses tend to be greatest at high temperatures.[8]

The persistence of organic contaminants in soils is determined by the balance between adsorption onto soil colloids, uptake by plants, and transformation or degradation processes (determined by the amount of contaminant, its form, and soil properties). In the case of pesticides, organochlorine molecules are regarded as being highly persistent with a duration in the soil of 2–5 years but some estimates put this up to 10 or more years.

6 METHODS OF SITE INVESTIGATION

In investigations of land suspected of being significantly contaminated, the type of contaminants suspected, their source, and the nature of the site are all important factors determining the choice of sampling procedure. For example, agricultural land suspected of being contaminated by fallout from a point source of contamination several kilometres away would be investigated differently from a former industrial site where the remaining buildings and other structures indicate the areas most likely to be heavily contaminated. It is convenient to divide the investigation of contaminated sites into two categories:

(a) Large areas, including agricultural land, contaminated by atmospheric fallout from distant or non-point sources : investigations on land of this type will usually need to use a grid pattern of sampling with most emphasis on topsoil (0–15 cm, especially 0–5 cm) but samples will need to be collected from greater depths (15–30, 30–45 cm) to determine whether contaminants have been translocated down the profile. The normal practice for agricultural land in the UK is to take samples with a screw auger in a W-pattern with 25 cores per 5 ha block. The cores for each depth sample are bulked together to give one sample. If the soil appears to differ

[16] M. Stieber, K. Bockle, P. Werner, and F. H. Frimmel, 'Contaminated Soil '90', ed. F. Arendt, M. Hinsenveld, and W. J. van den Brink, Kluwer Academic, Dordrecht, 1990, p.473.

in colour or texture in part of the area being considered, each distinct area should be (bulk) sampled separately.

Preliminary sampling surveys of areas affected by fallout from discrete sources are often carried out using transects where samples are collected at regular intervals in straight lines both up and downwind from the source. The siting of the transect may have to be modified to take account of the prevailing wind direction, accessibility of the area, and the constraints of coastlines and topography.

(b) Discrete areas of industrial or other obvious contaminating activity and land adjacent to them: in this situation it is vitally important to know the processes used on the site, the raw materials used, the products and byproducts, the wastes from those processes, and the waste disposal practices.[17] The location of potential areas of major contamination ('hot spots') within the site, such as storage and effluent tanks, bulk raw materials stores, and waste heaps should be determined in a preliminary desk study of old plans, or careful 'detective work' on the site. These potential 'hot spots' should be carefully examined and sampled separately from any systematic survey of the whole site. Any pre-existing pits or other subterranean structures which are found should be sampled. The normal practice is to take the samples for the analytical survey of the site on a grid pattern often with 25–50 metres between points—although a closer grid will provide a more reliable indication of the occurrence of contaminants within the site especially where the distribution is very heterogeneous.

Ideally, soil samples should be taken from trial pits (with appropriate safety precautions) rather than boreholes. This allows observations to be made about the stratification of materials and other relevant points. The excavation may release gases, vapours, or liquids and so vigilance is necessary and gas sampling may need to be carried out as a precaution.[17] Boreholes can be drilled more rapidly than excavated pits, thus allowing the site to be surveyed in a shorter time and they are the only feasible way of sampling below 3 or 4 metres at a large number of sites. In practice, a combination of pits and boreholes are often used. Although most sampling is done on some sort of a grid, supplemented by intensive sampling in potential 'hot spot' areas, it is advisable to take additional samples from areas between grid points in order to allow a more random approach.[17] Careful investigation of the ground before the sampling is essential; areas where the soil material differs in colour, texture, moisture content, and apparent organic content should be noted and included in the sampling.

Clearly, cost will be a major consideration in the planning of any site survey, but it has been demonstrated in many well publicized cases that inadequate surveys which fail to reveal major contamination usually result in many times greater costs of remediation and legal actions after the site has been developed for houses or wherever there is a high risk to human health (see Section 10.3).

[17] E. E. Finnecy and K. W. Pearce, 'Understanding our Environment', 1st Edn., ed. R. E. Hester, Royal Society of Chemistry, London, 1986, p.172.

Brief details of commonly used analytical methods are summarized in Table 1. Samples of soil are normally air-dried or oven dried <30 °C but care should be taken over the possibility of harmful vapours being released from the samples.

7 THE EFFECTS OF LAND CONTAMINATION

Soil contamination can restrict the options available for the use of land because of the potential hazards posed by the contaminants. Some examples of the hazards are:

Hazard	Examples of Contaminants
(1) Direct ingestion of contaminated soil (mainly children or animals)	As, Cd, Pb, CN^-, coal tars, phenols
(2) Inhalation of dusts and vapours from contaminated soil	toluene, benzene, xylene, various organic solvents, Rn, Hg, metal-rich particles
(3) Uptake by plants of contaminants hazardous to animals and people through food chain	As, Cd, Pb, Tl, PAHs
(4) Phytotoxicity	SO_4^{2-}, Cu, Ni, Zn, CH_4
(5) Deterioration of building materials and services	SO_4^{2-}, SO_3^{2-}, Cl^-, coal tar, phenol, mineral oils, organic solvents
(6) Fires and explosions	CH_4, S, coal dust, oils, petroleum, tar, rubber, high calorific organic wastes (old landfills)
(7) Contact of people with contaminants during demolition	coal tar, phenols, asbestos, radionuclides, PAHs, TCDDs, *etc.*
(8) Contamination of water	CN^-, SO_4^{2-}, soluble metals, solvents, pesticides

(Adapted from references 18 and 19.)

Contaminants in soil can affect humans by absorption into the body through oral, inhalation, or cutaneous pathways and their effects are usually related to the amount absorbed into the body. Inhalation is an important route of intake for volatile compounds such as organic solvents and mercurial compounds. Dust particles <7 μm can penetrate alveoli in the lungs and are a hazard to workers involved in moving contaminated soil and possibly to people using playing fields constructed on contaminated land. Oral intake is a particularly serious problem

[18] M. J. Beckett and D. L. Sims, 'Contaminated Land', ed. J. W. Assink and W. J. van den Brink, Martinus Nijhoff, Dordrecht, 1986, p.285.

[19] Interdepartmental Committee on the Redevelopment of Contaminated Land, 'Guidance on the Assessment and Redevelopment of Contaminated Land', Guidance Note 59/83, Department of the Environment, London, 1987.

Table 1 *Summary of the analytical methods used for soil contaminants*

Contaminant	*Method*
Heavy metals (in acid digests or partial extracts)	Flame atomic absorption spectrophotometry (FAAS) or inductively coupled plasma–atomic emission spectrometry (ICP–AES)
As, Bi, Hg, Sb, Se, Sn, and Te (in acid digests or partial extracts)	Hydride generation atomic absorption spectroscopy (HGAAS) using sodium borohydride in NaOH
Borate (water soluble)	ICP–AES (using quartz or plastic apparatus instead of glass)
Organic pollutants (in organic solvents)	Gas chromatography (GC) or combined with Mass Spectrometry (GC–MS)
Cyanides	Colorimetrically—pyridine pyrazalone (blue) reaction
Sulfates (water soluble)	Elution through cation exchange resin, followed by titration of eluate with standard NaOH
Sulfates (total)	Dissolution in HCl, precipitation of Al and Fe, followed by gravimetric determination using $BaCl_2$
Chlorides (in HNO_3)	Volhard's Method, back titration with ammonium thiocyanate after initial precipitation with $AgNO_3$
pH (in distilled water or $CaCl_2$ or KCl)	Electrometrically using glass electrode

with young children playing on contaminated land when they lick their fingers. Oral intake in adults is mainly through the consumption of soil on vegetables (and also the contaminants accumulated in the plants). Cutaneous absorption is only a problem with lipophyllic organic solvents.

8 CRITICAL CONCENTRATIONS FOR CONTAMINANTS IN SOILS

Having established that an area of soil is contaminated, it is necessary to make a decision about the action that needs to be taken in order to avoid unnecessary risk of health effects or of damage to structures. For this purpose, various sets of critical concentrations are in use around the world. For practical purposes, readers are advised to obtain the most up to date values used in their own country or region. The ranges of critical concentrations used in two European countries (UK and The Netherlands) for the interpretaion of contaminated land analytical data are given as examples in Tables 2 and 3.

In the Netherlands, a system has been used which involves three indicative values: A—the 'normal' reference value, B—the test value to determine the need for further investigations, and C—the value above which the soil definitely needs

Table 2 *UK Department of the Environment Trigger Concentrations for Environmental Contaminants*[17,19] *(total concentrations except where indicated)*

Contaminant	*Proposed uses*	*Threshold* (Trigger Concentrations $\mu g\ g^{-1}$)	*Action*
5 *Contaminants which may pose hazards to health*			
As	Gardens, allotments	10	†
	parks, playing fields, open space	40	
Cd	Gardens, allotments	3	
	parks, playing fields, open space	15	
Cr (hexavalent*)	Gardens, allotments	25	
	parks, playing fields, open space	1000	
Cr	Gardens, allotments	600	
	parks, playing fields, open space	1000	
Pb	Gardens, allotments	500	
	parks, playing fields, open space	2000	
Hg	Gardens, allotments	1	
	parks, playing fields, open space	20	
Se	Gardens, allotments	3	
	parks, playing fields, open space	6	
(*hexavalent Cr extracted by 0.1 M HCl adjusted to pH 1 at 37.5 °C)			
Contaminants which are phytotoxic but not normally hazardous to health			
B (water soluble)	Any uses where plants grown	3	
Cu (total)	Any uses where plants grown	130	
(extractable**)		50	
Ni (total)	Any uses where plants grown	70	
(extractable)		20	
Zn (total)	Any uses where plants grown	300	
(extractable)		130	

**Extracted in 0.05 M EDTA
†Action concentration yet to be specified.

cleaning-up. Examples of these indicative values for three typical contaminants are (in $\mu g\ g^{-1}$): for Cd, A = 1, B = 5, and C = 20; for total complex cyanides, A = 5, B = 50, and C = 500; and for total PCBs, A = 0.05, B = 1.0, and C = 10.[20]

9 RECLAMATION OF CONTAMINATED LAND

Once an area of land has been identified as being contaminated, it becomes necessary to decide what action ought to be taken with regard to restrictions on its

[20] G. Gieseler, 'Contaminated Land in the EC', Report of Contract No. 85—B 6632-11-006-11-N (EC), Dornier System GMBH, Friedrichshafen, 1987.

Table 3 *UK Department of the Environment Trigger Concentrations for Contaminants associated with former coal carbonization sites*[19]

Contaminant	*Proposed use*	*Threshold* (Trigger Concentrations $\mu g\ g^{-1}$)	*Action*
PAHs	Gardens, allotments	50	500
	Landscaped areas	1000	10 000
Coal tar	Gardens, allotments	200	—
	Landscaped areas, open space	500	—
	Buildings, hard cover	5000	—
Phenols	Gardens, allotments	5	200
	Landscaped areas	5	1000
Free cyanide	Gardens, allotments, landscaped areas	25	500
	Buildings, hard cover	100	500
Complex cyanides	Gardens, allotments	250	1000
	Landscaped areas	250	5000
	Buildings, hard cover	250	Nil
Thiocyanate	All uses	50	Nil
Sulfate	Gardens, allotments, landscaped areas	2000	10 000
	Buildings	2000	50 000
	Hard cover	2000	Nil
Sulfide	All uses	250	1000
Sulfur	All uses	5000	5000
Acidity	Gardens, *etc.*	pH < 5	pH < 3

use and, or, requirements for the amelioration or 'clean-up' of the soil. There are several options available for the remediation of contaminated sites. The choice of option will depend on the nature of the contaminants, the type of soil, the characteristics of the site, its intended use, the relative costs of the appropriate options, and the regulations which apply in the country or region where the site is located. The remediation options can vary from the minimum of reducing the bioavailability of the contaminants, to the maximum of either complete clean-up of the soil, or its removal from the site.

9.1 Removing Contaminated Soil from the Site

This requires stripping off the contaminated depth of soil and either sending it to a licensed landfill, or cleaning it by thermal or chemical means. Landfilling is expensive and the costs of transporting soil are very high. The landfilling option would probably be used for relatively small urban sites contaminated with hazardous materials, or derelict industrial sites where discrete areas have been contaminated with highly toxic materials (*e.g.* PAHs, PCBs, radioisotopes, cyanides, Cd).

Although in its infancy, soil cleaning is being used increasingly in the Netherlands, Germany, and the United States where there is a requirement to restore sites to allow any future use, including the growing of food crops. Technological developments have enabled mobile equipment to be taken to large areas of contaminated land so that clean-up can be done on site. Nevertheless, the operation does involve the stripping of the contaminated soil so that it can be processed. Techniques of soil cleaning include:

(i) superheated steam to remove organic micropollutants and hydrocarbons;
(ii) incineration to remove all organic pollutants—leaving ash and mineral particles;
(iii) washing the soil with selected extractants (such as chelating agents or dilute acids) to remove certain inorganic or organic contaminants.

These cleaning techniques destroy the soil organic matter and biomass. Further expense and time is needed to allow this material to develop into a soil by its integration with humic material and the activities of soil micro-organisms.

9.2 Covering the Contaminated Land with Barren Soil

Covering the contaminated land with a layer of clean soil, provides an uncontaminated rooting medium for most plants (except deep rooting shrubs and trees) and prevents loss of contaminated soil by wind or water erosion. The method is expensive in terms of obtaining and transporting the clean soil (cover loam) and placing it on top of the contaminated land, usually to a depth of at least one metre. A membrane of a plastic material is usually placed between the cover loam and the underlying contaminated soil. This method is unlikely to be permanent owing to the possibility of soluble contaminants moving upwards through the cover soil by capillarity in dry conditions, or roots reaching the contaminated soil. Gases and vapours from the contaminated layer may also escape to the surface through pores and cracks in the cover soil.

9.3 Remediation *in situ* to Remove Contaminants

The *in situ* methods can include:

(i) irrigation of the soil with chemical reagents to extract the harmful contaminants and remove them by an underdrainage system;
(ii) the use of cultures of micro-organisms to decompose persistent organic pollutants, such as at derelict gas works sites;
(iii) abandoning a site until radioactive contaminants have decayed through one or more half-lives.

The irrigation of relatively permeable soils, such as sands and sandy loams, with dilute acid or chelating agents can be effective if an underdrainage system is installed to collect the leachate. A Cd contaminated site in the Netherlands has been successfully treated in this way with dilute HCl. The bioavailability of the remaining metals in acid leached soils will probably be increased and so the soils will need liming to pH 7 to counteract this.

Biotreatment of contaminated land involves creating conditions to promote the growth of micro-organisms which have the ability to degrade persistent organic contaminants, such as PAHs and chlorinated hydrocarbons. The technique is relatively inexpensive but slow. The rate at which degradation occurs depends on: the structure of the contaminant molecules, their toxicity to soil micro-organisms, their water solubility, the organic matter content of the soil, soil pH and temperature, and the supply of oxygen and nutrients.[21] Bacteria of the *Pseudomonas* spp. have been found to degrade TCDDs (dioxins) and several other species of bacteria have been found to degrade PAHs.

9.4 Remediation by Reducing Bioavailability

This is the most widely practiced form of remediation in many humid regions of the world and is particularly appropriate for trace metal contaminants, such as Cd, Cu, Cr, Pb, Ni, and Zn. Liming the soil to pH 7 or higher (with $CaCO_3$) renders these metals less mobile and bioavailable.[10] This method is regularly used in vineyards where Cu toxicity occurs as a result of the accumulation of Cu from fungicides used for up to a century. The most practicable method of keeping the bioavailability of heavy metals to a minimum in sewage sludge-amended soils is by keeping the pH at around 6.5 or 7. Other methods, often used together with liming, are: to add relatively large amounts of organic matter with the aim of locking-up metals as stable complexes with organic colloids and, or, to apply phosphatic fertilizer to stimulate crop growth and precipitate insoluble metal phosphates.

Permanently waterlogged soils have strong reducing conditions and many metals are present as insoluble sulfides. If possible, heavy metal polluted marsh areas should be left in a waterlogged state. Methylation of mercury can occur under these conditions and some of this metal may be released as volatile methyl compounds. Aeration of waterlogged soils leads to oxidation of the sulfides and a flush of highly available metal ions in a more acid soil solution.

Policies on cleaning-up contaminated sites differ between countries. Some, including the Netherlands require sites to be cleaned up to a specification which enables the land to be used for almost any purpose. This is technologically demanding and consequently very expensive. The UK policy is that the condition of a site determines its suitability for the use to which it is put; therefore, clean-up may not be required in all cases. However, where the contaminants are left in the soil they may remain for a long time (thousands of years in some cases, *e.g.* Pb)[10] and may affect later uses of the land or pose a hazard to groundwater quality.

[21] R. J. F. Bewley, 'Contaminated Land', ed. J. W. Assink and W. J. van den Brink, Martinus Nijhoff, Dordrecht, 1986, p.759.

10 CASE STUDIES

10.1 Heavy Metal Contamination from Metalliferous Mining

Zinc was mined around the village of Shipham in Somerset, England, during the Eighteenth and Nineteenth Centuries and very high concentrations of Pb, Zn, and Cd were left in the soils and waste heaps around the village. The significance of the contamination was not fully realized until the late 1970s when a nationwide geochemical survey and associated follow-up studies revealed that people occupying houses built on the contaminated land between 1951 and 1980 could be at risk from excessive concentrations of Cd and Pb in vegetables and fruit grown in their own gardens. Some of the highest Cd concentrations reported for agricultural or horticultural soils were found ($<470\ \mu g\ g^{-1}$ Cd) together with Pb108–6540 $\mu g\ g^{-1}$ and Zn 250–37 200 $\mu g\ g^{-1}$. Although the mean Cd concentration in nearly 1000 samples of fresh vegetables was almost 17 times higher than the national average of 0.015 $\mu g\ g^{-1}$ Cd, no obvious adverse health effects were found in the group of 500 volunteers investigated. Nevertheless, the soils will remain indefinitely contaminated with these metals and the residents of affected houses have been advised not to consume produce from their own gardens.[22]

In contrast to Shipham, elderly women living in the mining-polluted Jinzu Valley in Toyama Province, Japan, during World War II were found to be suffering from severe skeletal deformation ('Itai-itai' disease) due to exposure to Cd and other heavy metals. The concentrations of Cd in the soils were much lower ($<5\ \mu g\ g^{-1}$) than in Shipham, but the bioavailability of the metal was much higher. The people lived mainly on a diet of rice grown in contaminated paddy fields with a low intake of protein and calcium, the water supply was also contaminated, and all the women affected had given birth to more than one child.[10] From this, it can be seen that many other factors have to be considered apart from the type and amount of contamination in order to assess the impacts on human health.

10.2 Heavy Metal Contamination of Domestic Garden Soils in Urban Areas

Thornton *et al.*[23] reported concentrations ($\mu g\ g^{-1}$) of Pb $<$ 14 125 with a geometric mean (GM) of 230, Cd $<$17 (GM 1.2), Cu $<$ 16 800 (GM 53), and Zn $<$ 14 568 (GM 260) in samples from 3550 urban gardens in the UK. The geometric mean concentrations ($\mu g\ g^{-1}$) of these metals in 579 gardens within Greater London were higher than the national values quoted above for: Pb (647), Cd (1.3), Cu (73), and Zn (424). Heavy metal concentrations were higher in the gardens of older houses. For 97 inner city gardens in Birmingham, it was found that Pb levels were significantly higher in the gardens of houses more than 35 years old, in gardens within a 500 m radius of a commercial garage, those in close

[22] H. Morgan, *Sci. Total Environ.* (Special Issue: The Shipham Report), 1988, **75**, No. 1, 143.
[23] I. Thornton, E. Culbard, S. Moorcroft, J. Watt, M. Wheatley, and M. Thompson, *Environ. Technol. Letters*. 1983, **6**, 137.

proximity to: demolition sites, tips, metallurgical industries, and within 10 m of a road. The sources of the heavy metals in urban domestic garden soils include: atmospheric fallout of metals from car exhausts and industrial processes, paint scrapings, coal ash and soot, bonfires, buried rubbish, manures, and fertilizers.

10.3 Soil Contaminated by Landfill and Waste Disposal Operations

A section of a cutting originally intended to be the Love Canal at Niagara Falls in New York State, USA, had been used for the disposal of chemical waste in the 1940s by the Hooker Chemical Company. About 20 000 tonnes of waste, including intermediates from the manufacture of chlorinated pesticides had been dumped. The waste pit was eventually covered with soil and the site later used for the construction of a school and houses.[24]

Although there had been reports of school children being affected by irritation of eyes and respiratory tract, the full extent of the contamination problem was not appreciated until heavy precipitation in the winter of 1977–78 caused the water table to rise to the surface and into the basements of adjacent houses. This brought many noxious chemicals with it and some buried drums broke through the softened covering soil. More severe health effects were reported and the US Environmental Protection Agency (EPA) carried out a series of investigations. Analysis of the groundwater at the site revealed 82 different chemicals of which 27 were on the EPA's priority list and 11 were carcinogenic. The site was declared a disaster area and 239 families were evacuated initially followed by many more later.[24] The EPA has documented 170 000 waste disposal sites in the USA, of which 2100 have been used for industrial wastes; interestingly, Love Canal only ranked 25th in a hazard rating of these sites.[24]

The village of Lekkerkirk, 20 km east of Rotterdam, in the Netherlands, was constructed in 1970–71 beside the River Lek on reclaimed land which had been raised <3.5 m by a layer of household and demolition waste covered by 0.7 m of sand. However, this elevation was still lower than that of the river and hence the groundwater flowed upwards carrying pollutants from the underlying waste towards the ground surface. In 1978, the contamination of the surface soil was manifested by deterioration of plastic drinking water pipes, noxious odours in the houses, and growth abnormalities in garden plants. By 1980, it was found that the pollutants had entered the drinking water through the deteriorated pipes and 75% of the houses in the village had to be evacuated. The empty houses were supported on hydraulic jacks while 87 000 m^3 of contaminated fill was excavated from beneath them and subsequently disposed of by incineration. The main contaminants in the waste were paint solvents and resins, giving concentrations in the soil of: toluene ($<1000\ \mu g\ g^{-1}$), lower boiling point solvents ($<3000\ \mu g\ g^{-1}$), and metals: Sb ($230\ \mu g\ g^{-1}$), Cd $<97\ \mu g\ g^{-1}$), Hg ($<8.2\ \mu g\ g^{-1}$), Pb ($<740\ \mu g\ g^{-1}$), and Zn ($<1670\ \mu g\ g^{-1}$).[17]

[24] S. E. Manahan, 'Environmental Chemistry', Brooks/Cole (Wadsworth), Monterey, California, USA, 1984, pp.612.

10.4 Former Gasworks Sites

From the late Nineteenth Century, most towns in industrialized countries with indigenous coal supplies produced coal gas for industrial and domestic use. As a result, there are now thousands of sites of former gasworks posing a major contamination problem in urban areas where derelict land is needed for housing, schools, and shops.

The dry heating of coal in an oven produced crude coal gas, coke, ash, and cinders. The crude gas was subsequently purified by being passed through tar separators, condensors, and wet purification to remove NH_3. HCN, phenols, and creosols, and finally dry purification with ferric oxide to remove sulfur and cyanide compounds. These processes resulted in several highly toxic compounds being produced which accumulated on gas works sites. The ash, cinders, and spent ferric oxide contain Pb, Cu, As, CN, sulfate, and sulfide, and the tars contain: hydrocarbons, phenols, benzene, xylene, naphthalene, and PAHs.[25]

A survey of eight former gas works sites in the UK showed the following maximum values ($\mu g\ g^{-1}$) for contaminants in the soil: sulfate <250 000, elemental S <250 000, free CN <64, total CN <8000, total phenols <1000, toluene extract 250 000, sulfide <4000, and total metals: As <250, Cd <64, Pb <4000, and Cr <250.[17] In some cases, old landfills near gas works may have received large amounts of spent oxides and other easily transported solid wastes.

10.5 Other Derelict Industrial Sites

Aldred and Lord[26] reported surveys of three derelict industrial sites in Manchester, England. The site of a former bleaching and dye works was contaminated with metals, sulfate, and phenols with the highest concentrations found in the areas of effluent storage and disposal. Total metal concentrations ($\mu g\ g^{-1}$) included: Cr <10 700, Cu <1350, Pb <950, and As <100 but bioavailable levels were also high. Sulfate concentrations were <10 500 $\mu g\ g^{-1}$, phenols <100 $\mu g\ g^{-1}$, and the pH ranged from 3.2–9.2.

A former petroleum storage site contained total oil concentrations in the soil of <0.4% with paraffins totalling <2300 $\mu g\ g^{-1}$ and benzene, toluene, and xylene present at <150 $\mu g\ g^{-1}$. The major hazard of the site was the danger of flash ignition; the low flash points (closed cup) of the contaminated soil were 5.6 °C–25.6 °C. The options for the future development of the site were either to abandon any building development, or to remove the soil causing the flash risk and carry out appropriate follow-up remediation.

The soil on the site of a solvent recovery works contained high concentrations of solvents ($\mu g\ g^{-1}$): toluene <2100, xylene <4000, white spirit <1700, with hundreds of minor organic compounds, giving a total concentration of quantifiable solvents of <7600. Local experience in Manchester had shown that less than

[25] J. M. Roels and R. Kabos, 'Contaminated Land', ed. J. W. Assink and W. J. van den Brink, Martinus Nijhoff, Dordrecht, 1986, p.337.

[26] J. B. Aldred and D. W. Lord 'Experiences in the Investigation of Contaminated Land' preprint for 'Symposium on Hazardous Waste Disposal and the Re-Use of Contaminated Land', Society of Chemical Industry, London, 1984.

50 μg g^{-1} total solvents caused no odour problem, 50–500 μg g^{-1} is potentially hazardous, and concentrations of above 500 μg g^{-1} pose a high risk. The options being considered for this site were either to remove all the solvent-contaminated soil, replace with clean soil and use it for a recreational amenity, or to level and cap it with clay and tarmac for use as a car park.

CHAPTER 6

Integrative Aspects of Pollutant Cycling

ROY M. HARRISON

1 INTRODUCTION: BIOGEOCHEMICAL CYCLING

The earlier chapters of this book have followed the traditional sub-division of the environment into compartments (*e.g.* atmosphere, oceans, *etc.*). Whilst these sub-divisions accord with human perceptions and have certain scientific logic, they encourage the idea that each compartment is an entirely separate entity and that no exchanges occur between them. This, of course, is far from the truth. Important exchanges of mass and energy occur at the boundaries of the compartments and many processes of great scientific interest and environmental importance occur at these interfaces. A physical example is that of transfer of heat between the ocean surfaces and the atmosphere, which has a major impact upon climate and a great influence upon the general circulation of the atmosphere. A chemically-based example is the oceanic release of dimethyl sulfide to the atmosphere, which may, through its decomposition products, act as a climate regulator (see Chapter 4).

Pollutants emitted into one environmental compartment will, unless carefully controlled, enter others. Figure 1 illustrates the processes affecting a pollutant discharged into the atmosphere.[1] As mixing processes dilute it, it may undergo chemical and physical transformations before depositing in rain or snow (wet deposition) or as dry gas or particles (dry deposition). The deposition processes cause pollution of land, freshwater, or the seas, according to where they occur. Similarly, pollutants discharged into a river will, unless degraded, enter the seas. Solid wastes are often disposed into a landfill. Nowadays these are carefully designed to avoid leaching by rain and dissemination of pollutants into groundwaters, which might subsequently be used for potable supply. In the past, however, instances have come to light where insufficient attention was paid to the potential for groundwater contamination, and serious pollution has arisen as a result.

[1] W. H. Schroeder and D. A. Lane, *Environ. Sci. Technol.*, 1988, **22**, 240.

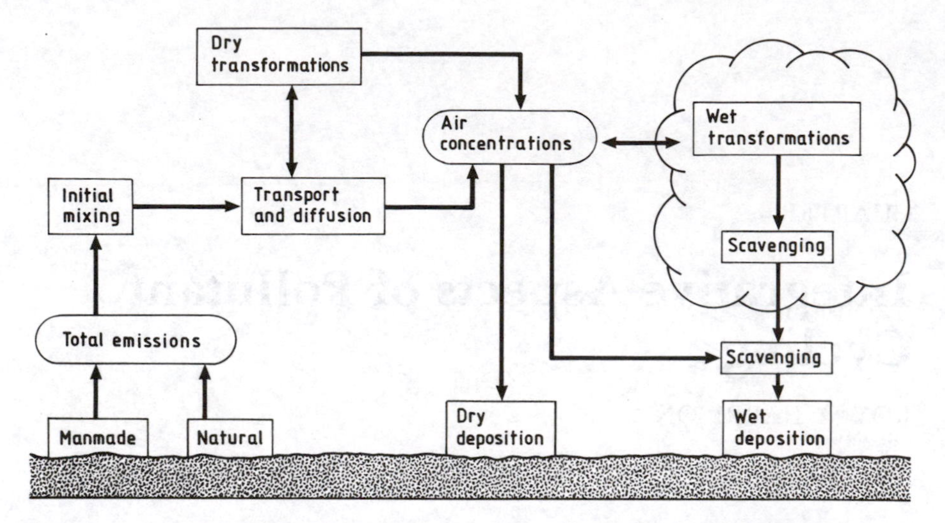

Figure 1 *Schematic diagram of the atmospheric cycle of a pollutant*[1] (Reprinted from Environmental Science and Technology by permission of the American Chemical Society)

Another important consideration regarding pollutant cycling is that of degradability, be it chemical or biological. Chemical elements (other than radioisotopic forms) are, of course, non-degradable and hence once dispersed in the environment will always be there, although they may move between compartments. Thus, lead, for example, after emission from industry or motor vehicles, has a rather short lifetime in the atmosphere, but upon deposition causes pollution of vegetation, soils, and waters.[2] On a very long timescale, lead in these compartments will leach out from soils and transfer to the oceans, where it will concentrate in bottom sediments.

Some chemical elements undergo chemical changes during environmental cycling which completely alter their properties. For example, nitrate added to soil as fertilizer can be converted to gaseous nitrous oxide by biological denitrification processes. Nitrous oxide is an unreactive gas with a long atmospheric lifetime which is destroyed only by breakdown in the stratosphere. As will be seen later, nitrogen in the environment may be present in a wide range of valence states, each conferring different properties.

Some chemical compounds are degradable in the environment. For example, methane (an important greenhouse gas) is oxidized via carbon monoxide to carbon dioxide and water. Thus, although the chemical elements are conserved, methane itself is destroyed and were it not continuously replenished, would disappear from the atmosphere. The breakdown of methane is an important source of water vapour in the stratosphere, illustrating another, perhaps less obvious, connection between the cycles of different compounds.

[2] R. M. Harrison and D. P. H. Laxen, 'Lead Pollution: Causes and Control', Chapman and Hall, London, 1981.

Table 1 *Size and vertical mixing of various reservoirs (from Brimblecombe[3])*

	Mass (kg)	*Mixing time* (years)
Biosphere*	4.2×10^{15}	60
Atmosphere	5.2×10^{18}	<0.2
Hydrosphere	2.4×10^{21}	1600
Crust	2.4×10^{22}	$>3 \times 10^{7}$
Mantle	4.0×10^{24}	$>10^{8}$
Core	1.9×10^{24}	

* Plants, animals, and organic matter are included but coal and sedimentary carbon are not. The mixing time of carbon in living matter is about 50 years

Degradable chemicals which cease to be used will disappear from the environment. PCBs are no longer used industrially to any significant degree, having been replaced by more environmentally acceptable alternatives. Their concentrations in the environment are decreasing, although because of their slow degradability (*i.e.* persistence), it will take many years before their levels decrease below analytical detection limits.

The transfer of an element between different environmental compartments, involving both chemical and biological processes is termed biogeochemical cycling. The biogeochemical cycles of the elements lead and nitrogen will be discussed later in this chapter.

1.1 Environmental Reservoirs

To understand pollutant behaviour and biogeochemical cycling on a global scale, it is important to appreciate the size and mixing times of the different reservoirs. These are given in Table 1. The mixing times are a very approximate indication of the timescale of vertical mixing of the reservoir.[3] Global mixing can take very much longer as this involves some very slow processes. These mixing times should be treated with considerable caution as they oversimplify a complex system. Thus, for example, a pollutant gas emitted at ground level mixes in the boundary layer (*ca.* 1 km) on a timescale typically of hours. Mixing into the free troposphere (1–10 km) takes days, whilst mixing into the stratosphere (10–50 km) is on the timescale of several years. Thus, no one timescale describes atmospheric vertical mixing and the same applies to other reservoirs. Such concepts are useful, however, when considering the behaviour of trace components. For example, a highly reactive hydrocarbon emitted at ground level will probably be decomposed in the boundary layer. Sulfur dioxide, with an atmospheric lifetime of days, may enter the free troposphere but is unlikely to enter the

[3] P. Brimblecombe, 'Air Composition and Chemistry', Cambridge University Press, Cambridge, 1986.

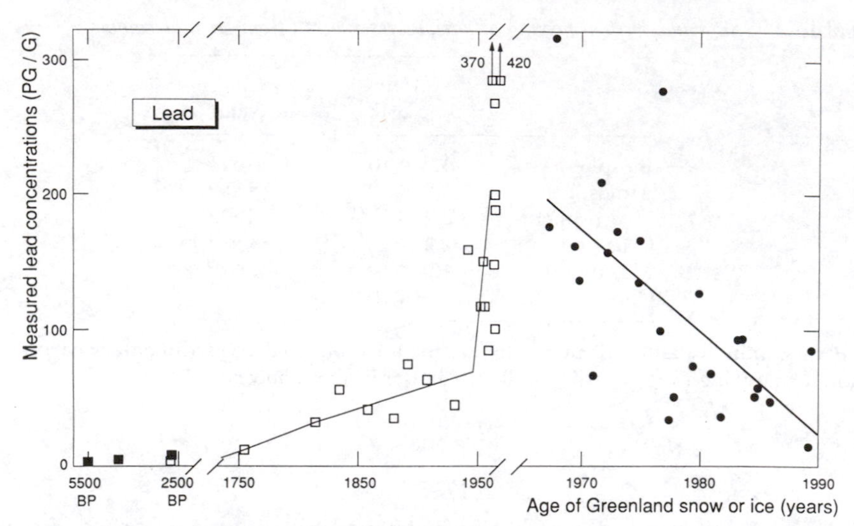

Figure 2 *Changes in lead concentrations in Greenland ice and snow from 5500 BP to present (from Boutron* et al.[4]*)* (Reprinted by permission from *Nature (London)*, **353**, p. 153; Copyright ©1991 Macmillan Magazines Ltd.)

stratosphere. Methane, with a lifetime of several years, extends through all of the three regions.

It should be noted from Table 1 that the atmosphere is a much smaller reservoir in terms of mass than the others. The implication is that a given pollutant mass injected into the atmosphere will represent a much larger proportion of total mass than in other reservoirs. Because of this, and the rather rapid mixing of the atmosphere, *global* pollution problems have become serious in relation to the atmosphere before doing so in other environmental media. The converse also tends to be true, that once emissions into the atmosphere cease, or diminish, the beneficial impact is seen on a relatively short time-scale. This has been seen in relation to lead, for instance, where lead in Antarctic ice (derived from snow) has shown a major decrease resulting from diminishing use of leaded petrol[4] (Figure 2). Improved air quality in relation to CFCs will take longer to achieve because of the much longer atmospheric lifetimes (>100 years) of some of these species.

1.2 Lifetimes

A very useful concept in the context of pollutant cycling is that of the lifetime of a substance in a given reservoir. We can think in terms of substances having sources, magnitude S, and sinks, magnitude R. At equilibrium:

$$R = S$$

[4] C. F. Boutron, U. Gorlach, J.-P. Candelone, M. A. Bolshov, and R. J. Delmas, *Nature (London)*, 1991, **353**, 153.

An analogy is with a bath; the inflow from a tap (S) is equal to the outflow (R) when the bath is full. An increase in S is balanced by an increase in R. If the total amount of substance in the reservoir (analogy = mass of water in the bath) is A, then the lifetime, τ is defined by:

$$\tau = \frac{A\ (\text{kg})}{S\ (\text{kg s}^{-1})} \tag{1}$$

If the removal mechanism is a chemical reaction, its rate may be described as follows:

$$R' = \frac{\mathrm{d}[A]}{\mathrm{d}t} = k[A] \tag{2}$$

(In this case $\mathrm{d}[A]/\mathrm{d}t$ describes the rate of loss of A if the source is switched off; obviously with the source on, at equilibrium $\mathrm{d}[A]/\mathrm{d}t = 0$). The latter part of equation (2) assumes first order decay kinetics, *i.e.* the rate of decay is equal to the concentration of A, termed $[A]$, multiplied by a rate constant, k. As discussed later this is often a reasonable approximation.

Taking equation (1) and dividing both numerator and denominator by the volume of the reservoir, allows it to be re-written in terms of concentration. Thus:

$$\tau = \frac{[A]\ (\text{kg m}^{-3})}{S'\ (\text{kg m}^{-3}\ \text{s}^{-1})} \tag{3}$$

since $S' = R'$

$$\tau = \frac{[A]}{k[A]} = k^{-1} \tag{4}$$

Thus the lifetime of a constituent with a first order removal process is equal to the inverse of the first order rate constant for its removal. Taking an example from atmospheric chemistry, the major removal mechanism for many trace gases is reaction with the hydroxyl radical, OH. Considering two substances with very different rate constants[5] for this reaction, methane and nitrogen dioxide:

$$CH_4 + OH \rightarrow CH_3 + H_2O \tag{5}$$

$$\frac{-\mathrm{d}}{\mathrm{d}t}[CH_4] = k_2[CH_4][OH] \quad k_2 = 8.4 \times 10^{-15}\ \text{cm}^3\ \text{molec s}^{-1} \tag{6}$$

$$NO_2 + OH \rightarrow HNO_3 \qquad k_2 = 1.1 \times 10^{-11}\ \text{cm}^3\ \text{molec s}^{-1} \tag{7}$$

Making the crude assumption of a constant concentration of OH radical (more justifiable for the long-lived methane, for which fluctuations in OH will average out, than for short-lived nitrogen dioxide),

$$\frac{-\mathrm{d}}{\mathrm{d}t}[CH_4] = k_2[CH_4][OH]$$

$$= k_1'[CH_4]$$

[5] B. J. Finlayson-Pitts and J. N. Pitts Jr., '*Atmospheric Chemistry*', John Wiley and Sons, Chichester, 1986.

where
$$\begin{aligned} k'_1 &= k_2[\text{OH}] \\ &= 8.4 \times 10^{-15} \times 1 \times 10^6 \\ &= 8.4 \times 10^{-9}\ \text{s}^{-1} \end{aligned}$$

assuming an OH concentration[6] of 1×10^6 molec cm^{-3}.
Then from equation (4)

$$\begin{aligned} \tau &= k^{-1} \\ &= (8.4 \times 10^{-9})^{-1}\ \text{s} \\ &= 3.8 \text{ years for } CH_4 \end{aligned}$$

By analogy, for nitrogen dioxide, the lifetime,

$$\tau = 25 \text{ hours}$$

This general approach to atmospheric chemical cycling has proved useful in many instances. For example, measurements of atmospheric concentration, $[A]$, for a globally mixed component may be used to estimate source strength, since

$$S' = R' = \frac{\mathrm{d}[A]}{\mathrm{d}t} = k_2[A][\text{OH}]$$

and

$$S = S' \cdot V$$

where V is the volume of atmosphere in which the component is mixed. Source strengths estimated in this way, for example for the compound methyl chloroform, CH_3CCl_3, known to destroy stratospheric ozone, may be compared with known industrial emissions to deduce whether natural sources contribute to the atmospheric burden.

There is an interesting relationship between the spatial variability in the concentration of an atmospheric trace species and its atmospheric lifetime.[3] Compounds such as methane and carbon dioxide with an atmospheric lifetime of years show little spatial variability around the globe, as the timescale of mixing of the entire troposphere is of the order of a year. Short-lived species are removed before they can mix thoroughly and show far greater spatial variation. By analogy, they also show far more temporal variation at a given measuring point.

2 TRANSFER FLUXES BETWEEN ENVIRONMENTAL COMPARTMENTS

2.1 Air–Land Exchange

The land surface is an efficient sink for many trace gases. These are absorbed or decomposed on contact with plants or soil surfaces. Plants can be particularly

[6] C. N. Hewitt and R. M. Harrison, *Atmos. Environ.*, 1985, **19**, 545.

Table 2 *Some typical values of deposition velocity*

Pollutant	*Surface*	*Deposition velocity* ($cm\ s^{-1}$)
SO_2	Grass	1.0
SO_2	Ocean	0.5
SO_2	Soil	0.7
SO_2	Forest	2.0
O_3	Dry grass	0.5
O_3	Wet grass	0.2
O_3	Snow	0.1
HNO_3	Grass	2.0
CO	Soil	0.05
Aerosol (<2.5 μm)	Grass	0.1

active because of their large surface area and ability to absorb water-soluble gases. The deposition process is crudely described by the deposition velocity, v_g,

$$v_g\ (\text{cm s}^{-1}) = \frac{\text{Flux}\ (\mu\text{g m}^{-2}\ \text{s}^{-1})}{\text{Atmospheric concentration}\ (\mu\text{g m}^{-3})}$$

Since the deposition process itself causes a gradient in atmospheric concentration, v_g is defined in relation to a reference height, usually 1 metre, at which the atmospheric concentration is measured. For reasons described later, v_g is not a constant for a given substance, but varies according to atmospheric and surface conditions. However, some typical values are given in Table 2, which exemplify the massive variability.

For some trace gases, for example, nitric acid vapour, dry deposition represents a major sink mechanism. In this case the process may have a major impact upon atmospheric lifetime. Taking as another example, ozone in the boundary layer, dry deposition is believed to be the main sink mechanism. Assuming a typically dry deposition velocity of 1 cm s^{-1} and a boundary layer height of 1000 metres, (H),

$$\frac{-\text{d}}{\text{d}t}[\text{O}_3]\ (\mu\text{g m}^{-3}\ \text{s}^{-1}) = \frac{\text{Flux}\ (\mu\text{g m}^{-2}\ \text{s}^{-1})}{\text{Mixing depth (m)}}$$

$$= \frac{v_g \cdot [\text{O}_3]}{H}$$

$$= k[\text{O}_3] \quad \text{where } k = v_g/H$$

By analogy with equation (4),

$$\tau = \frac{H}{v_g}$$

$$= 1000/0.01\ \text{s}$$

$$= 28\ \text{hours}$$

Thus, taking the boundary layer as a discrete compartment, the lifetime of ozone with respect to dry deposition is around 1 day. The lifetime in the free troposphere (the section of the atmosphere above the boundary layer) is longer, being controlled by transfer processes in and out, and chemical reactions. The stratospheric lifetime of ozone is controlled by photochemical and chemical reaction processes.

Dry deposition processes are best understood by considering a resistance analogue. In direct analogy with electrical resistance theory, the major resistances to deposition are represented by three resistors in series. These are as follows:

(i) r_a, the aerodynamic resistance describes the resistance to transfer towards the surface through normally turbulent air;
(ii) r_b, the boundary layer resistance describes the transfer through a laminar boundary layer (approximately 1 mm thickness) at the surface;
(iii) r_s, the surface (or canopy) resistance is the resistance to uptake by the surface itself. This can vary enormously, from essentially zero for very sticky gases such as HNO_3 vapour which attaches irreversibly to surfaces, to very high values for gases of low water solubility which are not utilized by plants (*e.g.* CFC's).

Since these resistances operate essentially in series, the total resistance, R, which is the inverse of the deposition velocity, is equal to the sum of the individual resistances.

$$R = \frac{1}{v_g} = r_a + r_b + r_s \tag{8}$$

Some trace gases have a net source at the ground surface and diffuse upwards; an example is nitrous oxide.

Whether the flux is downward or upward, it is driven by a concentration gradient in the vertical, dc/dz. The relationship between flux, F, and concentration gradient is:

$$F = K_z dc/dz$$

where K_z is the eddy diffusivity in the vertical (a measure of the atmospheric conductance). Fluxes, and thus deposition velocities can be estimated by measurement of a concentration gradient simultaneously with the eddy diffusivity.[7] It is usually assumed that trace gases transfer in the same manner as sensible heat (*i.e.* convective heat transfer, not radiative or latent heat) or momentum. Thus the eddy diffusivity for either of these parameters is measured, usually from simple meteorological variables (gradients in temperature and wind speed).

A few substances are capable of showing both upward and downward fluxes. An example is ammonia. Ammonium in the soil, NH_4^+, is in equilibrium with ammonia gas, NH_{3g},

[7] J. A. Garland, *Proc. R. Soc. London, Ser. A*, 1977, **354**, 245.

$$NH_4^+ + H_2O \leftrightharpoons NH_{3g} + H_3O^+ \tag{9}$$

when atmospheric concentrations of ammonia exceed equilibrium concentrations at the soil surface, the net flux of ammonia is downwards. When atmospheric concentrations are below the equilibrium value, ammonia is released into the air.[8]

2.2 Air–Sea Exchange

The oceans cover some two-thirds of the earth's surface and consequently provide a massive area for exchange of energy (climatologically important) and matter (an important component of geochemical cycles).

The seas are a source of aerosol, (*i.e.* small particles) which transfer to the atmosphere. These will subsequently deposit, possibly after chemical modification, either back in the sea (the major part) or on land (the minor part). Marine aerosol comprises largely unfractionated seawater, but may also contain some abnormally enriched components. One example of abnormal enrichment occurs on the eastern coast of the Irish Sea. Liquid effluents from the Sellafield nuclear fuel reprocessing plant in west Cumbria are discharged into the Irish Sea by pipeline. At one time, permitted discharges were appreciable and as a result radioisotopes such as ^{137}Cs and several isotopes of plutonium have accumulated in the waters and sediments of the Irish Sea. A small fraction of these radioisotopes were carried back inland in marine aerosol and deposited predominantly in the coastal zone.[9] Whilst the abundance of ^{137}Cs in marine aerosol was reflective only of its abundance in seawater (an enrichment factor—see Chapter 4—of close to unity), plutonium was abnormally enriched due to selective incorporation of small suspended sediment particles in the aerosol. This has manifested itself in enrichment of plutonium in coastal soils on the west Cumbria coast,[10] shown as contours of $^{239+240}Pu$ deposition (pCi cm^{-2}) to soil in Figure 3.

The seas may also act as a receptor for depositing aerosol. Deposition velocities of particles to the sea are a function of particle size, density, and shape, as well as the state of the sea. Experimental determination of aerosol deposition velocities to the sea is almost impossible and we have to rely upon data derived from wind tunnel studies and theoretical models. The results from two such models appear in Figure 4, in which particle size is expressed as aerodynamic diameter, or the diameter of an aerodynamically equivalent sphere of unit specific gravity.[11,12] If the airborne concentration in size fraction of diameter d_i is c_i, then

$$\text{Total flux} = \sum^{i} v_g(d_i) \cdot c_i$$

[8] R. M. Harrison, S. Rapsomanikis, and A. B. Turnbull, *Atmos. Environ.*, 1989, **23**, 1795.
[9] R. S. Cambray and J. D. Eakins, *Nature (London)*, 1982, **300**, 46.
[10] P. A. Cawse, UKAEA Report No. AERE—9851, 1980.
[11] S. A. Slinn and W. G. N. Slinn, *Atmos. Environ.*, 1980, **14**, 1013.
[12] R. M. Williams, *Atmos. Environ.*, 1982, **16**, 1933.

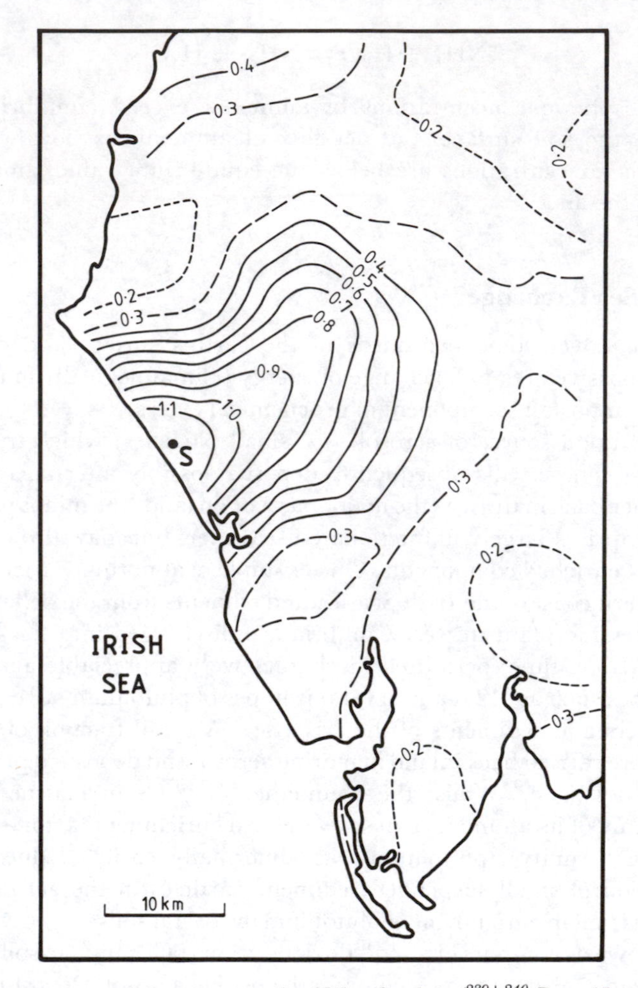

Figure 3 *Concentrations of plutonium in soils of West Cumbria ($^{239+240}$Pu to 15 cm depth; pCi cm^{-2}) (from Cawse[10]). The point marked S indicates the position of the Sellafield reprocessing works*

where $v_g(d_i)$ is the mean value of deposition velocity appropriate to the size fraction d_i. Measurements show that whilst most of the lead is associated with small, sub-micrometer particles, the larger particles comprise the major part of the flux.

Airborne concentrations of particulate pollutants are not uniform over the sea. The spatial distribution of zinc over the North Sea[13] averaged over a number of measurement cruises appears in Figure 5. Spatial patterns of other metals and many man-made pollutants are similar, reflecting the impact of land-based source regions, with concentrations falling towards the north and centre of the sea.

[13] C. R. Ottley and R. M. Harrison, Eurotrac ASE Annual Report, 1990.

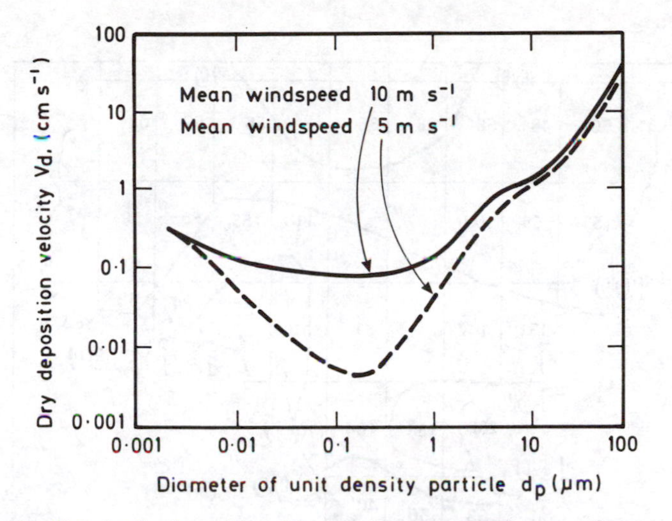

Figure 4 *Calculated values of deposition velocity to water surfaces as a function of particle size and wind speed*

Because of its position and relatively high pollution loading, the North Sea is a focus of considerable interest. An inventory of inputs of trace metals (*e.g.* Pb, Cd, Zn, Cu, *etc.*) accords similar importance to riverine inputs and atmospheric deposition.[14] Controls have now been applied to many source categories and total inputs of the metals indicated have in general declined appreciably. One particular example is lead, for which most European countries introduced severe controls on use in gasoline during the 1980s and atmospheric concentrations have fallen accordingly. Although the data are less clear, it might be anticipated that concentrations in river water will also decline as a result of reduced inputs from direct atmospheric deposition and in run-off waters from highways and land surfaces.

As explained in Chapter 4, the sea may be both a source and a sink of trace gases. The direction of flux is dependent upon the relative concentration in air and seawater.[15] If the concentration in air is C_a, the equilibrium concentration in seawater, C_w (equ) is given by

$$C_w\,(\text{equ}) = C_a H^{-1} \tag{10}$$

where H is the Henry Law constant.

If C_w is the actual concentration of the dissolved gas in the surface seawater and

$$C_w = C_w\,(\text{equ})$$

the system is at equilibrium and no net transfer occurs. If, however, there is a concentration difference, ΔC, where

[14] R. F. Critchley, *Proc. Int. Conf. Heavy Metals in the Environment*, Heidelberg, Germany; CEP Consultants, Edinburgh, 1983, 1108.

[15] P. S. Liss and L. Merlivat, 'The Role of Air–Sea Exchange in Geochemical Cycling', ed. P. Buat-Menard, Reidel, 1986, p. 113.

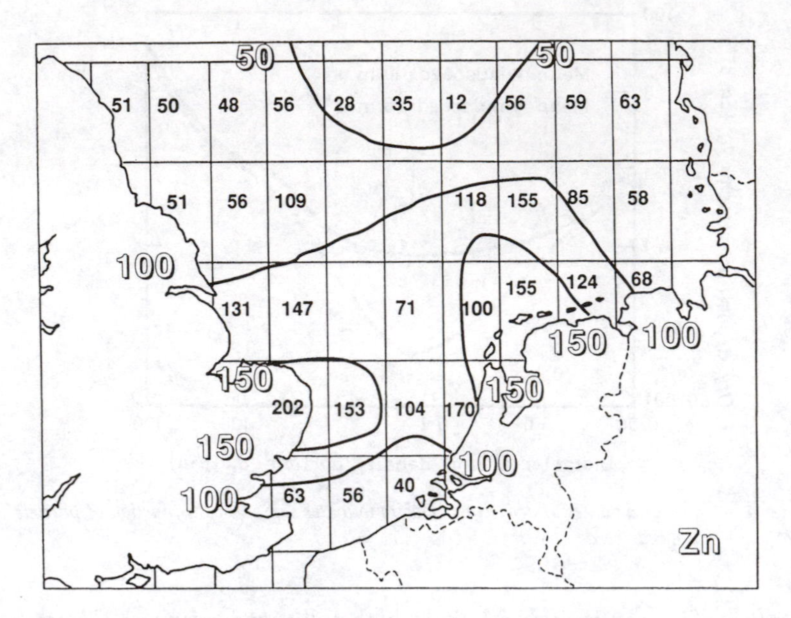

Figure 5 *Spatial distribution of zinc concentrations in air over the North Sea during 1989 (from Ottley and Harrison*[13]*) (in* ng m^{-3}*)*

$$\Delta C = C_a H^{-1} - C_w \qquad (11)$$

there will be a net flux. If

$$C_a H^{-1} > C_w$$

the water is sub-saturated with regard to the trace gas and transfer occurs from air to water. Conversely, gas transfers from supersaturated water to the atmosphere if

$$C_a H^{-1} < C_w$$

The rate at which gas transfer occurs is expressed by

$$F = K_{(T)w} \Delta C \qquad (12)$$

where $K_{(T)w}$ is termed the total transfer velocity. This can be broken down into component parts as follows:

$$\frac{1}{K_{(T)w}} = \frac{1}{\alpha k_w} + \frac{1}{Hk_a} = r_w + r_a \qquad (13)$$

where k_a and k_w are the individual transfer velocities for chemically unreactive gases in air and water phases, respectively and α (= $k_{reactive}/k_{inert}$) is a factor which quantifies any enhancement of gas transfer in the water due to chemical

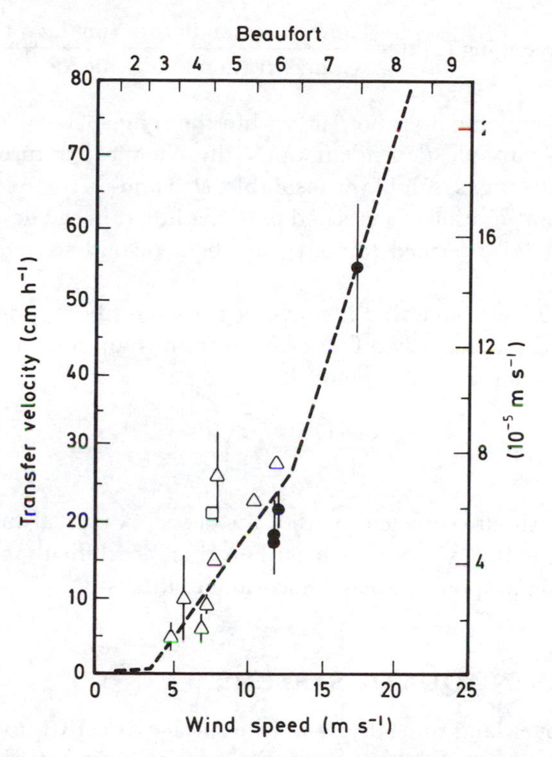

Figure 6 *Air–sea transfer velocities for carbon dioxide at 20 °C as a function of wind speed at 10 metres (m s^{-1} or Beaufort Scale). The graph combines experimental data (points) and a theoretical line (from Watson et al.*[16]*)* (Reprinted by permission from *Nature (London)*, **349**, 145; Copyright © 1991 Macmillan Magazines Ltd.)

reaction. The terms r_w and r_a are the resistances to transfer in the water and air phases respectively and are directly analogous to the resistance terms in equation 8. For chemically reactive gases, usually $r_a \gg r_w$ and atmospheric transfer limits the overall flux. For less reactive gases the inverse is true and $K_{(T)w} \cong k_w$; the resistance in the water is the dominant term.

Much research has gone into evaluating k_w and $K_{(T)w}$, both in theoretical models, wind tunnel, and field studies. The results are highly wind-speed dependent due to the influence of wind upon the surface state of the sea. The results of some theoretical predictions and experimental studies[16] for CO_2 (a gas for which k_w is dominant) are shown in Figure 6.

In addition to dry deposition, trace gases and particles are also removed from the atmosphere by rainfall and other forms of precipitation (snow, hail, *etc.*), entering land and seas as a consequence. Wet deposition may be simply described in two ways. Firstly,

[16] A. J. Watson, R. C. Upstill-Goddard, and P. S. Liss, *Nature (London)*, 1991, **349**, 145.

$$\text{Scavenging ratio} = \frac{\text{Concentration in rain (mg kg}^{-1})}{\text{Concentration in air (mg kg}^{-1})}$$

Typical values of scavenging ratio[17] lie within the range 300–2000. Scavenging ratios are rather variable dependent upon the chemical nature of the trace substance (particle or gas, soluble or insoluble, *etc.*) and the type of atmospheric precipitation. Incorporation of gases and particles into rain can occur both by in-cloud scavenging (also termed rainout) and below-cloud scavenging (termed washout).

Numerical modellers often find it convenient to describe wet deposition by a scavenging coefficient, actually a first order rate constant for removal from the atmosphere. Thus, for trace substance A,

$$\frac{d[A]}{dt} = -\Lambda[A]$$

where Λ is the washout coefficient, with units of s^{-1}. A typical value of Λ for a soluble substance is $10^{-4}\ s^{-1}$ although actual values are difficult to measure and are highly dependent upon factors such as rainfall intensity.

3 TRANSFERS IN AQUATIC SYSTEMS

When rain falls over land some drains off the surface directly into surface water courses in surface run-off. A further part of the incoming rainwater percolates into the soil and passes more slowly either into surface waters, or into underground reservoirs. Water held in rock below the surface is termed groundwater, and a rock formation which stores and transmits water in useful quantities is termed an aquifer. Water which passes through soil or rock on its way to a river is chemically modified during transit, generally by addition of soluble and colloidal substances washed out of the ground. Some substances are removed from the water; for example river water often contains less lead than rainwater; one mechanism of removal is uptake by soil.

River waters carry both dissolved and suspended substances to the sea. The concentrations and absolute fluxes vary tremendously. The suspended solids load is largely a function of the flow in the river, which influences the degree of turbulence and thus the extent to which solids are held in suspension and resuspended from the bed, once deposited. Table 3 shows a comparison of 'average' riverine suspended particulate matter and surficial rock composition[18] for the major elements. Elements resistant to chemical weathering or biological activity (*e.g.* aluminium, titanium, iron, phosphorus) show some enrichment in the riverine solids, whilst more soluble elements are subject to weathering and are depleted in the solids, being transported largely in solution (sodium, calcium). Some pollutant elements such as the metals lead, cadmium, and zinc tend to be

[17] R. M. Harrison and A. G. Allen, *Atmos. Environ.*, 1991, **25A**, 1719.
[18] J. M. Martin and M. Meybeck, *Mar. Chem.*, 1979, **7**, 177–206.

Table 3 *A comparison of the concentration of major elements in 'average' riverine particulate material and surficial rocks (adapted from Martin and Meybeck[18])*

	Concentrations (g kg^{-1})	
Element	*Riverine particulate material*	*Surficial rocks*
Al	94.0	69.3
Ca	21.5	45.0
Fe	48.0	35.9
K	20.0	24.4
Mg	11.8	16.4
Mn	1.1	0.7
Na	7.1	14.2
P	1.2	0.6
Si	285.0	275.0
Ti	5.6	3.8

Table 4 *Average concentrations of the major constituents dissolved in rain and river water (adapted from Garrels and Mackenzie[19])*

	Concentrations (mg dm^{-3})	
Constituent	*Rain water*	*River water*
Na^+	1.98	6.3
K^+	0.30	2.3
Mg^{2+}	0.27	4.1
Ca^{2+}	0.09	15
Fe		0.67
Al		0.01
Cl^-	3.79	7.8
SO_4^{2-}	0.58	11.2
HCO_3^-	0.12	58.4
SiO_2		13.1
pH	5.7	

highly enriched in the solids relative to surficial rocks or soils due to man-made inputs.

The dissolved components of river water typically exhibit significantly higher concentrations than in rainwater[19] (Table 4), due to leaching from rocks and soils. Some insight into the processes governing river water composition may be

[19] R. M. Garrels and F. T. MacKenzie, 'Evolution of Sedimentary Rocks', ed. W. W. Norton, New York, 1971.

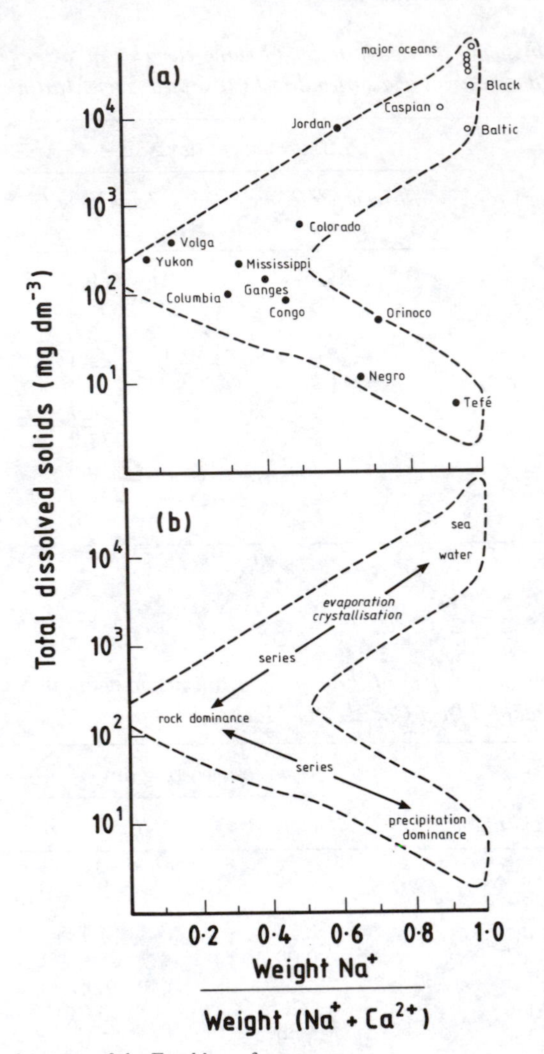

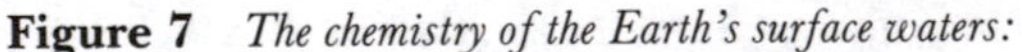

Figure 7 *The chemistry of the Earth's surface waters:*
(a) *typical values of the ratio $Na^+/(Na^+ + Ca^{2+})$ as a function of dissolved solids concentration for various major rivers and oceans;*
(b) *the processes leading to the observed ratios (from Gibbs[20]).* Copyright © 1970, American Association for the Advancement of Science

gained from Figure 7. Starting from the point of lowest dissolved salts concentrations, the ratio of Na/(Na + Ca) approaches one. This is similar to rainwater, and is termed the precipitation dominance regime. It is typified by rivers in humid tropical area of the world with very high rainwater inputs and little evaporation. As the suspended solids concentration increases the ratio Na/(Na + Ca) declines, indicating an increasing importance for calcium in the rock dominance regime. Here, increased weathering of rock provides the major source of dissolved solids. As dissolved solids increase further, the abundance of calcium decreases relative to sodium as the water becomes saturated with respect to

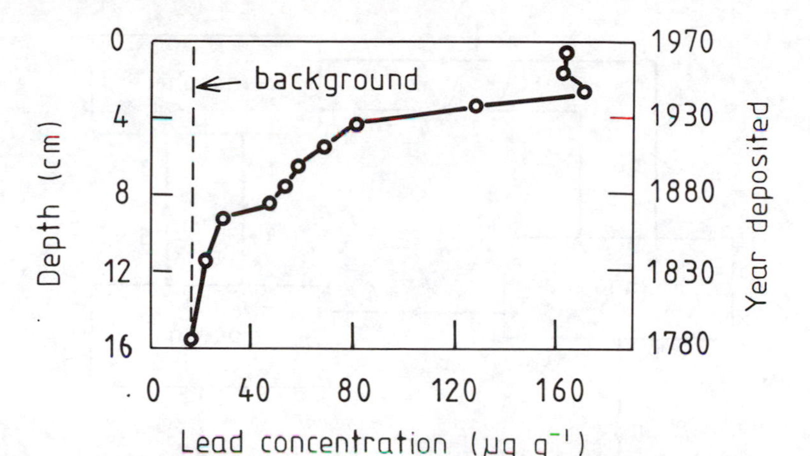

Figure 8 *Lead profile in a lake sediment in relation to depth and the year of incorporation (from Galloway* et al.[22]*)*

$CaCO_3$, and this compound precipitates. Waters in the evaporation/precipitation regime are typified by rivers in very arid parts of the world (*e.g.* River Jordan) and the major seas and oceans of the world.[20,21]

The flux of material in a river to the sea is expressed by:

$$\text{Flux (g s}^{-1}) = \text{Volumetric discharge (m}^3\text{ s}^{-1}) \times \text{Concentration (g m}^{-3})$$

In total, the rivers of the world carry around 4.2×10^{12} kg per year of dissolved solids to the oceans and 18.3×10^{12} kg per year of suspended solids.

In slow-moving water bodies such as lakes and ocean basins, suspended solids falling to the bottom produce a well stratified layer of bottom sediment. This is stratified in terms of age with the oldest sediment at the bottom (where when suitably pressurized it can form rock) and the newest at the top, in contact with the water. If burrowing organisms do not provide too much disturbance (termed bioturbation), the sediment can preserve a record of depositional inputs to the water body. An example is provided by Figure 8 in which lead is analysed in a sediment core dated from its radioisotope content.[22] The concentration rises from a background in around the year 1800, corresponding to the onset of industrialization. Considerably increased deposition is seen after 1930 due to the introduction of leaded petrol. Whilst some of the lead input is via surface waters, the majority probably arises from atmospheric deposition.

4 BIOGEOCHEMICAL CYCLES

A general model of a biogeochemical cycle appears in Figure 9. Although biota are not explicitly included, their role is a very important one in mediating

[20] R. J. Gibbs, *Science*, 1970, **170**, 1088.

[21] R. M. Harrison, S. J. de Mora, S. Rapsomanikis, and W. R. Johnston, 'Introductory Chemistry for the Environmental Sciences', Cambridge University Press, Cambridge, 1991.

[22] A. O. Davies and J. N. Galloway, 'Atmospheric Pollutants in Natural Waters', ed. S. J. Eisenreich, Ann Arbor, 1981, p. 401.

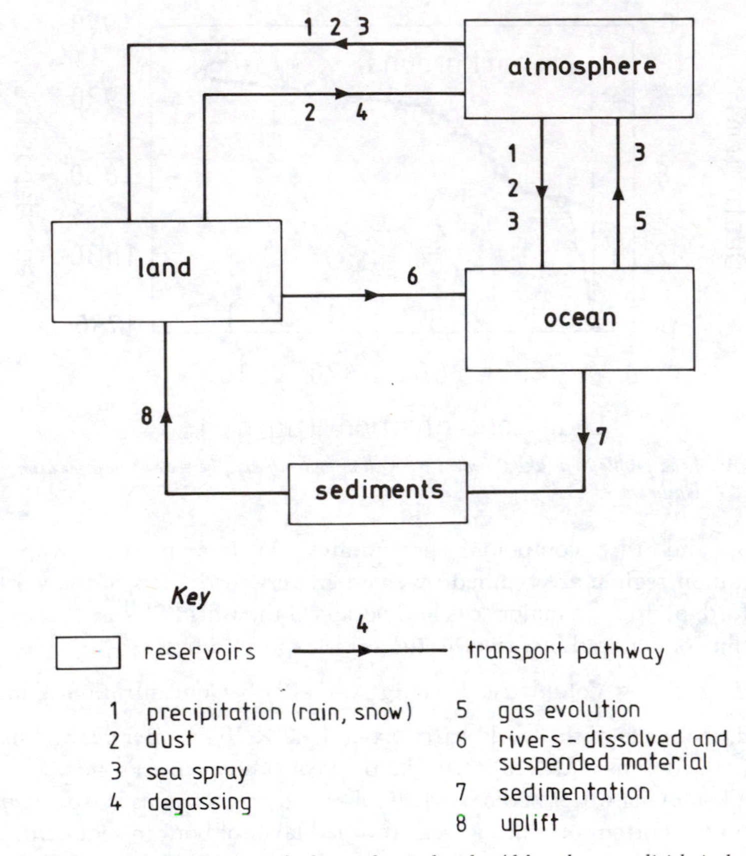

Figure 9 *Schematic representation of a biogeochemical cycle. Although not explicitly included, biota play a major role in many of the processes (after O'Neill[23])*

transfers between the idealized compartments of the model. For example, the role of marine phytoplankton in transferring sulfur from the ocean to the atmosphere in the form of dimethyl sulfide has been highlighted in Chapter 4. Biota play a major role in determining atmospheric composition. Photosynthesis removes carbon dioxide from the atmosphere and replenishes oxygen. In a world without biota, lightning would progressively convert atmospheric oxygen into nitrogen oxides and thence to nitrate which would reside in the oceans. Biota also exert more subtle influences. In aquatic sediments, micro-organisms often deplete oxygen more quickly than it can be replenished from the overlying water, producing anoxic conditions. This leads to chemical reduction of elements such as iron and manganese, which has implications for their mobility and bioavailability.

Biological reduction processes in sediments may be viewed as the oxidation of carbohydrate (in its simplest form CH_2O) with accompanying reduction of an oxygen carrier. In the first instance, dissolved molecular oxygen is used. The reaction is thermodynamically favoured, as reflected by the strongly negative ΔG.

$$CH_2O + O_2 \rightarrow CO_2 + H_2O \qquad \Delta G = -125.5 \text{ kJ mol}^{-1} \text{ e}^-$$

When all of the dissolved oxygen is consumed, anaerobic organisms take over. Initially, nitrate-reducing bacteria are favoured

$$2CH_2O + NO_3^- + 2H^+ \rightarrow 2CO_2 + H_2O + NH_4^+ \qquad \Delta G = -82.2 \text{ kJ mol}^{-1} \text{ e}^-$$

Once the nitrate is utilized, sulfate reduction takes over

$$SO_4^{2-} + H^+ + 2CH_2O \rightarrow HS^- + 2H_2O + 2CO_2 \qquad \Delta G = -25.6 \text{ kJ mol}^{-1} \text{ e}^-$$

Finally, methane-producing organisms dominate in a sediment depleted in oxygen, nitrate, and sulfate

$$2CH_2O \rightarrow CH_4 + CO_2 \qquad \Delta G = -23.5 \text{ kJ mol}^{-1} \text{ e}^-$$

Thus highly anoxic waters are commonly sources of hydrogen sulfide, H_2S from sulfate reduction and of methane (marsh gas). The formation of sulfide in sediments has led to precipitation of metal sulfides over geological time, causing accumulations of sulfide minerals of many elements, *e.g.* PbS, ZnS, HgS, *etc.*

4.1 Case Study 1: The Biogeochemical Cycle of Nitrogen

Nitrogen has many valence states available and can exist in the environment in a number of forms, depending upon the oxidizing ability of the environment. Figure 10 indicates the most important oxidation states and the relative stability (in terms of free energy of formation).[23] The oxides of nitrogen represent the most oxidized and least thermodynamically stable forms. These exist only in the atmosphere. Ammonia can exist in gaseous form in the atmosphere but rather rapidly returns to the soil and waters as ammonium, NH_4^+. Fixation of atmospheric N_2 by leguminous plants leads to ammonia, NH_3. In aerobic soils and aquatic systems, NH_3 and NH_4^+ are progressively oxidized by micro-organisms via nitrite to nitrate. The latter is taken up by some biota and used as a nitrogen source in synthesizing amino acids and proteins, the most thermodynamically stable form of nitrogen. After the death of the organism, microbiological processes will convert organic nitrogen to ammonium (ammonification) which is then available for oxidation or use by plants. Conversion of ammonia to nitrate is termed nitrification, whilst denitrification involves conversion of nitrate to N_2.

Figure 11 shows an idealized nitrogen cycle. The numbers in boxes represent quantities of nitrogen in the various reservoirs, whilst the arrows show fluxes.[23] It is interesting to note that substances involving relatively small fluxes and burdens can have a major impact upon people. Thus nitrogen oxides, NO, NO_2, and N_2O are very minor constituents relative to N_2 but play major roles in photochemical air pollution (NO_2), acid rain (HNO_3 from NO_2), and stratospheric ozone depletion (N_2O). Nitrate from fertilizers represents a very small flux but has major implications in terms of eutrophication of surface waters.

[23] P. O'Neill, 'Environmental Chemistry', George, Allen, and Unwin, 1985.

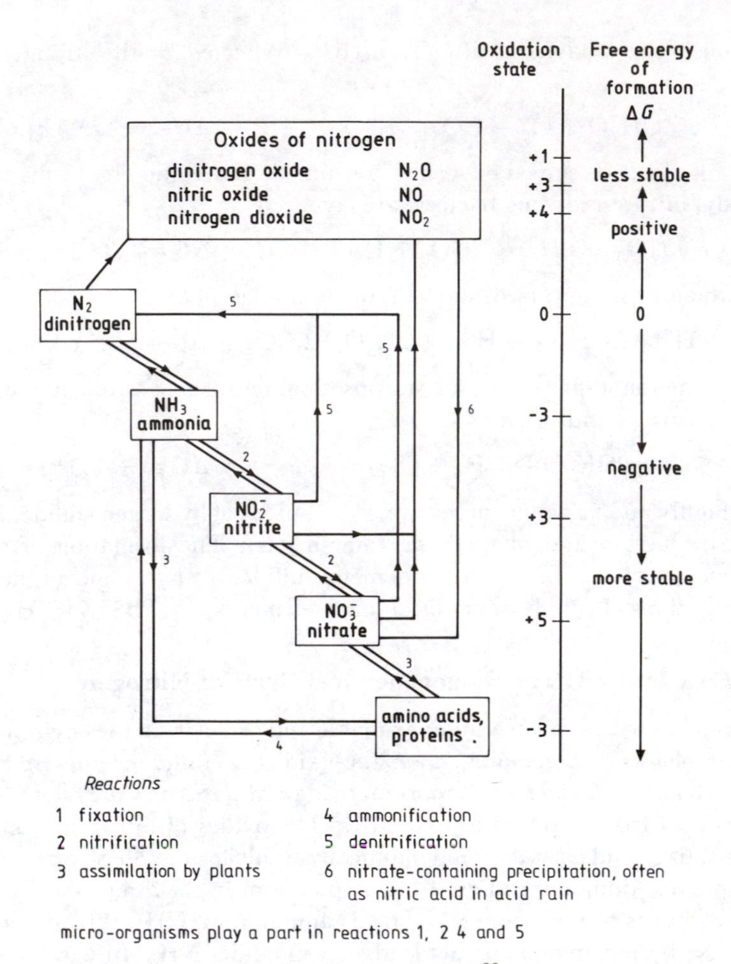

Figure 10 *Chemical forms and cycle of nitrogen from O'Neill[23])*

4.2 Case Study 2: Aspects of the Biogeochemical Cycle of Lead

Lead is a simpler case to study than nitrogen due to the small number of available valence states. The major use of lead until recently was as tetraalkyl lead gasoline additives in which lead is present as Pb^{IV}. The predominant compounds used are tetramethyl lead, $Pb(CH_3)_4$, and tetraethyl lead, $Pb(C_2H_5)_4$. These are lost to the atmosphere as vapour from fuel evaporation and exhaust emissions from cold vehicles, but comprise only about 1–4% of lead in polluted air.[2] Leaded gasoline also contains the scavengers 1,2-dibromoethane, CH_2BrCH_2Br and 1,2-dichloroethane, CH_2ClCH_2Cl which convert lead within the engine to lead halides, predominantly lead bromochloride, PbBrCl, in which lead is in the Pb^{II} valence state, its usual form in environmental media. About 75% of lead alkyl burned in the engine is emitted as fine particles of inorganic lead halides. Atmospheric emissions of lead arise also from industry; both these and vehicle-emitted lead are declining. Figure 12 shows trends in United States emissions of

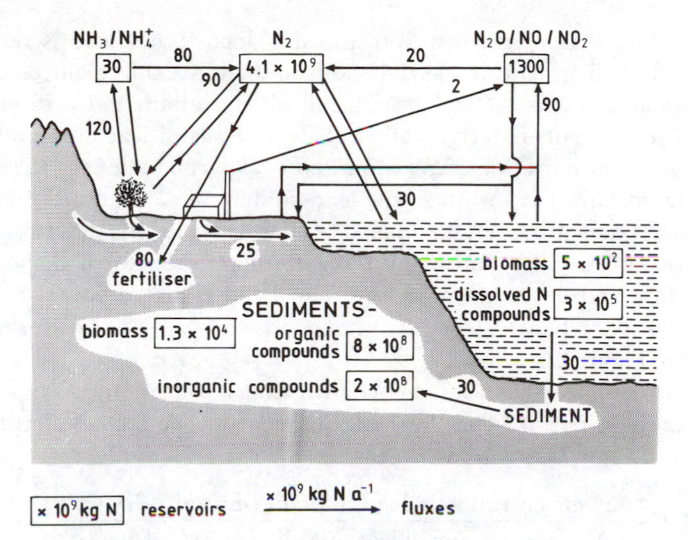

Figure 11 *Schematic representation of the biogeochemical cycle of nitrogen, indicating the approximate magnitude of fluxes and reservoirs (after O'Neill[23])*

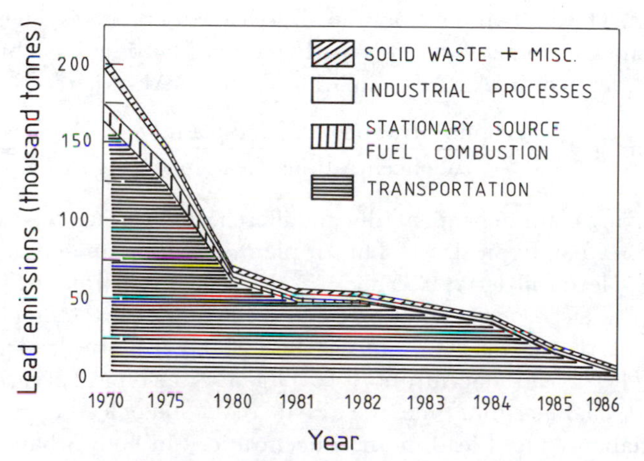

Figure 12 *Trends in United States lead emissions to the atmosphere, 1970–1986 (from USEPA[24])*

lead to atmosphere.[24] Lead emitted to the atmosphere has a lifetime of around 7–30 days and hence may be subject to long-range transport. Concentrations of trace elements in polar ice provide a historical record of atmospheric deposition. Measurements (Figure 2) have shown a marked enhancement in lead accompanying the increase in leaded gasoline usage, and a major decline in recent years attributable to reduced emissions to atmosphere.[4]

[24] U. S. Environmental Protection Agency, 'National Air Pollutant Emission Estimates, 1940–1986' EPA-450-4-87-024, 1988.

Atmospheric lead is deposited in wet and dry deposition. Lead is relatively immobile in soil, and agricultural surface soils in the UK exhibit concentrations approximately double those of background soil which contain *ca.* 15–20 mg kg^{-1} derived from soil parent materials, other than in areas of lead mineralization where far greater concentrations can be found. Local perturbations to the cycle of lead can be important. For instance, the lead content of garden soils correlates strongly with the age of the house. This is probably due to the deterioration of leaded paintwork on older houses and the former practices of disposing of household refuse and fire ashes in the garden. Lead is also of low mobility in aquatic sediments and hence the sediment may provide a record of historical lead deposition (see Figure 8).

Plants can take up lead from soil, hence providing a route of human exposure. Careful research in recent years has established transfer factors, termed the Concentration Factor, CF, where

$$\mathrm{CF} = \frac{\Delta\text{Concentration of lead in plant (mg kg}^{-1}\text{ dry wt.)}}{\Delta\text{Concentration of lead in soil (mg kg}^{-1}\text{ dry wt.)}}$$

The value of CF for lead is lower than for most metals and is typically within the range 10^{-3} to 10^{-2}. Much higher values have been estimated from earlier studies which ignored the importance of direct atmospheric deposition as a pathway for contamination. The direct input from the air to leaves of plants is often as large, or greater than soil uptake.[24,25] This pathway may be described by another transfer factor, termed the Air Accumulation Factor, AAF, where

$$\mathrm{AAF}\ (\mathrm{m^3\,g^{-1}}) = \frac{\Delta\text{Concentration of lead in plant (}\mu\text{g g}^{-1}\text{ dry wt.)}}{\Delta\text{Concentration of lead in air (}\mu\text{g m}^{-3}\text{)}}$$

Values of AAF are plant-dependent, due to differences in surface characteristics, but values of 5–40 are typical.[25,26] Thus a plant grown on an agricultural soil with 50 mg kg^{-1} lead will derive 0.25 mg kg^{-1} dry weight lead from the soil ($\mathrm{CF} = 5 \times 10^{-3}$), whilst airborne lead of 0.1 μg m^{-3} will contribute 2.0 μg g^{-1} ($\equiv$ mg kg^{-1}) of lead ($\mathrm{AAF} = 20\ m^3\,g^{-1}$). Thus in this instance airborne lead deposition is dominant. The air lead concentration of 0.1 μg m^{-3} was typical of rural areas of the UK until 1985. Since that time, the drastic reduction of lead in gasoline has led to appreciably reduced lead-in-air concentrations in both urban and rural localities.

Human exposure to lead arises from four main sources:[2,27]

(i) inhalation of airborne particles. The adult human respires approximately 20 m^3 of air per day. Thus for an urban lead concentration of 0.5 μg m^{-3}, *intake* is 10 μg per day. This is rather efficiently absorbed (*ca.* 70%) and therefore *uptake* is around 7 μg per day in this instance.

[25] R. M. Harrison and M. B. Chirgawi, *Sci. Total Environ.*, 1989, **83**, 13.
[26] R. M. Harrison and M. B. Chirgawi, *Sci. Total Environ.*, 1989, **83**, 47.
[27] Royal Commission on Environmental Pollution, '*Ninth Report: Lead in the Environment*', HMSO, London, 1983.

(ii) ingestion of lead in foodstuffs. The concentrations of lead in food obviously vary between different foodstuffs and even between different batches of the same food. Typical freshweight concentrations (much of the weight of some foods is water) are from <50 to 250 μg Pb kg^{-1}. Thus a food consumption of 1.5 kg per day represents an *intake* of around 150 μg per day and an *uptake* (10–15% efficient) of around 18 μg per day.

(iii) drinking water and beverages. Concentrations of lead in drinking water vary greatly, related particularly to the presence or absence of lead in the household plumbing system. Most households in the UK now conform to the EC standard of 50 $\mu g\ l^{-1}$ and a concentration of 15 $\mu g\ l^{-1}$ may be taken as representative. Gastrointestinal absorption of lead from water and other beverages is highly dependent upon food intake. After long fasting, absorptions of 60–70% have been recorded, 14–19% with a short period of fasting before and after the meal, and only 3–6% for drinks taken with a meal. If 15% is taken as typical, for a daily consumption of 1.5 litres, *intake* is 23 μg and *uptake* 4 μg.

(iv) cigarette smoking exposes the individual to additional lead.

Whilst both individual exposure to lead and the uptake efficiencies of individuals are very variable, it is evident that exposure arises from a number of sources and control of human lead intake, if deemed to be desirable, requires attention to all of those sources. An additional pathway of exposure, not easily quantified, and not included above is ingestion of lead-rich surface dust by hand to mouth activity in young children.

CHAPTER 7

Environmental Monitoring Strategies

C. NICHOLAS HEWITT AND ROBERT ALLOTT

1 OBJECTIVES OF MONITORING

The gathering of information on the existence and concentration of substances in the environment, either naturally occurring or from anthropogenic sources, is achieved by **measurement** of the substance or phenomenon of interest. However, single measurements of this type made in isolation are virtually worthless, since temporal and spatial variations cannot be deduced. Rather, it is necessary to *monitor* the parameter of interest by repeated measurements made over time (and often over space), with sufficient sample density, temporally and spatially, that a realistic assessment of variations and trends may be made.

Monitoring of the environment may be undertaken for a number of reasons and it is important that these be defined before sampling takes place. The generalization that 'monitoring is done in order to gain information about the present levels of harmful or potentially harmful pollutants in discharges to the environment, within the environment itself, or in living creatures (including ourselves) that may be affected by these pollutants'[1] may be expanded as follows:

(a) Monitoring may be carried out to assess pollution effects on man and his environment, and so to identify any possible cause and effect relationship between pollutant concentrations and, for example, health effects or climatic changes.

(b) Monitoring may be carried out in order to study and evaluate pollutant interactions and patterns. For example, source apportionment[2] and pollutant pathway studies usually rely on environmental monitoring.

(c) Monitoring may be carried out to assess the need for legislative controls on emissions of pollutants and to ensure compliance with emission standards.

[1] Department of the Environment, 'The Monitoring of the Environment in the United Kingdom,' Report by the Central Unit on Environmental Pollution, HMSO, London, 1974, pp. 66

[2] P. K. Hopke, 'Receptor Modelling in Environmental Chemistry,' John Wiley and Sons, New York, 1985.

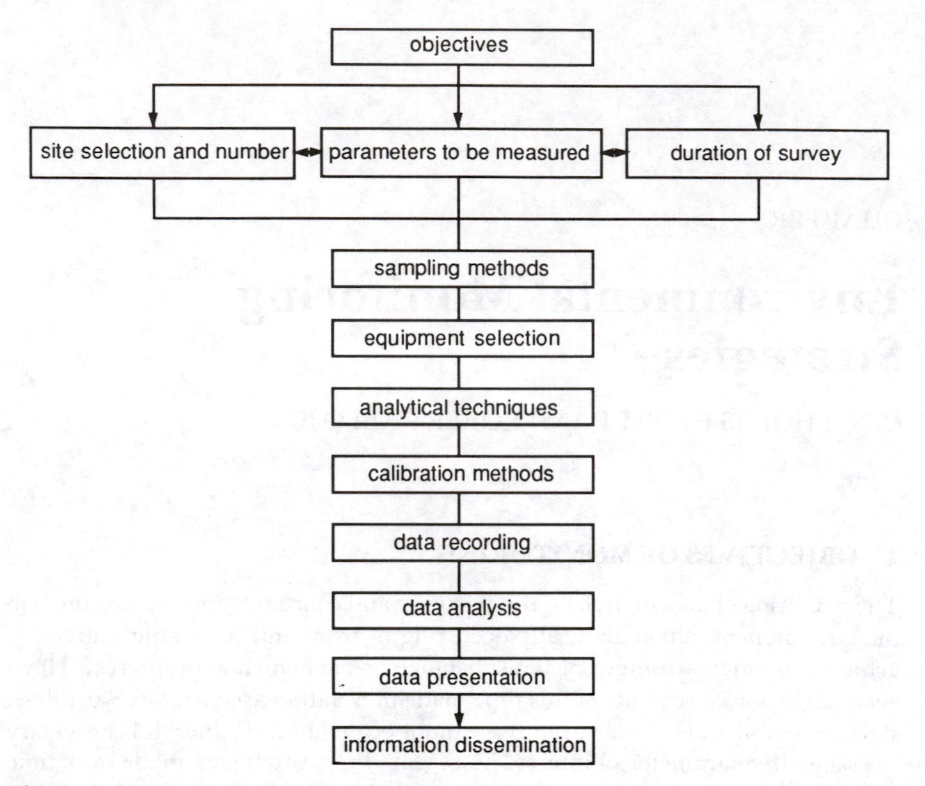

Figure 1 *Steps in the design of a monitoring programme*

An assessment of the effectiveness of pollution legislation and control techniques also depends upon subsequent monitoring.

(d) In areas prone to acute pollution episodes, monitoring may be carried out in order to activate emergency procedures.

(e) Monitoring may be carried out in order to obtain a historical record of environmental quality and so provide a database for future use in, for example, epidemiological studies.

(f) Monitoring may also be necessary to ensure the suitability of water supply for a proposed use (industrial or domestic) or to ensure the suitability of land for a proposed use (for example, for housing).

A basic problem in the design of a monitoring programme is that each of the above reasons for carrying out monitoring demands different answers to a number of questions. For example, the number and location of sampling sites, the duration of the survey, and the time-resolution of sampling will all vary according to the use to which the collected data are to be put. Decisions on what to monitor, when and where to monitor, and how to monitor are often made much easier once the purpose of monitoring is clearly defined. Therefore it is most important that the first step in the design of a monitoring programme should be to set out the objectives of the study. Once this has been done then the programme may be designed by consideration of a number of steps in a systematic way (see Figure 1) such that the generated data are suitable for the intended use. It is important also that the data produced by a monitoring programme should be continuously appraised in the light of these objectives. In this way, limitations in the design, organization, or execution of the survey may be identified at an early stage.

The aim of this chapter is to present and discuss the most important and relevant considerations that must be taken into account in the design and organization of a monitoring exercise. It is not intended to be a manual or practical guide to monitoring; rather it is hoped that it highlights the types of approaches that may be used and some of the problems likely to be encountered. The inclusion of case studies and references direct the reader to the more specific practical information which is available elsewhere.

2 TYPES OF MONITORING

The earth is comprised of three distinct media; the atmosphere, the hydrosphere, and the land. Pollutants can occur in any or all of the solid, liquid, or gaseous phases. However, the environment is not a simple system and consequently each of the three media may contain pollutants in each of the three phases. Monitoring may therefore be required for a particular pollutant in a specific phase in a particular environmental compartment (*e.g.* sulfur dioxide in air) or it may encompass two or more phases and/or media (*e.g.* dissolved and particulate phase metals in water). Pollutants in the environment originate from a multitude of different types of sources and the identification of these is a necessary prerequisite to the design of a monitoring programme. First, pollutant sources may be classified by their spatial distribution as point sources, line sources, or area sources. Point sources include industrial chimneys, liquid waste dishcarge pipes, and localized toxic waste dumps on land. Line sources may include highways, airline routes, and run-off from agricultural land, while area emissions may arise from extensive urban or industrial complexes.

Sources may be classified as either statutory or mobile, motor vehicles being the obvious example of the latter. Classification may also be made for air pollutant sources on the basis of the height of discharge, *i.e.* at street level, building level, stack level, or above the atmospheric boundary layer level.

A further important distinction may be made between 'planned', 'fugitive', and 'accidental' emissions to the environment:

(a) Planned emissions arise when (as is invariably the case) it is economically or technically impossible to completely remove all the contaminants in a discharge and hence the process operation allows pollutants to be discharged to the environment at known and controlled rates. Obvious examples of planned emissions include sulfur dioxide from power generation plants and low-level radioactive effluent during nuclear fuel reprocessing.

(b) Fugitive emissions arise when pollutants are released in an unplanned way, normally without first passing through the entire process. They therefore occur at a point sooner in the process that the stack or duct designed for 'planned' emissions. They generally originate from operations which are uneconomic or impractical to control, have poor physical arrangements for effluent control, or are poorly maintained or managed. An example is the escape of heavy metal contaminated dust from a lead works on vehicle tyres, arising from poor dust control and wheel washing arrangements.

(c) Accidental emissions result from plant failure, such as a burst filter bag or faulty valve, or from an accident involving either equipment or operator error (*e.g.* the Chernobyl reactor accident). Accidental emissions can give rise to very high concentrations but they normally occur only infrequently.

Classification of the sources of pollutants in this way allows the distinction of two differing approaches to their monitoring. On the one hand, samples may be taken of the effluent before discharge to, and dispersion in, the environment without consideration of source strengths and rates. Alternatively, samples may be taken from the ambient environment without consideration of source strengths and rates. Obviously neither one of these approaches alone can necessarily provide all the data required to resolve a particular problem and often it is desirable to complement one with the other.

2.1 Source Monitoring

2.1.1 General Objectives. Source monitoring may be carried out for a number of reasons:

(a) Determination of the mass emission rates of pollutants from a particular source, and assessment of how these are affected by process variations.
(b) Evaluation of the effectiveness of control devices for pollution abatement.
(c) Evaluation of compliance with statutory limitations on emissions from individual sources.

2.1.2 Stationary Source Sampling for Gaseous Emissions. A common feature of many industrial processes is that effluent output exhibits cyclical patterns. These may be related to working shift arrangements or be a function of the operations involved, but both require that source testing or monitoring be planned accordingly. Process operations should be reviewed so that discharges during the period of sampling are representative of the plant output in order to ensure that the samples themselves are representative of the effluent, and that the final pollutant analysis will be a representative measure of the entire output.

Two requirements have been specified for valid source monitoring.[3] First, the sample should accurately reflect the true magnitude of the pollutant emission at a specific point in the stack at a specific instant of time. This requirement is met by adequate sampling instrument design. Secondly, enough measurements should be obtained over time and space so that their combined result will accurately represent the entire source emission. This requires consideration of the emissions both in time and in space, across the entire cross-section of the stack.[4]

[3] W. C. Aschinger and R. T. Shigehara, *J. Air Pollut. Control Assoc.*, 1968, **18**, 605–9.
[4] J. R. Hodkinson, 'Air Sampling Instruments for Evaluation of Atmospheric Constituents,' American Conference of Governmental Industrial Hygienists, Cincinnati, Ohio, 1972.

In a circular flue, sampling at the centroids of equal-area annular segments will ensure that emission variations across the stack cross-section are quantified. In a rectangular flue sample points should be located at the centroids of smaller equal-area rectangles. Generally eight or twelve such sampling points are adequate to compensate for any deficiencies in the location of the sampling site with respect to the length of the stack and to non-ideal flow conditions at the site caused by bends, inlets, or outlets.[5] If it is a particulate pollutant which is being sampled within the stack, it is important that an isokinetic sampling regime is maintained (see Section 3.1.1 below).

2.1.3 *Mobile Source Sampling for Gaseous Effluents.* Vehicle and aircraft emissions are heavily dependent upon the engine operating mode (*i.e.* idling, accelerating, cruising, or decelerating) and the results obtained by sampling must be considered specific to the type of operating cycle used during the test.[6] Emission tests are usually performed with the vehicle on a dynamometer equipped with inertia fly wheels to represent the vehicle weight and brake loading on a level road.

2.1.4 *Source Monitoring for Liquid Effluents.* Liquid wastes and effluents often tend, like gaseous effluents, to be inhomogeneous and care is needed in selecting sampling positions. Having considered the location of the site in relation to plant operation (*e.g.* should the site be before or after a particular stage of the process of treatment) it is desirable that a region of high turbulence and/or good mixing be chosen. As for gaseous emissions, several samples may have to be taken across the cross-section of a pipe or channel. Sampling from vertical pipes is less liable to be affected by deposition of solids than sampling from horizontal pipes, and a distance of approximately 25 pipe-diameters downstream from the last inflow should ensure that mixing of the two streams is essentially complete.[7] If suitable homogeneous regions for sampling cannot be found, particularly where suspended materials are present, samples may have to be taken from several positions along the effluent stream.

Where the composition of a liquid effluent is known to vary with time, grab samples may be collected at set intervals, either manually or by use of an automatic sampler. An alternative approach is to sample at intervals varying with the flow rate so that a more representative composite may be obtained.

2.1.5 *Source Monitoring for Solid Effluents.* Solid effluents may arise from a number of different processes, including sludge after sewage treatment, ash residue from municipal incinerators, or low-grade gypsum from desulfurization plants attached to coal fired power stations. In general, solid wastes are even less homogeneous than either liquid or gaseous effluents. Therefore, great effort must be made to ensure that samples are representative of the bulk waste (see Section

[5] Pollution Control Branch, British Columbia Water Resources Service, 'Source Testing Manual for the Determination of Discharges to the Atmosphere,' 3rd Edn., Victoria, British Columbia, 1974.

[6] J. P. Soltan and R. J. Larbey, 'Symposium of Institution of Mechanical Engineers,' London, 1971, 218.

[7] A. L. Wilson, 'Design of Sampling Programmes', in 'Examination of Water for Pollution Control,' ed. M. J. Suess, Vol. 1, Pergamon, London, 1982.

3.3). Monitoring of sewage sludge is particularly common due to sludge acting as an efficient sorption material for heavy metals. Typically, 80–100% of the input lead in a sewage treatment plant is incorporated into the sludge, resulting in sludge lead concentrations of 120–3000 $\mu g\ g^{-1}$. Consideration must therefore be given to the concentrations of pollutants in the material before it is used as fertilizer, incinerated, dumped at sea, or used as land-fill. It may also be mentioned that the determination of the metal balance of a sewage treatment works is an interesting monitoring exercise which may be necessary when considering the fate of the treated effluent and solid waste.

In most countries guidelines exist to control the disposal of sewage sludges to land, usually based primarily upon the zinc, copper, and nickel content of the sludge. Hence considerable quantities of other metals, including lead, may be added to land over a normal 30 year disposal period. In the UK the disposal of lead-rich sewage sludges to land is controlled where direct ingestion by animals of contaminated grass or soil can occur. Where such ingestion can occur only sludges of lead content $<2000\ mg\ kg^{-1}$ dry weight should be disposed. Otherwise, where the sludge is to be mixed with the soil the total lead disposed over the normal 30 year period should be limited to $1000\ kg\ ha^{-1}$, corresponding to about $450\ \mu g\ kg^{-1}$ of soil of 200 mm depth.

Until fairly recently most trace metal analysis of environmental samples was designed to give a measure of the total elemental concentration in the sample as it was felt that this gave an adequate measure of the pollution load for that metal. It is now realized however that total metal concentrations are often not sufficient and that information based upon some form of physico-chemical speciation scheme is required. This may include, for example, solubility of the pollutant in acids of different strengths, the size distribution of particles, and the association with organic compounds. This is because the physical, chemical, and biological responses to a metal input will vary according to its physical and chemical speciation. One disadvantage of this type of analysis is that it is complicated and time-consuming compared with total metal determinations. Thus speciation studies are invariably limited to a few samples where many (tens or even hundreds) would be taken in a total-metal study.

Case study 1: Organic solvent residues at a landfill site.[8] Landfill sites have become recognized during the 1980s as a source of toxic and explosive substances, including methane and organic chemicals. The contamination of groundwater by these toxic organic chemicals is of major environmental concern in Europe and North America. At a landfill site studied near Ottawa, Canada, disposal of chlorinated and non-chlorinated solvents, wood preservatives, and small amounts of other wastes occurred between 1969 and 1980. Groundwater supplies were collected from monitoring wells within the landfill site using either piezometers or multi-level samplers attached to peristaltic pumps. Analysis was carried out by gas chromatography–mass spectrometry (GC–MS) which

[8] S. Lesage, R. E. Jackson, M. W. Priddle, and P. G. Riemonn, *Environ. Sci. Technol.*, 1990, **24**, 559–566.

enabled the identification and quantification of a wide range of volatile organic compounds, including dioxane (~300–2000 μg l^{-1}), diethyl ether (<2–658 μg l^{-1}), trichloroethene (7.4–583 μg l^{-1}), and 1, 1, 2-trichloro-1, 2, 2-trifluoroethane (Freon F113) (<5–2725 μg l^{-1}). The contaminant of greatest concern was 1, 4-dioxane, due to its toxicity and persistence. Freon F113 was the organic chemical found in greatest concentration.

2.2 Ambient Environment Monitoring

2.2.1 General Objectives. Monitoring the environment may be carried out for a number of reasons, as outlined above in Section 1. However, whatever the purpose of the survey the over-riding consideration when designing a programme is to ensure that the samples obtained provide adequate data for the purpose intended. Invariably this means that samples should be representative of conditions prevailing in the environment at the time and place of collection. Thus, not only must the sampling location be carefully chosen but also the sampling position at the chosen location.

The selection of a specific monitoring site requires consideration of four steps: identify the purpose to be served by monitoring; identify the monitoring site type(s) that will best serve the purpose; identify the general location where the sites should be placed; finally identify specific monitoring sites.

2.2.2 Ambient Air Monitoring. Air pollution problems vary widely from area to area and from pollutant to pollutant. Differences in meteorology, topography, source characteristics, pollutant behaviour, and legal and administrative constraints mean that monitoring programmes will vary in scope, content, and duration, and the types of station chosen will also vary. However ambient monitoring sites may be divided into several categories:

(a) source-orientated sites for monitoring individual or small groups of emitters as part of a local survey (*e.g.* one particular factory);
(b) sites in a more extensive survey (such as the UK National Survey of Air Pollution for smoke and sulfur dioxide) which may be located in areas of highest expected pollutant concentrations, high population density, or in rural areas to give a complete nationwide coverage;
(c) base-line stations to obtain background concentrations, usually in remote or rural areas with no anticipated changes in land-use.

2.2.2.1 Location of source-orientated monitors. Occasionally the effects and impact of a specific pollutant source are of sufficient interest or importance to warrant a special survey. This will usually include a site at the point of anticipated maximum ground-level concentration, which can be estimated from dispersion calculations (see Section 4.1 below), and also a nearby site to characterize the 'background' conditions in the area. Examination of meteorological records will usually be necessary in order to choose suitable locations for the sites and several

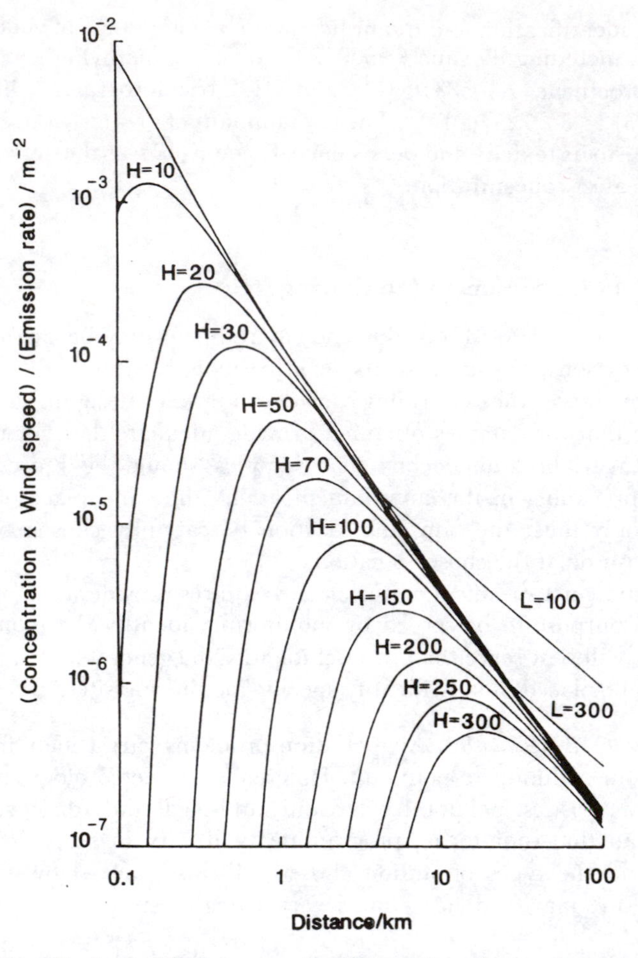

Figure 2 *Normalized ground-level concentrations from an elevated source for neutral stability. The effective stack height* (H) *is the sum of the release height* (e.g. *chimney height) and the height gained by the plume due to momentum and buoyancy*

computerized models are available for determining the areas of maximum average impact from a point source. Calculation of expected ground-level concentrations using the standard equations discussed in Section 4.1 below show that the concentration rises rapidly with distance from the source to a maximum and then falls gradually beyond the maximum,[9] as shown in Figure 2. This is for meteorological conditions of neutral stability and different heights of emission (H). The ordinates in this graph represent concentration normalized for emission rate (Q) and wind speed (U) and the various curves are for different source heights (H metres) and different limits to vertical dispersion (L). It is prudent

[9] 'WMO Operations Manual for Sampling and Analysis Techniques for Chemical Constituents in Air and Precipitation,' World Meteorological Organization, No. 299, Geneva, 1971, pp. 22.

therefore to locate the monitoring site somewhat beyond the distance where the maximum concentration is predicted. This allows some margin for error by placing the monitor in a region of relatively small concentration gradients. Obviously it is desirable to have an array of stations at differing distances and directions from the source and typically 4–6 samplers might be considered sufficient for monitoring a single point source.

In some cases pollutants are emitted to the atmosphere from a single source but in a more diffuse manner than from a single stack. Calculations of mass emission rates and distance of maximum ground-level concentration are more difficult to make for such diffuse or fugitive emissions which, in some cases, may have significant impact on the local air quality. In a study of cadmium and lead levels around a small secondary metal recovery factory[10] it was found that fumes created by the melting process were exhausted to the outside air by means of a roof-sited ventilation fan. Whilst lead-in-air values were not excessive (in the range 0.2–2.0 μg (Pb) m^{-3}, except for one 30 minute sample which gave 5.1 μg (Pb) m^{-3}, at 10 m from the source) high cadmium-in-air concentrations were found (maximum of 11.0 μg (Cd) m^{-3}), compared with typical urban concentrations of 0.001–0.003 μg (Cd) m^{-3}.

2.2.2.2 Location of monitors in larger-scale surveys. Often it is important to know the geographical extent of atmospheric pollution and to have localized information on source strengths or ground-level concentrations within a plume is not sufficient. For example the National Survey of Air Pollution (NSAP) monitoring network in the United Kingdom was established in 1961 following recognition of the need for the acquisition of a nationwide, day-to-day, and long-term bank of data of sulfur dioxide and smoke concentrations.[11] The original network of 1200 sites was based upon the assumption that it was necessary to monitor in rural areas (150 sites) and different types of urban areas such as high-density residential areas, industrial areas, commercial areas, and smoke-controlled areas. Since the application of this simple scheme in 1961 many of the original stations have ceased monitoring, others have been replaced, and additional stations have been added. Reasons for these changes include the need to monitor recently established smoke-control areas, new industrial estates and redeveloped areas, and surveys around new and projected power stations. In 1981 a rationalized long-term network of 150 NSAP stations was proposed, to be redesignated as the UK Smoke and Sulfur Dioxide Monitoring Network. However, about 400 existing sites were retained in the short-term to provide continuation of monitoring in urban areas where the EC air quality standards for smoke and SO_2 may be approached or exceeded. As concentrations fall to 'acceptable' levels in each urban area so these sites will be discontinued.

Several methods are available for the rationalization of an existing monitoring network, and in the case of the 106 NSAP stations in the Greater London area a spatial correlation analysis has been applied to determine which sites may be

[10] D. B. Turner, 'Workbook of Atmospheric Dispersion Estimates,' Nat. Air Poll. Cont. Admin., 1969, 84 pp.
[11] C. J. Muskett, L. H. Roberts, and J. Page, *Sci. Total Environ.*, 1979, **11**, 73–87.

discontinued without losing overall coverage of the area.[12] A high degree of correlation was found to exist between the majority of NSAP sites, and so rationalization may be achieved by considering all inter-station correlations and delimiting areas containing two or more stations which are highly correlated.

The design of monitoring networks for air pollution has been treated in several different ways. For example monitoring sites may be located in areas of severest public health effects, which involves consideration of pollutant concentration, exposure time, population density, and age distribution. Alternatively the frequency of occurrence of specific meteorological conditions and the strength of sources may be used to maximize monitor coverage of a region with limited sources. The Dosage Monitoring Survey Design procedure[13] is based upon the analysis of the ambient dosage (the product of the concentration by exposure time) which in turn is computed from source emission data and diffusion models. Potential air monitoring sites are then ranked according to their ability to represent the ambient dosage.

2.2.2.3 Location of regional-scale survey monitors. On the regional or global scale monitoring is usually concerned with long-term changes in background concentrations of pollutants and so the principal siting requirement is that truly representative baseline or background levels may be measured over a long time period without interference from local sources. It has been suggested[9] that baseline stations should be located in areas where no significant changes in land use practices are anticipated for at least 50 years within 100 km of the station, and should be away from population centres, major highways, and air routes.

One example of a network established to monitor distant sources for regional or global effects is the OECD 'Long-Range Transport of Air Pollutants' project which measured chemical components in precipitation and SO_2 and particulate sulfate in air.

2.2.2.4. Air monitoring case study. Case study 2: National and regional radon surveys in the UK.[15,16,17] Increasing awareness of the importance of human exposure to the naturally occurring radioactive gas radon has resulted in several national surveys being commissioned for the UK.[16,17] ^{222}Rn, commonly known as radon, is a colourless odourless gas which results from the decay of ^{238}U and in turn decays to a series of radioactive daughters. Two other isotopes of radon exist, ^{219}Rn and ^{220}Rn (thoron) within the ^{232}Th and ^{235}U radioactive decay series. However, it is ^{222}Rn which has the greatest health significance, since it has the largest proportion of alpha emitting, high activity progeny. The inhalation of radon and particularly its daughters is believed to be responsible for a major

[12] Warren Spring Laboratory, 'National Survey of Air Pollution 1961–1972,' Vol. 1, HMSO, London, 1972.
[13] C. M. Handscombe, and D. M. Elsom, *Atmos. Environ.*, 1982, **16**, 1061–1070.
[14] K. E. Nou and S. Mitsutomi, *Atmos. Environ.*, 1983, **17**, 2583–2590.
[15] C. N. Hewitt and M. Kelly, *Environ. Tech.*, 1990, **11**, 387–392.
[16] A. D. Wrixon, B. M. R. Green, P. R. Lomas, J. C. H. Miles, F. D. Cliff, E. A. Francis, C. M. H. Driscoll, A. C. James, and M. C. O'Riordon, 'Natural Radiation Exposure in UK Dwellings,' NRPB—R190, 1988.
[17] 'Radon, Report of the IEHO Survey on Radon in Homes 1987/8,' Institution of Environmental Health Officers, London, 1988.

Table 1 *Indoor radon concentration in UK dwellings (from Ref. 15)*

Area	Sample size	Mean concⁿ Bq m^{-3}	% of dwellings with radon concentration 100–200	200–400 Bq m^{-3}	>400
UK	2093	22.3	1.9	0.3	0.1
UK†	2583	30.2	3.3	1.2	0.4
SW	6500	200	22.2	15.9	12.7
NI*	174	20	1.7	0.6	0
NW**	674	20	1.3	0.3	0
NW†	133	—	13.5	6.8	0.8
UK	1900	55	5.8	3.2	1.6

* Systematic survey; † Winter period only; ** Charcoal detectors, not occupancy weighted; SW = SW England; NI = Northern Ireland; NW = NW England.

proportion of the overall annual dose of ionizing radiation received by the UK population.

Several methods exist for the measurement of radon in air and may be divided into active or passive techniques. The active techniques are generally only used for research or special survey purposes and normally consist of a pump which draws air through a filter trapping the radon decay products. The alpha radiation from these daughters is measured by scintillators or semi-conductor detectors.

Probably the most common method for monitoring radon is the passive alpha track detector and these have been used in UK national surveys.[16,17] Radon is allowed to diffuse into a small container, but radon daughters present in the ambient air are excluded. Inside the pot a piece of polycarbonate, cellulose nitrate, or allyl diglycol carbonate film is placed. Alpha particles, formed by the decay of radon, damage chemical bonds in the surface of the film. After a period of exposure, the surface of the film is etched with NaOH and tracks appear. The number density of these tracks is proportional to the average radon concentration and may be either counted using a computerized image analysing system under a microscope, or the tracks filled with a scintillant (ZnS–Ag) which fluoresces when exposed to an alpha source in proportion to the number of tracks present. The detectors may be calibrated by exposure to known radon concentrations for known periods of time. Exposure times of weeks to months are required with this method.

A further passive technique again involves the use of a container, through which radon, but not its daughters, may diffuse. However, the radon is subsequently adsorbed on to activated charcoal and air concentrations obtained by counting the γ-rays emitted by its daughters, after 3–5 days exposure.

In the National Radiological Protection Board (NRPB) survey of 2093 dwellings in the UK,[16] the distribution of observed radon concentrations was log normal, with an arithmetic mean of 22 Bq m^{-3} and a long tail of higher values (Table 1). About 0.4% of the sample was above 200 Bq m^{-3}, corresponding to about 50 000 houses over the whole country.

Figure 3 *Relative radon concentrations in homes throughout the UK* (Reproduced with permission from the National Radiological Protection Board)

The primary influence of bedrock geology on relative radon concentrations throughout the UK is exhibited in Figure 3. High levels of radon in homes (reaching 8000 Bq m^{-3}) were found in areas where the bedrock contained high concentrations of uranium. These included granites and mineralization in SW England and Scotland (up to 2000 p.p.m. U) and some shales and limestones in north and central England (about 800 p.p.m. U).

Other factors which influence radon levels were also identified, including season, time of day, meteorological conditions, ventilation (*e.g.* existence of double glazing), usage, and other occupancy habits. These can result in a factor of two differences between summer and winter concentrations. Furthermore, radon concentrations may vary between rooms, with the highest radon levels in basements and, on average, first floor bedrooms have concentrations two thirds those of living rooms.

In the UK, the limit for radon concentrations in homes, recommended by NRPB,[18] is currently 200 Bq m^{-3}. Parts of the country where there is a 1% probability or more of radon levels in houses exceeding this limit are to be regarded as Affected Areas and the NRPB recommends that control measures should be applied. This can be done by either preventing the gas entering the building (*e.g.* by sealing floors), increasing the ventilation rate (*e.g.* by extractor fan), or removing the radon decay products from the air (*e.g.* by electrostatic precipitator).

2.2.3 Environmental Water Monitoring. Pollutants enter the aquatic environment from the air (by dry deposition or in precipitation occurring either directly onto

[18] National Radiological Protection Board, Board Statement on Radon in Houses, Documents of the NRPB, 1.1, 1990.

the water surface or elsewhere within the catchment area), from the land (either in surface run-off or via sub-surface waters) and directly through effluent discharges (either domestic, industrial, or agricultural). The undesirable effects of pollutants in natural water may be due to:

(a) stimulation of water plant growth—eutrophication—which ultimately leads to deoxygenation of the water and major ecological change;
(b) their direct or indirect toxic effects on aquatic life;
(c) the loss of amenity and practical value of the water body, particularly as a source of water for public supply.

Apart from the monitoring of sources of pollutants in liquid effluents (Section 2.1.4 above) sampling may be carried out:

(a) in rivers, lakes, estuaries, and the sea in order to obtain an overall indication of water-quality;
(b) for rainwater, groundwater, and run-off water (particularly in the urban environment) to assess the influence of pollutant sources;
(c) at points where water is taken for supply, to check its suitability for a particular use;
(d) using sediments and biological samples in order to assess the accumulation of pollutants and as indicators of pollution.

Apart from the measurement of chemical and physical parameters the quantitative or qualitative assessment of aquatic flora and fauna is often used to give an indication of the presence or absence of pollution, and well recognized relationships are known to exist between the abundance and diversity of species and the degree of pollution. This is often used to assess the cleanliness of natural fresh waters and is known as biological monitoring.

2.2.3.1 Location of sampling sites. There are two main causes of heterogeneous distribution of quality in a water body. These are:

(a) if the system is composed of two or more waters which are not fully mixed (such as in themally stratified lakes or just below an effluent discharge in a river) and;
(b) if the pollutant distributes non-uniformly in a homogeneous water body (for example oil which tends to float, and suspended solids which tend to settle out of the water). Also chemical and/or biological reactions may occur non-uniformly in different parts of the system, so changing pollutant concentrations non-uniformly. When the degree of mixing is unknown it is advisable to conduct a preliminary survey before deciding on sampling locations. Rapidly obtainable measurements of water temperature, pH, dissolved oxygen, or electrical conductance may be used in this respect.

Sampling locations should generally be at points as representative of the bulk of the water body as possible, *e.g.* away from river or lake banks or the walls of channels or pipes, but often it will be desirable (and necessary) to take samples from several locations in order to obtain the required information.

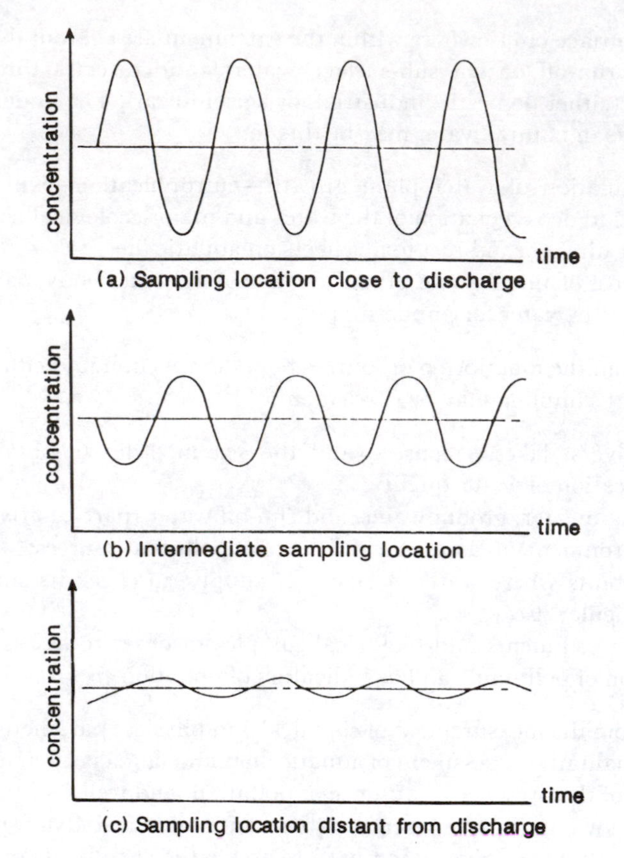

Figure 4 *Schematic diagram of the dependence of pollutant concentration on the distance downstream from a cyclically varying waste discharge*

When sampling from rivers and streams downstream of effluent discharges longitudinal, transverse, and vertical sampling arrays may be necessary to ensure that truly representative data are obtained. Studies of some pollutants require sampling at considerable distances downstream of effluent inputs, *e.g.* in investigating the sag in dissolved oxygen content. When a temporally-varying effluent discharge is under study it may be desirable to sample as close to the point of discharge as mixing allows in order to monitor short-term variations in concentration. However, if long-term average water quality is of interest then sampling should be carried out further downstream where longitudinal dispersion and mixing will have smoothed out the short-term variations (see Figure 4).

Sampling in estuaries presents special problems as great spatial and temporal variability may be exhibited. The appropriate locations for sampling will vary from estuary to estuary and will depend on the parameters of interest, but a minimum of 50 samples per survey might be appropriate. If one considers a compound which has an input at one end only of an estuary and which is not removed from or added to solution during its lifetime in the estuary then the concentration of that compound in the estuary will be solely dependent upon the dilution ratio between the river water and the seawater. Thus, for example, the concentration of chloride ion or salinity is dependent only upon the mixing of the fresh and the saline water bodies. This concept, 'conservative' behaviour, is an important one which must be taken into account when monitoring estuarine concentrations (see also Chapter 4). If a graph is drawn of the concentrations of the element of interest in an estuary against salinity then the data points will fall on a straight line (the theoretical dilution line) if physical mixing is the only process controlling the concentration of that element in the water. However, if the element of interest is added to or removed from the solution during mixing then the data will not plot on a straight line. In the case of lakes and reservoirs vertical stratification of pollutants may be very pronounced due to a reduction in dissolved oxygen from the surface downwards. A minimum of three samples is then considered necessary,[7] at 1 m below the surface, 1 m above the bottom, and at an intermediate point.

When water is abstracted from a river, lake, reservoir, or from an aquifer samples should be regularly taken at the point of abstraction and at the point where the water enters the distribution system. Several excellent handbooks with full descriptions of water sampling and analytical methods are available (*e.g.* References 19–22).

The determination of concentrations of trace metals in natural waters is a fundamental stage in the calculation of their budgets or cycles, but is subject to the same problems of sample contamination as occur for atmospheric samples from remote areas. All stages of the analysis, from sample collection, storage, and filtration to actual laboratory manipulation require care to prevent contamination occurring. Indeed for many years measured levels of many trace elements in seawater were purely an artefact of contamination during sampling and analysis.

2.2.3.2 Water monitoring case studies. Case study 3: The behaviour and variations of caesium and plutonium in estuarine waters.[23] Plutonium and caesium isotopes, in addition to other radionuclides, have been discharged into the Irish Sea in effluent from the nuclear fuel reprocessing plant at Sellafield since 1953. In this study,[23] the plutonium and caesium activity concentrations were monitored in the dissolved phase of estuarine waters, in order to investigate the behaviour of these radionuclides. Water samples were collected by boat at mid-channel, during spring and neap tides, along the Esk Estuary (10 km from Sellafield discharge point) and its tributaries (Irt and Mite). Samples of 1 or 2 dm^3 were collected using instantaneous isokinetic samplers at near surface (0.5 m) depth on high water surveys and mid-depth on tidal cycle surveys. In addition, a 10 dm^3

[19] 'Standard Methods for the Examination of Water and Wastewater,' 15th Edn., American Public Health Association, New York, 1980.

[20] M. J. Suess, 'Examination of Water for Pollution Control—A Reference Handbook,' 3 Vols, WHO/ Pergamon Press, Oxford, 1982.

[21] J. W. Clark, W. Viessman, and M. Hammer, 'Water Supply and Pollution Control,' 3rd Edn., Harper, New York, 1977, 841 pp.

[22] L. G. Hutton, 'Field Testing of Water in Developing Countries,' Water Research Centre, 1983, 120 pp.

[23] D. J. Assinder, M. Kelly, and S. R. Aston, *Environ. Technol. Lett.*, 1984, **5**, 23–30.

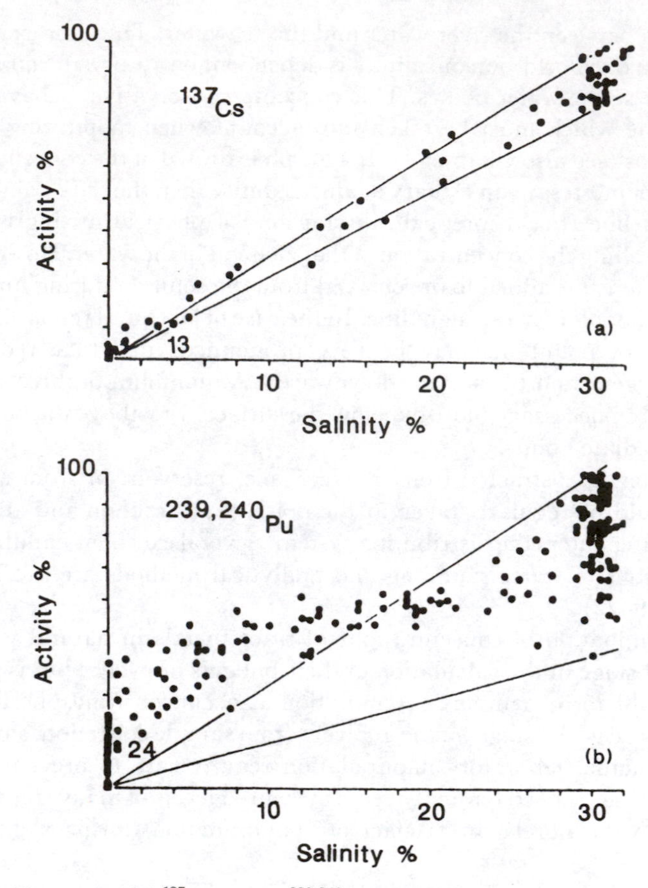

Figure 5 *Variation of (a) ^{137}Cs and (b) ^{239,240}Pu dissolved phase activity with salinity in the Esk Estuary, UK (Lines refer to level of confidence in theoretical mixing line)* (Reproduced with permission from *Environ. Tech. Lett.*, 1984, **5**, 27)

sample of river water was collected from the River Esk, well above the tidal limit. Fractionation into particulate and dissolved phases was achieved by filtration with 0.22 μm Millipore filters. Salinities were measured using a Goldberg refractometer and activity concentrations by γ-spectrometry and α-spectrometry following standard preparation and chemical separation procedures.[24]

Figure 5 shows the variation in plutonium and caesium dissolved phase activity (expressed as a percentage of the maximum 'seawater' value for each study) with salinity. Figure 5a clearly indicates that ^{137}Cs activities follow a theoretical dilution line between seawater and river water end members, exhibiting conservative behaviour. Therefore, caesium does not appear to re-equilibriate between the particulate and dissolved phases throughout estuarine

[24] RADREM, 'Sampling and Measurement of Radionuclides in the Environment,' HMSO, London, 1989.

waters of different salinities. However, dissolved plutonium activities do not follow a theoretical dilution line (Figure 5b). Higher activities are apparent in the dissolved phase, for salinities below 18‰, than predicted by conservative mixing. In addition, a wide variation of activities were recorded at high salinities. Thus monitoring of $^{239,240}Pu$ dissolved phase activity concentrations has shown that plutonium is apparently being released from the particulate phase at low salinities in the Esk Estuary.

Case study 4: Acidification of lakes.[25] The acidification of lakes in Northern Europe (Scotland, Norway, and Sweden) and North America has been directly linked to anthropogenic activities, in particular sulfur dioxide emissions from coal fired power stations. Much of the evidence for increase in acidity has been indirect, such as loss of fish populations, changes in aquatic plant and invertebrate communities, changes in sediment metal concentrations, and results from empirical models of lake acidification. A study of the Adirondack lakes in New York state[25] made direct comparisons of historical and recent lake survey data in order to determine if significant acidification had occurred in that region.

The historic data (1929–1934) included measurements of alkalinity (a measure of the acid neutralizing capacity of the water), pH, and CO_2 acidity. However, the pH data were considered to be unreliable due to the use of colorimetric indicator solutions which can alter the pH of a solution being measured. It was therefore decided to directly compare alkalinity values from this data set with extensive data collected between 1975–1985. Alkalinities for each lake were matched on the basis of a unique 'pond number' system adopted by the surveys. Data were excluded where lake identification was ambiguous, the lakes were known to have been treated with lime, or it was suspected that the electrode used for the alkalinity measurements was malfunctioning. The final data set consisted of 274 lakes.

Account was taken of the different analytical techniques employed to determine alkalinity. The historic method used titration to a fixed pH end point, determined by the 'faintest pink' colour of the methyl orange indicator. Modern surveys employed the Gran technique to determine the end point. As a result, the historic data set were corrected by an alkalinity subtraction of 54.6 μequiv. dm^{-3}.

The historic data had a mean 'corrected' alkalinity of 140.8 μequiv. dm^{-3} compared to a modern alkalinity of 99.8 μequiv. dm^{-3}. This indicates a mean estimated acidification of -41 μequiv. dm^{-3}. This loss of alkalinity was found to be significant ($p < 0.01$) by a Wilcoxon signed rank test. It was concluded that significant acidification had occurred in the Adirondack Mountain region since the 1930s. Inter lake differences in acidification were noted, probably due to factors such as atmospheric deposition, lake hydrology, geology, or soil type. Sensitivity of lakes to acidification has been related to crystalline bedrock and thin acid soils, which are generally associated with high elevation lakes. This study found a direct relationship between the mean lake elevation and mean acidification for lakes in each of the major catchments.

[25] C. E. Asbury, F. A. Vertucci, M. D. Mattson, and G. E. Likens, *Environ. Sci. Technol.*, 1989, **23**, 362–365.

2.2.4 Sediment, Soil, and Biological Monitoring. Soils and sediments may become polluted by a number of routes, including the disposal of industrial and domestic solid wastes, wet and dry deposition from the atmosphere, and infiltration by contaminated waters.

The main pollution hazards on land have been identified as follows:[1]

(*a*) Harmful substances may get into the soil or plants and so into the food supply.
(b) Substances may wash from the land and so pollute water supplies.
(c) Contaminants may be resuspended and subsequently inhaled.
(d) Substances polluting the land may make it potentially dangerous or unsuitable for future use (*e.g.* for housing or agriculture).
(e) Ecological systems may be damaged, with consequent loss to conservation and amenity.

Some potentially harmful substances, such as mercury or lead, are naturally present in soils but at concentrations which are not normally deleterious. Some activities however can cause elevated levels of these compounds. For example, mining may cause soils to be contaminated by metals, and the dumping of solid wastes on land will invariably introduce a wide variety of pollutants to the soil. On the other hand there are compounds which do not occur naturally, and their presence in soils and sediments is due entirely to man's activities. These substances include pesticides (particularly the organo-chlorine compounds such as DDT, toxaphene, aldrin, dieldrin) and artificial radionuclides (*e.g.* ^{137}Cs, ^{106}Ru).

As with the other types of media discussed above it is important that the background levels of pollutants be established in soils, sediments, and vegetation. One example of a large-scale investigation of this type is the geochemical survey of stream-bed sediments carried out in England and Wales.[26] Stream sediments are considered to represent a close approximation to a composite sample of the weathering and erosion products of rock and soil upstream of the sampling point and in the absence of pollution provide information on the regional distribution of the elements.

A second type of monitoring programme is required to establish actual levels of contamination in land or sediments known or believed to be affected by pollutants. In this case much more specific and localized monitoring may be required in order to quantify the degree of contamination. The contamination of sites often arises from their previous uses, particularly as coal-gas manufacturing plants, sewage works, smelters, waste disposal sites, chemical plants, and scrap-yards. A typical contaminated site may be found to contain variable concentrations of toxic elements and organic compounds, phenols, coal-tars, and oils, combustible material from undecomposed refuse, acidic or alkaline waste sludges, and sometimes methane accumulations. In addition, contaminated land

[26] J. S. Webb, I. Thornton, M. Thompson, R. J. Howarth, and P. L. Lowenstein, 'The Wolfson Geochemical Atlas of England and Wales,' Oxford University Press, Oxford, 1978.

is often formed by waste tipping and so may be poorly compacted and very unhomogeneous.

Some sites may contain underground pipework and structures from their previous uses and so present formidable sampling problems. Site investigation of this type is very expensive, involving bore holes or trial pits, and, in cases where a detailed site history is unavailable, determination of a large number of pollutants. However it is most important to be sure that the investigation is sufficiently rigorous as remedial measures are extremely costly and must be based upon adequate data. Common problems which have been identified[27,28] are:

—an inadequate number of samples;
—an inadequate range of determinands;
—bulking of samples when individual samples from specific locations are preferable;
—inappropriate analytical methods;
—inadequate referencing of sample locations;
—inadequate descriptions of samples;
—inadequate descriptions of trial pit strata;
—an ignorance of the nature of the required information.

Monitoring should be carried out following the application of sewage sludge or waste waters to agricultural land. Samples of surface water, groundwater, site soil, vegetation, and the sludge applied would normally be tested for faecal coliform, nutrients, heavy metals, and pH. Details of the necessary chemical and physical methods may be found elsewhere.[19] The results from the monitoring exercise may be compared to predicted levels derived from the application rates of sludges to land, soil type, nitrogen, phosphorus, and heavy-metal contents of the waste, and the nutrient-uptake characteristics of the cover crop.

When monitoring background levels and more specific pollution on land or in the sediments of a water body, measurements will often be made of levels in the plants or organisms that the soil or sediments support. In many cases flora or fauna provide excellent indicators of the degree of pollution as they may act as bioconcentrators (for example of heavy metals from suspended material in shellfish). Furthermore it is obviously important to monitor pollution levels in food and through the food chain. The simultaneous measurements of pollutant levels in soils and plants as well as in water, sediments, and aquatic biota are therefore often carried out. However the relationships between levels in these various media are often not simple and sampling and analysis of one of these is no substitute for a comprehensive monitoring programme. For example laboratory experiments indicated that metals are readily absorbed by the water hyacinth, *Eichornia crassipes*, but measurements of nickel, cadmium, zinc, cobalt, and lead in

[27] M. A. Smith, 'Redevelopment of Contaminated Land: Tentative Guidelines for Acceptable Levels of Elements in Soils,' ICRCL 38/80, DoE Central Directorate on Environmental Pollution, London, 1980.
[28] M. A. Smith, 'Controlling the Development of Contaminated Land,' Proc. Conf. Local Government and Environmental Pollution and Control, Coventry, 1980.

river water samples and water hyacinths revealed a lack of correlation between the corresponding pairs of concentrations.[29] Furthermore, high concentrations of nickel, lead, and zinc were found in the plants of rivers known to receive effluents containing metals, even when elevated levels were not detected in the water samples. Presumably this was due to water samples being taken during periods of low waste discharge and so showing low metal concentrations. Plant samples taken at the same time would show elevated concentrations having retained metal from periods of high water concentrations.

An example of the large-scale use of aquatic biota as a medium for the monitoring of pollutants in coastal waters is the US Mussel Watch Program.[30] This programme began in 1976 with the overall aim of providing strategies for pollutant monitoring in coastal waters using mussels and oysters collected on the west, east, and Gulf coasts of the USA. Analyses for trace metals, chlorinated hydrocarbons, petroleum hydrocarbons, and radionuclides (including ^{238}Pu, $^{239+240}$Pu, and ^{241}Am) were carried out at three laboratories with extensive intercalibration studies used to ensure compatibility of data. One conclusion of the initial report of this programme was that for metals the frequency of monitoring need not be yearly, but some small multiple of years for a given site. Thus for the same resources, human and financial, greater geographical coverage can be gained by increasing the sampling interval. On the basis of the data obtained in 1977–1978 national baseline concentrations for bivalves in unpolluted waters were suggested. The programme also established the importance of systematic repetitive sampling at the same sites over periods of years, which allowed elevated cadmium and plutonium concentrations in some areas to be attributed to coastal upwelling rather than as a consequence of localized anthropogenic sources.

Monitoring of concentrations of trace metals in crop plants grown on sewage sludge-amended soils indicates that the levels found will vary with crop species and properties of the soil substrate on which they have been grown. More recent studies have concentrated on the physico-chemical speciation of the metals in the applied material and the receiving soil and have demonstrated the importance of organic complexes in reducing free metal activity.

Case study 5: Metals in dusts.[31] A great number of studies have been conducted since the 1970s on dust and soil contamination with heavy metals, in particular lead, due to its predominant source as an anti-knock agent added to petrol. These studies have established the magnitude of different source of lead in the environment. However, it is still unclear whether or not the exposure of young children to relatively small amounts of lead has a detrimental effect on intelligence, behaviour, and educational and social attainment.

[29] J. E. L. Maddock, M. B. P. dos Santos, and R. S. Marinho, 'Monitoring of Metal Pollution in Tropical Rivers by Sampling and Analysis of *Eichomia Crassipes*,' Proc. Int. Conf. Environmental Pollution, London, CEP Ltd., Edinburgh, 1984.

[30] E. D. Goldberg, V. T. Bowen, J. W. Farrington, G. Harvey, J. H. Martin, P. L. Parker, R. W. Risebrough, W. Roberts, E. Schneider, and E. Gamble, *Environ. Cons.*, 1978, **5**, 101–125.

[31] D. J. A. Davies, I. Thornton, J. M. Watt, E. B. Culbard, P. G. Harvey, H. T. Delves, J. C. Sherlock, G. A. Smart, J. F. A. Thomas, and M. J. Quinn, *Sci. Tot. Environ.*, 1990, **90**, 13–29.

Although various studies have attempted to assess the significance of different pathways of lead to a child's total lead intake, the study described here[31] was a comprehensive environmental, biological, behavioural, and dietary investigation of 2 year old children in Birmingham. From a set of 183 randomly selected children, 97 completed the study. Various blood samples were taken at the beginning of the study and 56 children also provided a second sample about 5 months later.

Environmental monitoring was conducted at each of the children's homes. Indoor and outdoor air samples were obtained, over seven consecutive days, using aerosol monitors at a height of 1.5–2 m, connected via PVC tubing to a single vacuum pump. House dust samples were collected from the householders' own vacuum cleaner and from the child's playroom and bedroom, using a vacuum cleaner adapted to collect dust in a soxhlet thimble. Pavement and road dusts were sampled using a plastic dust pan and brush during dry conditions. A composite soil sample was also collected with a stainless steel trowel. On each of the seven days of the study, both hands were wiped thoroughly using a total of 3 wet wipes. A duplicate of the child's diet over the seven days was provided, along with a dietary record. Finally, mouthing behavioural patterns were analysed using a 30 item questionnaire or interviewing the parents and by recording the child on videotape in set situations. These situations included a period of free play, lunch, being read a 10 minute story, and watching a 15 minute video film. Each film was analysed to determine how much time a child touched the floor, objects, and the mouth.

The dust and soil samples were dried at 80°C for 24 h, after which the dust samples were passed through a 1 mm sieve, while the dust samples were sieved to <2 mm and reduced to a powder in an agate mill. The air, dust, and soil samples were analysed by inductively coupled plasma atomic emission spectroscopy. Handwipes, food, water, and blood were analysed by atomic absorption spectrophotometry.

The frequency distributions of lead concentrations in all the samples were found to be approximately log normal, consequently geometrical means were calculated (Table 2). These mean lead concentrations (*e.g.* mean indoor dust lead 424 $\mu g\ g^{-1}$) were similar to those found in previous studies in both Birmingham and elsewhere in the UK, although exposure to dietary lead was believed to be slightly elevated due to higher water lead concentrations. The indoor and outdoor air lead concentrations were strongly correlated, as observed in previous studies, with a mean indoor/outdoor air lead ratio of 0.61. An exception to this occurred in a house in which old paint was being removed by a machine sander. On these occasions the indoor lead concentration exceeded that outdoors. In addition, the dust lead concentration for the householder's own vacuum cleaner was 23 000 $\mu g\ g^{-1}$.

Relationships between blood lead, environmental lead, and behavioural measures were assessed using multiple regression analysis. This indicated the importance of the amount of lead in the house dust combined with a child's rate of hand touching activity to blood lead levels. Water lead and smoking habits of parents are also significant factors where blood lead is concerned. These results

Table 2 *Lead in blood, environmental samples, handwipes, and water (from Ref. 31)*

Sample	Units	N	Geometric mean	Percentiles 5th	Percentiles 95th
Blood	$\mu g\ (100\ ml)^{-1}$	97	11.7	6	24
Air					
Playroom	$\mu g\ m^{-3}$	607	0.27	0.08	0.88
Bedroom	$\mu g\ m^{-3}$	599	0.26	0.09	0.81
External	$\mu g\ m^{-3}$	605	0.43	0.12	1.53
Dust					
Playroom	$\mu g\ g^{-1}$	97	311	105	1030
Bedroom		96	464	109	2040
Average		94	424	138	2093
Vacuum[a]		92	336	97	1440
Doormat		42	615	120	4300
Pavement		97	360	127	1340
Road		97	527	195	1170
Dust 'loading'	$\mu g\ m^{-2}$	93	60	4	486
Soil	$\mu g\ g^{-1}$	86	313	92	1160
Handwipes	μg	704	5.7	1.9	15.1
Diet (food and beverages)	$\mu g\ week^{-1}$	96	161	82	389
Water	$\mu g\ l^{-1}$	96	19	5	100

[a] Householder's own cleaner

were used to predict a mean lead intake of 1 $\mu g\ day^{-1}$ via inhalation and 35 $\mu g\ day^{-1}$ via ingestion for the Birmingham children.

Case Study 6: Chernobyl derived radiocaesium in Cumbria, UK.[32] Following the Chernobyl reactor accident on 26th April 1986, a large number of monitoring studies have been conducted in the UK. Contamination of sheep by radiocaesium reared on upland fells in Cumbria and North Wales is of particular concern, due to the high levels of fallout received by these areas (20 kBq m^{-2} ^{137}Cs). Restrictions to movement and slaughter of those sheep with a ^{137}Cs muscle specific activity concentration greater than 1000 Bq kg^{-1} (fresh weight) have been imposed by UK authorities.

One study conducted at a farm in West Cumbria[64] investigated radiocaesium specific activities in local soils, vegetation, and sheep muscle. Three upland soils and one valley bottom soil were sampled within the same 1 km^2 area, using a 5 cm deep corer. The sampling programme continued from October 1986 to November 1989 with a sampling frequency of every two weeks for the first 5 months. Samples were analysed for total and ammonium exchangeable ^{134}Cs and ^{137}Cs by γ-spectrometry.

Vegetation samples were collected from improved pasture at the study farm and also from the nearby open fell, from August 1986 until June 1988 with a

[32] B. J. Howard, N. A. Beresford, and F. R. Livens, 'Transfer of Radionuclides in Natural and Semi-Natural Environments,' eds. G. Desmet, P. Nassimbeni, and M. Belli, Elsevier Science Publishers Ltd., New York, 1990.

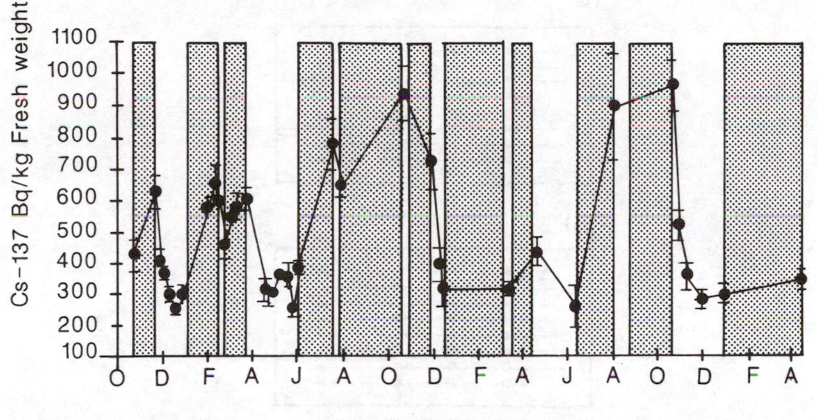

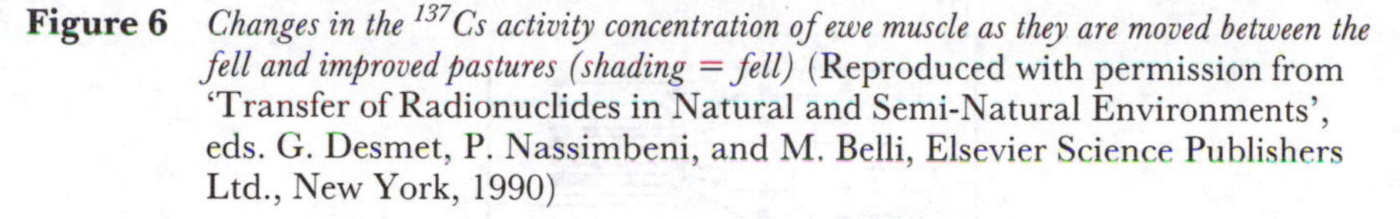
Figure 6 *Changes in the ^{137}Cs activity concentration of ewe muscle as they are moved between the fell and improved pastures (shading = fell)* (Reproduced with permission from 'Transfer of Radionuclides in Natural and Semi-Natural Environments', eds. G. Desmet, P. Nassimbeni, and M. Belli, Elsevier Science Publishers Ltd., New York, 1990)

frequency of once every two weeks. Live monitoring of the sheep for ^{137}Cs was achieved using a portable γ-spectrometry system which could be taken into the field. A similar monitoring programme to that for the soil and vegetation was adopted.

As a result of this survey it was discovered that a high percentage of the Chernobyl derived ^{137}Cs in upland soils was ammonium exchangeable (86–100%) and thus probably available for uptake by vegetation, compared to 0% for the valley-bottom soil. This was confirmed by the generally higher ^{137}Cs content in vegetation from the open fell (September 1987, 140–2080 Bq kg^{-1} dry weight) compared to the improved pasture (September 1987, $<$100–710 Bq kg^{-1} dry weight).

As a consequence of this radiocaesium behaviour, the ^{137}Cs activity concentrations of ewe muscle (Figure 6) declined when the sheep were brought onto enclosed pastures, but rose when they were returned to the open fell. However, despite an overall fall in the ^{137}Cs activity concentration in vegetation with time, the ^{137}Cs levels in sheep were higher in the summer/autumn of 1987 and 1988 than in the autumn of 1986. This appeared to be due, at least in part, to an increase in the availability of radiocaesium originating from Chernobyl fallout, as it had been incorporated into plant material, rather than when it was present as a direct deposit on vegetation surfaces as in 1986.

3 SAMPLING METHODS

3.1 Air Sampling Methods

Sampling systems for airborne pollutants usually consist of four component parts, the intake component, the collection or sensing component, the flow

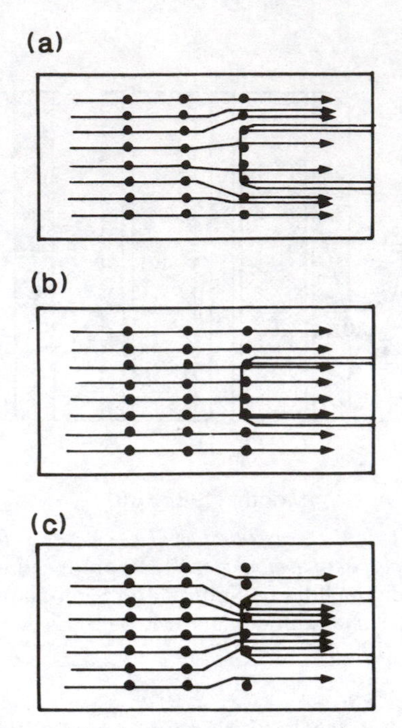

Figure 7 *Schematic diagram of (a) over-sampling of suspended particles; (b) isokinetic sampling; and (c) under-sampling of suspended particles*

measuring component, and the air moving device. All these must be constructed of materials which are chemically and physically inert to the sampled air (other than the collector or sensor itself).

3.1.1 Intake design. The nature of the intake is determined by the type and objective of the sampling technique, and may vary from a vertical opening for the passive collection of dustfall in a deposit gauge to a thin-walled probe used for source sampling of aerosols. Common problems which may require consideration are the non-reproducible collection of the sample portion from the air mass due to poor inlet design, adhesion of aerosols to tube walls, loss or change of analyte by chemical reaction with inlet materials, adsorption of gaseous components on inlet materials, and condensation of volatile components within the transfer lines.

The sampling of aerosols presents particular difficulties in inlet design. A basic requirement is that the velocity of the sample entering the system intake should be the same as the velocity of the gas being sampled. This is necessary because, if the streamlines of the sampled gas are disturbed by the intake probe, particles travelling in the gas flow and possessing inertia directed along the streamlines will continue into the probe while the 'carrier gas' will be diverted away from (if the probe intake velocity is too low) or into (if the intake velocity is too high) the inlet (Figure 7). Thus either a greater number of particles per unit volume of gas than exists in the actual gas flow, or a fewer number will be collected. Only when the intake velocity at the face of the probe is equal to the approach velocity of the gas stream will the streamline pattern remain unaltered and the correct number of particles per unit volume of gas enter the probe. This is known as isokinetic sampling.

Sampling of ambient air masses is seldom made isokinetically as sophisticated equipment is required to maintain the inlet facing into the wind and to adjust the sampling velocity to match changes in wind speed. This is feasible, but what is difficult is then to interpret the analysis of the collected sample as the flow rate is temporally variable. However isokinetic sampling is usually practicable, and indeed necessary, when sampling flue gases.

3.1.2 Sample Collection. The methods most commonly used for the collection of atmospheric particulate samples are:

—filtration;
—impingement: wet or dry impingers, cascade impactors;
—sedimentation: by gravity in stagnant air, thermal precipitators;
—centrifugal force, cyclones;

and for gaseous samples are

—adsorption;
—absorption;
—condensation;
—grab sampling.

3.1.2.1 Filtration. This is by far the most common technique. The type of filter medium chosen will depend upon a number of factors. These include the collection efficiency for a given particle size,[33,34] pressure drop and flow characteristics of the filter type,[35] background concentrations of trace constituents within the filter medium, and the chemical and physical suitability of the filter with regard to the sampling environment.

3.1.2.2 Impingement. Impingers consist of a small jet through which air stream is forced, so increasing the velocity and momentum of suspended particles, followed by an obstructing surface on which the particles will tend to collect. Wet impingers operate with the jet and collection surface under liquid and require high flow rates for optimum collection efficiency.

Cascade impactors use the aerodynamic impaction properties of particles to separate the sample into different size fractions by use of sequential jets and collection surfaces. Increasing jet velocity at each stage fractionates the sample. Figure 8 shows the principle of a commonly used cascade impactor. This consists of up to seven stages backed by a membrane filter, each stage containing accurately drilled holes which align over a solid portion of the adjacent plates. The holes in each successive stage are smaller than those in the preceding plate,

[33] M. Katz, 'Measurement of Air Pollutants, Guide to the Selection of Methods,' WHO, Geneva, 1969.
[34] R. K. Skogerboe, D. L. Dick, and P. J. Lamothe, *Atmos. Environ.*, 1977, **11**, 243–249.
[35] B. Y. H. Liu, D. Y. H. Pui, and K. L. Rubow, 'Aerosol in the Mining and Industrial Work Environments,' eds V. A. Marple and B. Y. H. Liu, Vol. 3, Ann Arbour Science, 1983, pp. 898–1038.

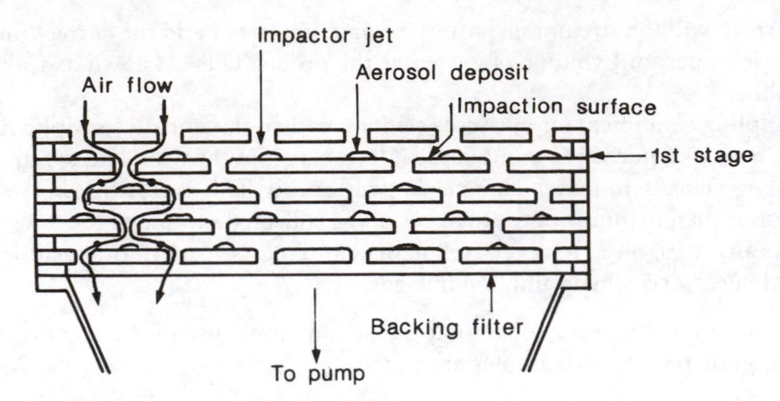

Figure 8 *Schematic representation of a cascade impactor*

and since air is drawn through the instrument at a constant flow rate the effective velocity at each stage increases. The largest particles are impacted on the first stage and the smallest are collected on the back-up filter. The range of particle diameters collected on each stage may be determined by laboratory calibration or by theoretical calculations. However impaction sampling at normal flow rates (0.01–0.04 m^3 min^{-1} or 0.6–1.1 m^3 min^{-1} for Hi-Vol cascade impactors) and atmospheric pressure is only efficient for particles with aerodynamic diameters $>0.3\,\mu m$. Also the collection efficiency of each stage will vary according to particle type, some being very 'sticky', others liable to bounce off. Other problems of cascade impactor sampling include wall losses and the aggregation of particles and the mechanical breaking of agglomerates which result in inaccurate size distribution measurements.

3.1.2.3 Sedimentation. The collection of particulate material by allowing it to deposit into a collection vessel is the simplest of all air pollution measurement techniques. However the presence of the bowl or cylinder in the path of the falling particles will change their flow pattern and it is not clear whether the collected materials is truly representative of actual conditions. The British Standard deposit gauge consists of a collection bowl connected to a bottle and supported by a galvanized steel stand. During wet weather dust is washed down from the bowl, but during dry weather high winds may blow dust out of or into the bowl, so producing erroneous dust loadings. At the end of the sampling period (usually one month) a measured volume of water is used to wash any dust in the bowl into the collection bottle, and the pH, total particulate mass, and water volume determined.

A major disadvantage of the standard deposit gauge is that no directional resolution of the source of particulate material is possible. The directional deposit gauge[36] consists of four cylinders mounted on a common post, with open slots facing the four quadrants of the compass (Figure 9). Each cylinder has a removable collection bottle at its base. After collection a suspension of the dust

[36] British Standard Method for the Measurement of Air Pollution. Part 5—Directional Deposit Gauges BS 1747, Part 5, 1972.

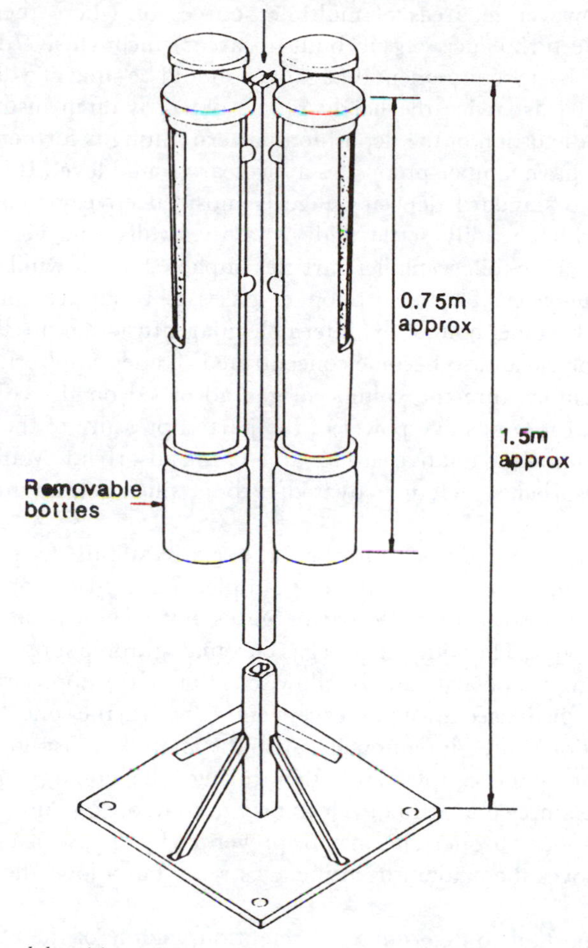

Figure 9 *Directional deposition gauge*

may be placed in a glass cell and a measure of the dust loading made by the amount of obscuration of a beam of light passing through the cell. Density fractionation of the collected materials is also possible by using a mixture of di-iodomethane and acetone which allows a density gradient of 0.8–3.3 g cm^{-3} to be achieved. By comparison of the density gradient fractionation of material collected in a directional deposit gauge with material collected from likely sources (*e.g.* pulverized fly ash storage heaps, slag heaps, cement works, *etc.*) simple source apportionment studies may be attempted. Other techniques, including dispersion staining and microscopic examination may also be used in this way, whilst the compilation of a reference library of dusts from known sources in a given area will greatly ease the practical difficulties of dust identification.

The siting of deposit gauges often presents problems. When attempting to monitor emissions from one major source the use of several directional gauges around the source may be successful in confirming that emissions occur from that source. However in areas of multiple sources or where there is significant atmospheric turbulence (*e.g.* in built-up areas) inconclusive data may be obtained. The basic requirement that gauges should be sited at a distance from any object of at least twice the height of the object is often insufficient to avoid significant distortion of the deposition pattern. Siting is further complicated by the need to have tamper-proof sites at or near ground level. It is worth pointing out that the standard deposit gauge is most effective for collection of large particles which readily settle under gravity, whilst the vertical slots of the directional gauge collect smaller particles impacted by the wind more effectively.

3.1.2.4 Adsorption. The adsorption of gases is a surface phenomenon. Gas molecules become bound by intermolecular attraction to the surface of a collection phase and so become concentrated. Under equilibrium conditions at constant temperature the volume of gas adsorbed on the collection phase is proportional to a positive power of the partial pressure of the gas, and is also dependent upon the relative surface area of the adsorbent. Materials commonly used as adsorbents include activated carbon, silica gel, alumina, and various porous polymers.

When selecting a suitable adsorbent the relative affinity for polar or non-polar compounds must be considered. For example, activated carbon is non-polar and therefore will absorb non-polar organic gases, but exclude polar compounds such as water vapour. The wide range of gas chromatographic supports available vary in their degree of polarity and so allow selection of the appropriate type.

The adsorbent used must not react chemically with the collected sample unless chemisorption is used intentionally. Also the analytes must not react with other constitutents of the sampled air, either during collection or storage. It has been found for example that tetraalkyl lead may decompose on Porapak Q by reaction with atmospheric ozone. This may be prevented by the use of a selective prefilter which removes the oxidant from the air stream but allows the analytes to pass through.[37]

It is important to determine the retention volume of the adsorbent (*i.e.* the volume of air which may be passed without breakthrough of the analyte) with respect to the species being collected. This should be high enough to allow sufficient of the analytes to be collected for analysis. The desorption properties of the material are also important to ensure quantitative recovery of the sample, preferably with regeneration of the adsorbent for subsequent use. Activated carbon is a very efficient adsorber, so making quantitative desorption difficult. Steam stripping may result in hydrolysis reactions with the analytes and vacuum distillation and solvent extraction are not without their problems. Compounds on support bonded porous polymers may conveniently be thermally desorbed by flushing with an inert carrier gas. In the case of tetraalkyl lead on Porapak Q a two-stage thermal desorption system utilizing an intermediate cryogenic trap cooled with liquid N_2 and then flash-heated to 90 °C allows quantitative recovery as well as direct injection of the sample in a very small volume of gas onto the gas chromatograph column.[37]

[37] C. N. Hewitt and R. M. Harrison, *Anal. Chim. Acta.*, 1986, **188**, 229–238.

Since adsorption is temperature-dependent, collection efficiency and an increase in retention volume may be achieved by cooling the adsorbent. However, problems with blockages by ice may then occur. With the increase in sophistication of detection systems in recent years, particularly by the interfacing of chromatographic separation techniques with mass detectors and spectrophotometers, the use of adsorbents as preconcentrators is also increasing. However, care must always be exercised to avoid non-quantitative collection, break-through effects due to exceeding the retention volume of the system, decomposition, and non-quantitative recovery of the sample.

3.1.2.5 Absorption. Gases may be collected by being dissolved in a liquid collection phase or by chemical reaction with the absorbent. The simple Dreschel bottle may be used or may be modified by the inclusion of a fritted diffuser to create small bubbles and so enhance the collection efficiency.

An example of a simple absorption technique which allows an estimate of NO_2 concentration to be made with relatively little capital outlay is the use of triethanolamine diffusion tubes.[38] An acrylic tube 7.5 cm long × 1 cm i.d. is fitted with a fixed cap at one end and removable cap at the other. A fine wire mesh coated in triethanolamine is placed in the closed end of the tube and adsorbs NO_2 as it diffuses from the open end. The NO_2 is determined spectrophotometrically at the end of the sampling period. These passive samplers are very cheap to construct and analyse and have been used as an effective primary survey technique before embarking upon a more expensive monitoring exercise based upon the standard chemical method or chemiluminescent techniques.[39] In this survey highest NO_2 concentrations were identified at main road junctions.

3.1.2.6 Condensation. By cooling an air stream to temperatures below the boiling point of the substance of interest it is possible to condense gases from the air and so concentrate them. However, a limitation of the method is that water vapour present in the air will also freeze and so progressively block the trap. This may be overcome by using a first trap of large volume designed to collect water and a second trap at a sufficiently low temperature to collect the analytes. Coolants of temperatures −183 °C or lower (*e.g.* liquid N_2) should not be used for this purpose as they will condense atmospheric oxygen and result in a serious combustion hazard.

3.1.2.7 Grab sampling. Rather than utilizing a concentration technique in the field, samples may be collected in an impermeable container and returned to the laboratory for analysis. Grab samples of this type have been collected in FEP-Teflon bags for hydrocarbon determination by GC for example. In this technique the deflated bag is contained within a rigid box which is slowly evacuated. Air is thus drawn into the flexible bag which may be sealed when inflated. Samples can then be drawn at a later stage from the bag by hypodermic gas-tight syringe.

3.1.2.8 Case studies. Two air sampling techniques are in such widespread use that they require separate consideration. These are:

[38] A. J. Apling, K. J. Stevenson, B. D. Goldstein, R. J. W. Melia, and D. H. F. Atkins, 'Air Pollution in Homes 2: Validation of Diffusion Tube Measurements of Nitrogen Dioxide,' Warren Springs Laboratory LR 311 (AP), 1979.

[39] C. N. Hewitt, *Atmos. Environ.*, 1991, **25B**, 429–434.

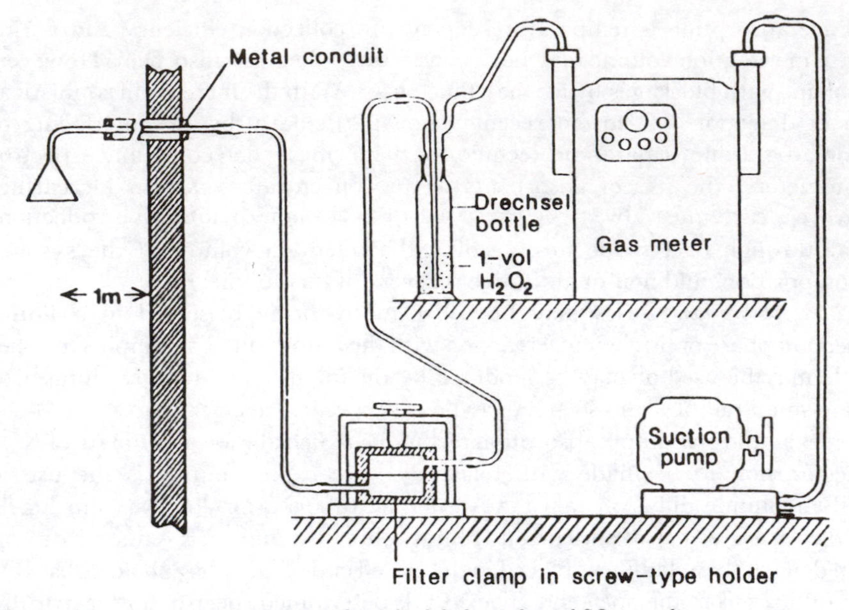

Figure 10 *Schematic arrangement of standard daily smoke and SO_2 sampling apparatus*

(a) the UK National Survey of Air Pollution smoke and SO_2 sampling apparatus, and
(b) the US National Air Sampling Network Hi-Vol method for total suspended particles (TSP) measurement.

(a) NSAP smoke and SO_2 method. The equipment used in the National Survey of Air Pollution in the UK is shown in Figure 10.[12] The pump draws about 2 m^3 of air per day through a filter paper held in a brass clamp which removes particulate material and then through a Dreschel bottle. This contains dilute H_2O_2 which removes SO_2, converting it to H_2SO_4, which is determined by titration. A simple sophistication is the addition of an eight-port sample changer which allows eight 24 hr samples to be collected sequentially in an array of eight filter papers and bottles, and so requires operator attendance only once a week. Although the H_2O_2 method will actually measure the net gaseous acidity of the air it is considered that for ordinary urban situations it will give a good estimate of the SO_2 concentration and hence results are expressed in terms of $\mu g\ SO_2\ m^{-3}$. The particulate loading of the exposed filter paper is estimated by measuring the reflectance of the paper, so obtaining a measure of the staining property of the air. A calibration curve is then used to convert the darkness of the stain to concentrations of equivalent standard smoke. Thus inaccuracies in these methods may become pronounced when there are acidic or basic gaseous components other than SO_2 present in the air (HCl giving a positive interference and ammonia a negative interference) or when particulates of 'non-standard' staining properties (*e.g.* light coloured ammonium compounds) are present. When the filter paper is extremely heavily loaded the reflectometer method may

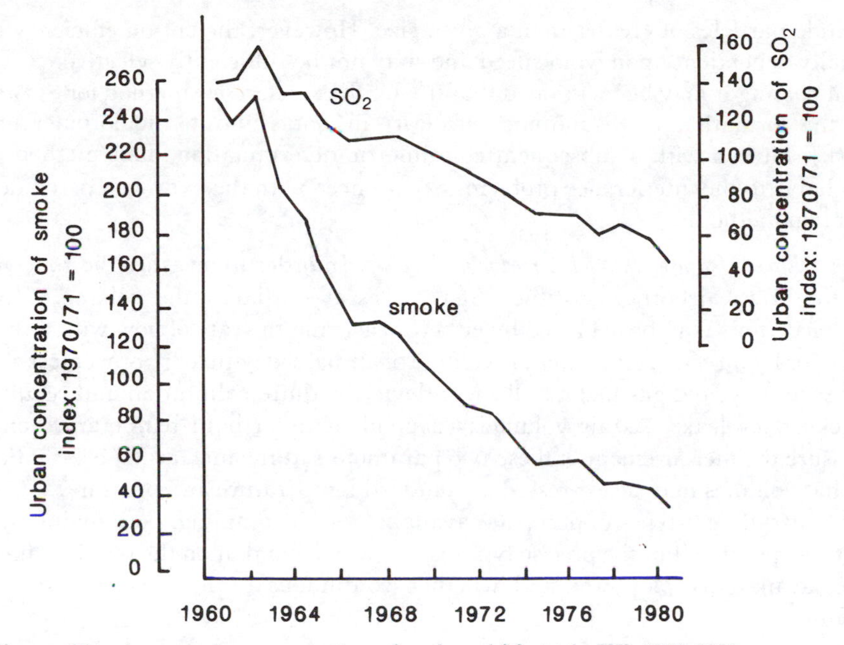

Figure 11 *Average urban concentrations of smoke and SO_2 in the UK, 1960–1981*

severely underestimate the actual smoke level. Particulate losses may also occur at the inlet to the apparatus and in the inlet tube. The smoke stain method gives concentrations of Standard Smoke which are not directly comparable with TSP levels determined by the US Hi-Vol method. Figure 11 shows the fall in smoke and SO_2 concentrations found by the National Survey during the period 1960–1980.

(b) Hi-Vol method for suspended particulates. This method is the current US EPA reference method for total suspended particulates (TSP) and is used in the US National Air Sampling Network,[40] although is being superseded by the use of a size selective inlet which excludes particles of aerodynamic diameter $>10\ \mu m$ (the PM_{10} instrument). A high flow rate blower draws the air sample into a covered housing and through a 20×25 cm rectangular glass fibre filter at 1.1–1.7 $m^3\ min^{-1}$. The mass of particles collected on the filter is determined gravimetrically and extraction techniques may then be used to remove the material for chemical analysis. However, when glass fibre filters are used, reaction with acidic components may result in artifact formation, for example sulfate from gaseous SO_2. Although 24 hr sampling periods are commonly used, timer devices are available to switch on and off the blower at predetermined intervals. However, passive sampling by the settlement of particulates onto the filter during periods when the pump is not operating may cause a positive error in the determination. Recent modifications include size-selective inlets which will

[40] EPA, Reference Method for the Determination of Suspended Particulates in the Atmosphere (High Volume Method), US Federal Register 36, No. 84, 1971.

exclude particles of greater than a given size. However, the cut-off efficiency is usually dependent upon wind speed and may not be sufficiently selective.

In passing it may be mentioned that the US EPA reference method for SO_2 is by the formation of disulfitomercurate(II) in potassium tetrachloromercurate(II) solution with a subsequent colorimetric determination. This method is less liable to the interference problems experienced with the hydrogen peroxide/titration method.

3.1.3 Flow Measurement and Air Moving Devices. In order to measure the concentration of an airborne constituent it is necessary to know the volume of air sampled. This may be achieved by either measuring the rate of flow with a rate meter or by directly measuring the volume of air passed with a dry or wet test gas meter or a cycloid gas meter. All these devices require calibration and regular checking for leaks. As air volume is dependent upon both temperature and pressure the measurement of these two parameters at the meter inlet is essential so that volumes may be expressed at standard temperature and pressure.

Many different types of pump are available for air sampling, both mains and battery operated, but the precise type chosen will depend upon the required flow rate, availability of power and whether continuous or intermittent flow is required.

3.2 Water Sampling Methods

For many applications no special water sampling system is required as an appropriate sample container immersed in the water may be adequate. The main requirement is that a portion of the material under investigation small enough in volume to be transported and handled conveniently but still accurately representing the bulk material should be collected. Typically a 0.5–2 dm^3 volume is sufficient. When samples are required from depth, two types of collection vessel may be used. The first consists of a cylinder with hinged lids at both ends. The container is lowered into the water with both lids opened and at the desired depth a messenger weight sent down the wire which closes them. This type is not suitable for trace metal work as contaminated surface waters may result in contamination of the vessel and the messenger may scour metallic particles from the wire. The second type consists of a sealed container filled with air which is lowered to the required depth. A messenger is again sent down to open, and another to close, the lid. These devices are discussed in further detail elsewhere.[19,41] Alternatively, pressure sensors may activate the lid.

Automatic sequential samplers are available which will collect a given volume of water into an array of bottles. They have been used, for example, in collecting stormwater run-off from roads and the sampling sequence may be triggered when the flow in a flume reaches a certain height.

A recent innovation in the field of water monitoring is the use of adsorption or filtration media to concentrate the species of interest *in situ*. Using a completely self-contained sealed unit of inert material housing a peristaltic pump and power

[41] J. P. Riley, 'Chemical Oceanography,' Vol. 2, Academic Press, New York, 1965.

supply with only the adsorption tube inlet and outlet open to the water, contamination of the sample can be completely avoided. Very large volumes of water may be processed. Future developments of this technique employing micro-processor controlled pumps and arrays of collection tubes in moored units is anticipated.

One aspect of water sampling that has attracted a great deal of research is the development of efficient methods of sampling the water surface microlayer, the top 100 μm or so. Early methods centred on the collection of lipids which were known to affect the transfer rates of gases across the air–water interface, and examples of these were the use of a stainless-steel mesh (efficient only for molecules of chain-length of 16 carbon atoms or more) and the Harvey Skimmer which picks up a continuous film of water 60–100 μm thick which is then scraped off the rotating drum by a blade. However both these devices collect microscopic surface organisms, known as neustron, as well as the abiotic molecules and lipids which may be leached from the neustron during the subsequent extraction procedure.

A recent development is the use of a germanium slide which will pick up a coherent layer of any surfactants present on the water surface when it is withdrawn perpendicularly. The use of infrared spectrophotometry may then be used to identify the functional groups present, germanium being transparent to infrared radiation. Another method utilizes a large container with an outlet at the bottom and equipped with a mechanical stirrer. As water drains out of the container 'floatables' adhere to the vessel walls and may then be rinsed off with solvent.[19]

When considering sampling methods for use on inland waters or in coastal waters and estuaries the sophisticated techniques developed for use at sea may be found to be impracticable due to their need for heavy lifting gear on the sampling vessel. Monitoring work must often be undertaken on such waters using small boats without such equipment. One ingenious method of collecting water samples at different depths using very limited resources on a small boat is to lower a weighted plastic tube to the desired depth and to use a small peristaltic pump to draw water up and into a collection bottle. In this way completely uncontaminated samples may easily be obtained from depths of 30 m or more. Whether the particulate fraction of material present in the water can be quantitatively collected in this manner is not clear.

Whichever method of collecting samples is used care must be taken to ensure that neither the sample storage containers nor any collecting vessels used contaminate or alter the sample. This may occur by:

(a) Leaching of contaminants from the surface of imperfectly cleaned containers;
(b) Leaching of organic substances from plastics or silica and sodium or other metals from glass;
(c) Adsorption of trace metals onto glass surfaces or organics onto plastic surfaces. In the case of metals this may be avoided by prior acidification of the container, but this may in turn exacerbate problem (a);

(d) Reaction of the sample with the container material, *e.g.* fluoride may react with glass;
(e) Change in equilibrium between pollutants in particulate and solution phases.

If a solvent extraction technique is used to concentrate the analytes prior to analysis, care must be taken to ensure that the reagents and containers used are themselves sufficiently clean. Some commonly used materials and techniques have been shown to cause severely elevated metal levels in water.

Different determinands require different methods of preservation in order to prevent significant changes between the time of sampling and of analysis. Generally, acidification to pH 2 and refrigeration to 4 °C will be adequate although complete stability of every constituent can never be guaranteed and, at best, chemical, physical, and biological processes affecting the sample can only be slowed down. Samples may be filtered directly after collection in the field to separate the particulate and solution phases. The solution phase may then be acidified to prevent adsorption of the pollutant to the container wall. Details of sample preservation methods for individual determinands and maximum storage times may be found elsewhere, as may details of the minimum satisfactory sample size.

3.3 Soil and Sediment Sampling Methods

Soils and sediments are typically very inhomogeneous media and large lateral and vertical variations in texture, bulk composition, water content, and pollutant content may be expected. For this reason large numbers of samples may be required to characterize a relatively small area. Although surface scrapings may be taken it is often necessary to obtain cores so that vertical profiles of the determinands may be obtained or cumulative deposition estimated. Plastic or chromium plated steel tubing of 2.5 cm internal diameter is often suitable, and if the samples are sealed into the tubes and air excluded they may be satisfactorily stored at low temperatures until required. Otherwise they may be extruded in the field and stored in plastic bags. Various core sampling devices are available for obtaining cores of bottom sediments from lakes, *etc.* (*e.g.* the Jenkin corer).

Grab samples of soils are easily obtained manually and stored in pre-cleaned plastic bags. Sometimes composite samples formed by the bulking together of a number of individual samples may be sufficient, but generally analyses of individual samples is to be preferred. In the case of sediments, grab samplers are available for operation at considerable depths, examples being the Ponar, Orange-peel, and Peterson grabs. Alternatively a dredge may be used to obtain a composite sample along a strip of the sediment surface.

Wet soils and sediments which are to be analysed while still wet should not be collected or stored in bags, but in rigid containers. The vessel should be filled as completely as possible leaving no airspace at the top and a bung inserted so as to displace excess water without admitting air.

Some determinands in soils and sediments are liable to change during storage and require the use of preservation techniques. For example nitrate in soil can be

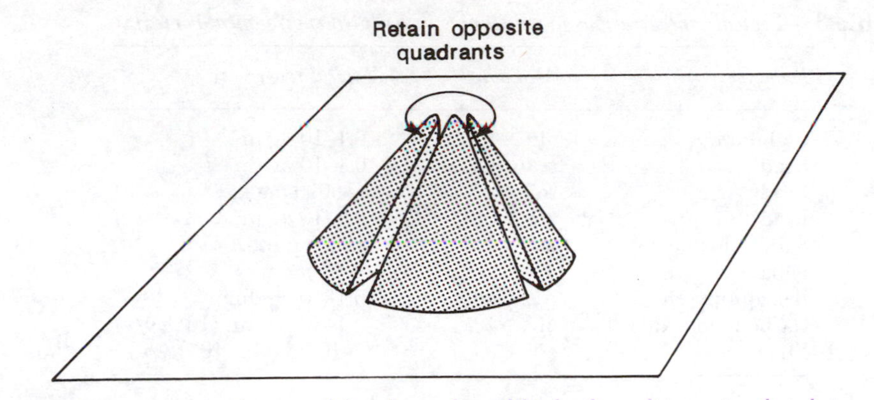

Figure 12 *Schematic diagram of the sub-sampling of dried soil or sediment using the technique of coning and quartering*

extracted into potassium chloride solution and preserved with toluene. Usually, however, air-dried soils and sediments may be disaggregated, sub-sampled by coning and quartering (see Figure 12) and stored in suitable containers, but as always sample contamination must be avoided at each stage.

There are important effects associated with grain size which should be considered in the analysis of soils or sediments. First, many pollutants are associated with particle surfaces and therefore occur in highest concentrations in the smaller grain sized material. Secondly sub-sampling from a bulk sample may be very difficult due to size segregation effects and it may be necessary to grind the sample to a very fine powder to ensure homogeneity prior to division of the sample.

4 MODELLING OF ENVIRONMENTAL DISPERSION

A characteristic feature of environmental monitoring studies is that substances may be found over very large ranges of concentrations, and therefore the analytical techniques employed must be extremely flexible. Some typical concentrations of substances in polluted environmental media are given in Table 3. Not only will large differences of concentrations be found from area to area but even small temporal and lateral changes can result in large changes in pollutant concentrations. The temporal and spatial variability of lead in air and dust samples is illustrative of the way in which pollutant concentrations vary and have recently been discussed in relation to the design of monitoring programmes.[43] In this study it was shown by collating data collected at several sites in London that short-term concentrations of atmospheric lead (of sampling period less than one week) can vary considerably at the same site, with a factor of up to 4 between the highest and lowest weekly averages at any one site. Traffic flow differences cannot entirely account for these variations, and local weather conditions are

[42] K. H. Mancy and H. E. Allen, 'Design of Measurement Systems', in 'Examination of Water for Pollution Control,' ed. M. J. Suess, WHO/Pergamon Press, Oxford, 1982.
[43] M. J. Duggan, *Sci. Total Environ.*, 1984, **33**, 37–48.

Table 3 *Typical concentrations of substances in polluted environmental media*

Pollutant	*Medium*	*Typical ranges*
Cadmium	air	0.1–10 ng m^{-3}
Lead	seawater	0.1–10 ng dm^{-3}
Lead	soil	15–5000 mg kg^{-1}
Lead	air	0.1–10 $\mu g\ m^{-3}$
Sulfur dioxide	air	1–200 p.p.b.
Sulfate	air	1–25 $\mu g\ m^{-3}$
Benzo(*a*)pyrene	freshwater	0.1–10 ng dm^{-3}
Carbon monoxide	air	0.1–50 p.p.m. (10^{-6} v/v)
Ethene	air	1–100 p.p.b. (10^{-9} v/v)

probably the dominant factor. However, when longer averaging periods are used the temporal variability at any one site is much reduced, but a seasonal effect becomes apparent. Thus monitoring periods of at least a month or two are required in order to obtain representative results with possibly some adjustment made for the season. The spatial variability of urban lead-in-air concentrations was found in this study to be rather slight and although lead concentrations decreased away from the carriageway small changes in sampling position (*e.g.* at the ground floor as opposed to the third floor windows of a building facing the street) made up to 30% difference to lead concentration. Thus the precise choice of sampling position at a site may not be critical unless the concentrations are very close to some pre-determined guideline or limit value, in which case compliance may be found in one position but not at another close by.

Dust samples collected weekly at the same pavement sites were found to have lead concentrations which varied by about the same relative amount as air–lead concentrations (*i.e.* weekly samples had a coefficient of variation of *ca.* 30%). However, spatial variability of lead-in-dust concentrations was found to be high over short distances (*i.e.* a few metres). It was suggested therefore that this can be overcome by taking dust samples over a large area (*e.g.* 5 m^2) and that in this way representative dust–lead values may be obtained.

In order to appreciate the variability of pollutant levels, and hence appreciate the complexity of designing an adequate monitoring programme, it is necessary to have some understanding of environmental dispersal, mixing and sink processes, and of the time scales on which these processes act. Rather than discussing in detail the physical, chemical, and biological processes responsible for changes in pollutant concentrations after discharge to the environment which have been extensively presented elsewhere the salient features of some of the techniques by which these changes can be anticipated will be shown.

4.1 Atmospheric Dispersal

Material discharged into the atmosphere is carried along by the wind and mixed into the surrounding air by turbulent diffusion. In the vertical plane the

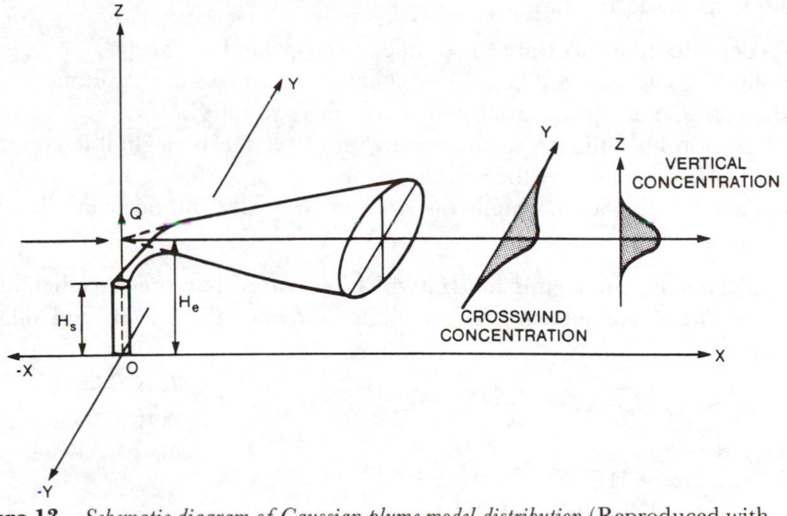

Figure 13 *Schematic diagram of Gaussian plume model distribution* (Reproduced with permission from 'Pollution Causes, Effects, and Control', 2nd Edn., ed. R. M. Harrison, Royal Society of Chemistry, Cambridge, 1990)

dispersion continues until the turbulent boundary layer is uniformly filled whilst in the horizontal plane dispersion is theoretically unlimited and usually proceeds more rapidly than in the vertical plane. In the simplest models of plume dispersion the degree of turbulence, and hence of mixing, is described by an atmospheric stability classification which is dependent upon the amount of incoming solar radiation, wind speed, and cloud cover, but surface roughness is also important in producing turbulence, especially in the case of large buildings or topographic features.

The most commonly used model of plume dispersion[44] is that described by a Gaussian distribution characterized by standard deviations σ_y and σ_z in the vertical and horizontal directions respectively (see Figure 13).

The basic equation for Gaussian plume dispersion is:

$$\chi(x,y,z) = \frac{Q}{2\pi\sigma_y\sigma_z u_{(z)}} \exp\left[-\frac{1}{2}\left(\frac{y}{\sigma_y}\right)^2\right]\left\{\exp\left[-\frac{1}{2}\left(\frac{z-H_e}{\sigma_z}\right)^2\right] + \exp\left[-\frac{1}{2}\left(\frac{z+\mathrm{H_e}}{\sigma_z}\right)^2\right]\right\}$$

where χ is the pollutant concentration at point (x,y,z) (μg m^{-3}); Q is the pollutant emission rate (μg s^{-1}); $u_{(z)}$ is the wind speed (m s^{-1}) at the effective emission height, H_e; σ_z is the standard deviation of the plume concentration in the vertical at distance x; σ_y is the standard deviation of the plume concentration in the horizontal at distance x; x,y,z are the lateral, transverse, and vertical directions (m), downwind with the base of the stack as the co-ordinate origin; H_e is the effective height of the plume (m).

[44] F. Pasquill, 'Atmosphere Diffusion,' Ellis Horwood, Chichester, 1974.

Assumptions made are that:

— no deposition occurs from the plume at the ground surface;
— pollutant levels are not altered by chemical processes in the plume;
— there is no effect from surface obstructions (*e.g.* buildings);
— dispersion by diffusion in the downwind direction is negligible compared with bulk transport by the wind;
— the constituents are normally distributed vertically and horizontally across the plume.

The effective height of emission H_e (which is greater than the stack height due to the momentum and buoyancy of the plume) may be calculated from Holland's equation;[45]

$$H_e = H + \left\{\frac{V_s d}{u}\left(1.5 + 2.68 \times 10^{-3} p\, d \frac{T_s - T_a}{T_s}\right)\right\}$$

where V_s = stack gas exit velocity (m s^{-1})
d = internal stack diameter (m)
u = wind speed (m s^{-1})
p = atmospheric pressure (mb)
T_s = stack gas temperature (K)
T_a = air temperature (K)
H = stack height (m)

and 2.68×10^{-3} is a constant with units of mb^{-1} m^{-1}.
The wind speed at height z (u_z) may be obtained from the power law:

$$u_z = u_{10}\,(z/10)^n$$

where u_{10} is the speed at the reference height of 10 m, and n varies according to surface roughness, but may be taken to equal 0.2.

The next step is to determine the appropriate Pasquill stability class and then evaluate the estimate of σ_y and σ_z as a function of downwind distance.[44] In order to obtain the ground level concentration below the plume centre line (*i.e.* $y = z = 0$) the general equation reduces to:

$$\chi(x) = \frac{Q}{\pi \sigma_y \sigma_z u} \exp\left[-\frac{1}{2}\left(\frac{H_e}{\sigma_z}\right)^2\right]$$

or for a ground level source with no effective plume rise ($H_e = 0$):

$$\chi(x) = \frac{Q}{\pi \sigma_y \sigma_z u}$$

Full treatments of these equations and their applications are available elsewhere, including modifications to include line and area sources.[46,47]

[45] J. Z. Holland, 'A Meteorological Survey of the Oak Ridge Area,' Atomic Energy Comm. Report ORO-99, Washington DC, 1953, pp. 554–559.
[46] D. B. Turner, 'Workbook of Atmospheric Dispersion Estimates,' US Environmental Protection Agency, Research Triangle Park, North Carolina, 1970.
[47] R. H. Clarke, 'A Model for Short and Medium Range Dispersion of Radionuclides Related to the Atmosphere,' National Radiological Protection Board Report NRPB-R91, Harwell, 1979.

One useful simplification which gives an approximation (with an estimated error of 30%) of ground-level concentration from a ground-level release is that of Clancey.[48] The equation is applicable only to 'average' meteorological conditions but may provide a useful first-order approximation:

$$C_x = 32Q/1.75\,\bar{u}x$$

where C_x = concentration (g m^{-3}) on centre line of mean wind direction at distance x(m)
Q = source strength (g s^{-1})
$\bar{u}$ = mean wind speed (m s^{-1}).

Another useful equation is the modification of Sutton's classic formula by Lutzke[49] which provides a simple prediction of the downwind distance at which a certain limiting concentration will be reached. Again assuming 'average' meteorological conditions and at distances greater than the first few tens of metres from the source and assuming that significant turbulence from buildings *etc.* is absent this is given by:

$$\chi_{\text{critical}} = KQ^c/s\bar{u}$$

where χ_{critical} = the critical distance on the centreline of the dispersing plume
K = a constant (9.23)
c = a constant (0.55)
Q = source strength (kg h^{-1})
s = critical concentration (g m^{-3})
$\bar{u}$ = average wind speed (m s^{-1})

The critical concentration, s, might be the threshold limit value (TLV) or lower explosive limit (LEL) or some other value.

Another type of calculation which is frequently required is that to determine the minimum height of a chimney which will give adequate dispersion to ensure that critical ground level concentrations are not exceeded. A simple algorithm has been developed which allows this estimate to be easily made.[50]

All attempts to model atmospheric diffusion and mixing processes are liable to be, at best, only good estimates of the real situation and care must be taken to:

(a) understand the physical, chemical, and mathematical limitations of the model, and;
(b) avoid treating the output from models as providing definite answers.

Although accurate measurements may be made of wind speed it is in the determination of the turbulence characteristics of the atmosphere that uncertainties arise, which in turn lead to uncertainties in the model. At short distances downwind, with steady winds, uniform terrain and no local obstructions the Gaussian plume model may be expected to give good estimates, but in any but ideal conditions they soon become order-of-magnitude estimates only.

[48] V. J. Clancey, 'The Evaporation and Dispersion of Flammable Liquid Spillages,' 5th Symposium on Process Hazards, Institute of Chemical Engineering, London, 1974.
[49] K. Lutzke, *Staub Reinhalt. Luft*, 1974, **34**, 33–39.
[50] Department of Environment, 'Chimney Heights 1956 Clean Air Act Memorandum, 3rd Edn., HMSO, London, 1981.

4.2 Aquatic Mixing

The physical transfer and transport of pollutants in the aquatic environment is determined by the same two processes which determine the mixing of pollutants in the atmosphere. These are:

(a) advection, caused by the large-scale movement of water, and;
(b) mixing or diffusion, due to small-scale random movements which give rise to a local exchange of the pollutant without causing any net transport of water.

The combined effect of advection and diffusion is known as dispersion. These two processes occur over a very wide range of scales, both of spatial extent and frequency, which necessitates the use of averaging procedures when defining their role in pollutant transfer. As in the lower atmosphere, there is usually a constraint on the vertical dispersion component, induced either by water depth, or in deeper waters by thermal or density stratification. On a large scale in the oceans there will be a vertical circulation driven by density differences but on a small scale vertical motion due to turbulence or eddy diffusion will tend to be suppressed by stratification.[51] Similar restrictions on vertical mixing occur in rivers due to limited depth, but here horizontal mixing in the cross-channel direction is also constrained.

A large number of models have been developed to describe the movement of pollutants in the aquatic environment, many being analogous to those used in air pollution studies. The fact that such models are necessary is due to limitations in our detailed knowledge of the velocity field, and this in turn may lead to uncertainties in the prediction of dispersion patterns.

The dispersion of pollutants in rivers has attracted a great deal of study, particularly in the context of effluent discharges and the ability of rivers to dilute them.[52] Unlike the oceans which have traditionally, but unreasonably, been considered to have an infinite capacity for dilution, the deterioration in water quality due to pollutant discharges is often manifestly apparent in rivers.

Various models have been applied to the problem of modelling river dispersion with varying levels of complexity. An extremely simple model of change in water quality downstream from a discharge into a river would be to assume that an exponential decay in concentration occurs, *i.e.*

$$C_x = C_0 \, e^{-kt}$$

where C_x is the concentration at point x, C_0 is the concentration at the point of discharge, k is the decay rate, and t is the time taken for flow from the point of discharge to point x.

A more recent approach to river modelling is to consider the river to be divided into a number of reaches, with each reach being considered as a continuously

[51] G. Kullenburg, 'Physical Processes in Pollutant Transfer and Transport in the Sea,' Vol. 1, ed. G. Kullenberg, CRC Press, Boca Raton, Florida, 1982.

[52] A. James, 'Water Quality Modelling in Water Quality in Catchment Ecosystems,' ed. A. M. Gower, J. Wiley, Chichester, 1980.

stirred tank reactor, and between each reach an appropriate time delay is placed. It is assumed that the major dispersive mechanism can be explained by the aggregated effect of all 'dead-zone' phenomena in the river between the two sampling points. This 'aggregated dead-zone' model is thus a combination of the continuously stirred tank reactor and a factor to account for the advection component of dispersion.[53]

4.3 Variability in Soil and Sediment Pollutant Levels

Obviously, the same mechanisms of dispersion as operate in fluid or gaseous media do not occur in soils and sediments. Physical mixing may occur, such as during agricultural practices, dredging of estuaries, or bioturbation by burrowing organisms, but usually only on a fairly limited scale. The level of contamination in a soil or sediment will depend upon the deposition rate of the pollutant and its subsequent rate of movement through the soil or sediment column. The rate of movement of a contaminant through these solid media is dictated by the degree of adsorption to or leaching from the particles and the flux rate of pore water, to transfer the pollutant to deeper horizons. The physico-chemical properties of the water, soil, or sediment particles and the pollutant all influence the rate of adsorption or leaching. For example, conditions which favour the adsorption of radiocaesium and lead in soils include low rainfall and high clay (particularly illite) content, whereas mercury and copper tend to accumulate in soils with a high organic content. Contaminant concentrations are generally always higher in soils or sediments with finer grain sizes, due to the increased total surface area available for adsorption. This effect is illustrated in Figure 14. All these factors lead to great variability in pollutant concentrations, which simply emphasizes the need for carefully designed monitoring programmes.

5 DURATION AND EXTENT OF SURVEY

5.1 Duration of Survey and Frequency of Sampling

The duration of a pollution monitoring programme is entirely dependent upon the purpose of the study, and can vary from the time taken to collect a limited number of, for example street dust samples in an urban area to tens of years for long-term surveillance projects such as the UK National Survey of Air Pollution. The choice of the frequency of sampling, *i.e.* the duration of each sample period and the interval between successive measurements, is also dependent upon the objectives of the study. Pollutant concentrations in air and water fluctuate with varying degrees of rapidity and in order to characterize their behaviour it is necessary to measure these changing levels; long-term mean data may be sufficient for some purposes but will not be adequate where information of short-

[53] P. C. Young, 'Quantitative Systems Methods in the Evaluation of Environmental Pollution Problems,' in 'Pollution: Causes, Effects, and Control,' 2nd Edn., ed. R. M. Harrison, Royal Society of Chemistry, Cambridge, 1990, p. 367.

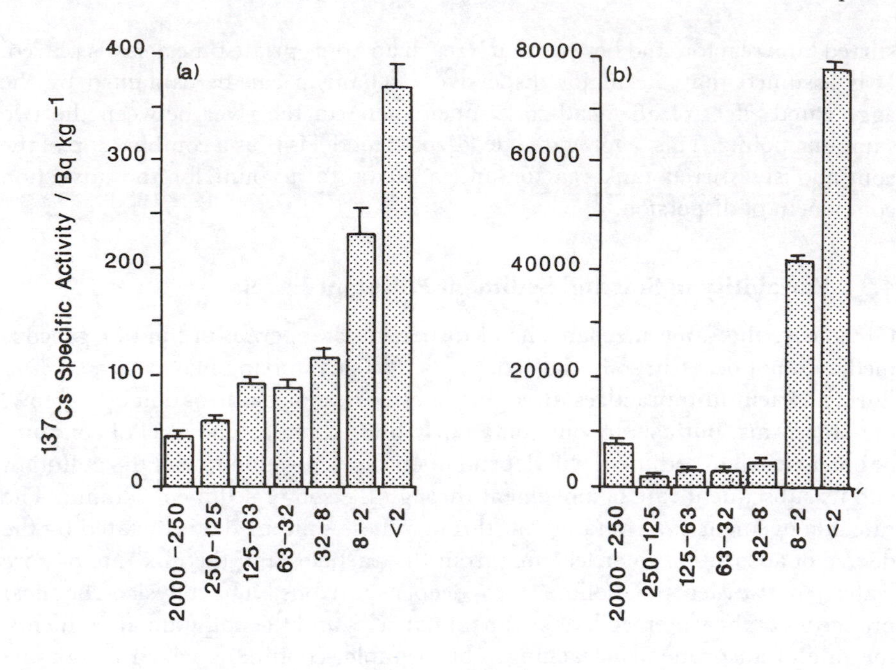

Figure 14 *The distribution of* ^{137}*Cs activity concentrations in different grain size ranges for Cumbrian (a) soil and (b) intertidal sediment*

term high-level episodes is required. Generally it has been found that if random sampling techniques are used, the number of samples required will increase as the standard geometric deviation of the pollutant concentrations increases, *i.e.* the greater the fluctuation of the pollutant level, the more numerous the samples that must be taken to accurately assess the variations. If the variations in levels during the period of interest are essentially random, independent, and normally distributed then the number of samples which must be taken in order to estimate the period mean within certain limits and prescribed confidence limits may be calculated.[7] However, in order to do this a reasonable estimate of the standard deviation of the data is required: it is assumed that the random variations follow a normal distribution and that the results of successive samples are not serially correlated but are independent. These criteria are rarely met. As a more general guideline it may be assumed to be necessary to have a sampling interval at least ten times shorter than the fluctuation cycle time. For example the entire variance structure of a diurnally fluctuating pollutant concentration may be obtained from roughly 12 samples, each with a two-hour averaging period. If the annual trend of levels is required then probably 12 monthly samples each year would be adequate. Thus, although a continuous and instantaneous record of pollutant level may be required for some purposes, *e.g.* to monitor very short-term changes in air quality due to the oscillating passage of a plume over the sampling station, it is not always necessary. Further, if a short sampling period is chosen it rapidly becomes necessary to efficiently record and store the large volumes of data generated, which will themselves often be averaged over a longer period for analysis.

Fast-response continuous monitors are now available for the more common gaseous pollutants and for many determinations of water quality. The term 'fast-response' implies a response time measured as a 90% rise of less than about 2 mins, but in most cases is of the order of seconds. Thus very rapid temporal variations are measurable and with the advent of on-line microprocessor data handling and reduction systems the large amounts of data produced are more easily handled. Fast-response continuous monitors do not generally have a predictable response and cannot be calibrated solely in terms of chemical stoichiometry and hence need calibration with a standard atmosphere or solution. Some of the commonly employed fast-response methods of gaseous air pollutant analysis are shown in Table 4, and some of the methods used for water analysis are shown in Table 5.

One further consideration when deciding on sampling frequency is that if the measurements are being made in order to assess whether a given environmental quality standard is being satisfied, then the data resolution must be sufficient for that purpose. As an example the European Community air quality standard for smoke and sulfur dioxide specifies that daily (*i.e.* 24 h) values are required, and it is therefore necessary to take 24 h samples. It would be pointless to collect data with a continuous fluorescent analyser with a response time of two minutes, if the only purpose of the exercise is to see whether this standard is being met.

5.2 Methods of Reducing Sampling Frequency

Once the desired sampling frequency has been selected it may be found to be impracticable with the resources available and some means of reducing the number of samples will then be required. This may be done by:

(a) reducing the number of sampling locations;
(b) reducing the sampling frequency; or
(c) reducing the number of determinands.

It has already been shown above that there are several methods available for the rationalization of an existing monitoring network where the quality at one location is correlated sufficiently well with that at another location, and these methods of analysis may be applied to all types of monitoring networks. Similarly, statistical analysis of past data may show that one determinand is sufficiently well correlated with one or more others, such that it may be used as an indicator of quality.

When sampling water, soils, sediment, and flora and fauna the use of composite samples may be of value in reducing sample numbers. Composite samples are formed by mixing together individual samples to give an indication of the average quality over an area or during a sampling period. They may also be formed by continuous or intermittent collection of samples into one container over a given period, as, for example, the collection of atmospheric particulate material on a filter. Individual samples collected at different locations may be

Table 4 *Summary of commonly employed methods for measurement of gaseous air pollutants*

Pollutant	*Measurement technique*	*Sample collection period*	*Response time*[a] *(continuous techniques)*	*Minimum concentrations*
Total hydrocarbons	Non-dispersive infra-red (NDIR)		5 s	1 p.p.m. (as hexane)
	Flame ionization analyser		0.5 s	10 p.p.b. (as methane)
Specific hydrocarbons	GLC/flame ionization detection (FID)	[b]		1 p.p.b.
Carbon monoxide	NDIR		5 s	0.5 p.p.m.
	Catalytic methanation/FID	[c]		10 p.p.b.
	Electrochemical cell		25 s	1 p.p.m.
Sulfur dioxide	Absorption in hydrogen peroxide/titration	24 h		2 p.p.b.
	Absorption in hydrogen peroxide/conductivity determination		3 min	10 p.p.b.
	Absorption in tetrachloromercurate/spectrophotometry	15 min		10 p.p.b.
	Flame photometric analyser		25 s	0.5 p.p.b.
	Fluorescent analyser		2 min	0.5 p.p.b.
Oxides of nitrogen	Conversion to nitrite/azo dye formation	30 min		5 p.p.b.
	Chemiluminescent reaction with ozone		1 s	0.5 p.p.b.
Ozone	Oxidation of potassium iodide/spectrophotometry	30 min		10 p.p.b.
	Oxidation of potassium iodide/electrolytic cell		30 s[d]	10 p.p.b.
	Chemiluminescent reaction with ethene		3 s	1 p.p.b.
	UV absorption		30 s	3 p.p.b.
Peroxyacetyl nitrate	GLC/electron capture detection	[c]		1 p.p.b.

[a] Time taken for a 90 per cent response to an instantaneous concentration change
[b] Grab samples of air collected in an inert container and concentrated prior to analysis
[c] Instantaneous concentrations measured on a cyclic basis by flushing the contents of a sample loop into the instrument
[d] Time for a 75 per cent response

Table 5 *Summary of some methods of analysis of water*

Pollutant or determinand	*Measurement technique*	*Response time*
pH	Electrometric	10 s
Biochemical oxygen demand	Dilution/incubation	5 days
Chemical oxygen demand	Dichromate oxidation	2 h
Metals	AAS	—
Organometallics	GC–AAS	—
Nitrate	Colorimetric	5 min
Nitrate	UV spectrophotometric	1 min
Formaldehyde	Photometric	6 min
Phenols	GC	30 min

mixed together in proportion to the volumes of the sampled bodies, as in the case of a non-homogeneous water body, and so give a better indication of average quality.

In the case of atmospheric pollutants it is often desirable to estimate the likely daily maximum as well as the daily mean concentration, and several methods have been proposed which allow this to be done on the basis of a few discrete samples of short duration. For example the same statistical concentration frequency distribution of SO_2 levels as is provided by continuously recorded data can be obtained from a limited number of randomly collected short term samples.

5.3 Number of Sampling Sites

The choice of the number of sampling sites to be used in a particular survey is very dependent, as are so many other design parameters, on the objectives of the study. In the simplest case of source-orientated monitoring of atmospheric emissions from a single stack or site then 4–6 sampling locations might be considered sufficient. This may be thought to be too few but operational constraints may prevent this number being increased. In the case of the UK National Survey of Air Pollution it was previously thought necessary to obtain daily data from about 1200 sites, although this number has now been greatly reduced.

When too few sampling sites are used in a source-orientated monitoring programme it is possible that atmospheric (or aqueous) emissions may pass between them without being detected at all. The probability of a fixed number of sample stations detecting a release is a function of the quantity released, the number of samplers, the distance of the samplers from the source, the plume dimensions, the height of release, and its duration. This type of analysis has been applied to the environmental monitoring of atmospheric releases.

6 PREREQUISITES FOR MONITORING

Before monitoring begins certain information, techniques, and methodologies must be available in order for the survey to be successfully carried out. As has already been stressed, a prime requirement is that the objectives of the study should be defined, but as well as this the following need consideration:

—availability of meteorological or hydrological data;
—availability of emission data;
—likely pollutant concentrations to be expected;
—availability of suitable monitoring equipment;
—availability of sensitive and specific analytical techniques;
—definition of suitable environmental quality standards.

6.1 Meteorological Data

When carrying out air pollution measurements it is desirable, and often essential, to have access to meteorological data. Care must be taken to ensure that the data used are meaningful and representative of the area of study, wind data in particular being very susceptible to local interference. Light, robust anemometers and wind vanes are now generally available and when mounted on a dismountable 10 m mast may be used in the field, thus obviating the need to rely on data obtained from another, possibly less representative, source. Details of the type of instruments and advice on siting requirements are available from an authoritative source,[54] although data-logging systems have recently been greatly advanced. Less easily obtained parameters which may be required are the lapse rate and the height of any atmospheric temperature inversions. These are rather difficult to measure, requiring accurate measurements of temperature at increasing heights or acoustic radar observations, but such data are usually obtainable from large meteorological stations.

Measurements of low-level atmospheric turbulence are made using a bivane and anemometer on a 10 m mast. This instrument measures the horizontal and vertical fluctuations of the wind; siting of the mast is obviously critical and a data logging system is required to cope with the large amounts of data generated.

Meteorological data may be required for several reasons. Firstly, when fast response measurements are unavailable it may be desirable to construct time-weighted pollution roses which show how pollutant concentrations vary with surface wind directions. For this the hourly wind direction at 10 m height is required, the record divided into sixteen $22\frac{1}{2}°$ sectors and the duration in each sector is tabulated. The rose may then be calculated using:

$$(\mathrm{TWMC})_n = \frac{\sum_{i=1}^{m} (t_{i,n}\, C_i)}{\sum_{i=1}^{m} \mathrm{t}_{i,n}}$$

[54] 'Handbook of Meteorological Instruments,' HMSO, London, 1956.

where $(TWMC)_n$ is the time-weighted mean concentration of the pollutant for the nth section, $t_{i,n}$ is the number of hours for which the wind was in sector n during the ith sampling period, C_i is the concentration of the pollutant for the ith period, m is the number of sampling periods, and n takes values from 1 to 16 (sector $n = 1$ being $0°–22\frac{1}{2}°$, sector 2 being $22\frac{1}{2}°–45°$, *etc.*).

Alternatively the trajectory of a parcel of air over synoptic-scale distances may be required for source-apportionment or dispersion studies. For this the surface pressure field over a very large area is needed and the geostrophic wind vector estimated from the isobar spacing and alignment.[55,56]

The analogous hydrographic data may be required for dispersion studies in the sea, and flow data for rivers and lakes may be needed although may not be so readily available as meteorological data.

6.2 Source Inventory

One, often very cost-effective, method of identifying the likely occurrence of pollutants prior to actual monitoring is by collecting and collating detailed information on the pollution emissions in a given area or to a particular river. Such an emission or source inventory should contain as much information as possible on the types of source as well as the composition of emissions and the rates of discharge of individual pollutants. Supplementary information describing the raw materials, processes, and control techniques used should also be collected. Detailed discussion on the test procedures to adopt in compiling an emission inventory is available,[57] and the applications of the completed inventory were identified by these authors to be:

(a) guiding emission—reduction efforts;
(b) helping to locate monitoring stations and alerting networks;
(c) indicating the seasonal and geographic distribution of the pollution burden;
(d) assisting in the development of implementation strategies;
(e) pointing out the priority of air or water quality problems;
(f) aiding regional planning and zoning;
(g) air and water quality diffusion modelling;
(h) predicting future air and water quality trends;
(i) determining cost–benefit ratios for air and water pollution control;
(j) community education and information programmes.

Although a very useful tool, the emission inventory is no substitute for actual measurements of pollutant levels and should be considered as a complementary technique to monitoring, not as an alternative.

[55] E. F. Danielsen, *J. Meteorol.*, 1961, **18**, 479–486.
[56] R. I. Sykes and M. Hatton, *Atmos. Environ.*, 1976, **10**, 925–934.
[57] A. T. Rossano and T. A. Rolander, 'The Preparation of an Air Pollution Source Inventory,' in 'Manual on Urban Air Quality Management', 1976.

6.3 Suitability of Analytical Techniques

As was shown in Table 3, pollutants may be found in environmental media over very wide concentration ranges. Often there are several procedures available by which a pollutant may be analysed, but they may have widely differing sensitivity and specificity. As an example, sulfur dioxide in air may be determined by absorption in hydrogen peroxide and the resultant acid determined by acid–base titration or by conductivity measurement. Although acidic or basic compounds interfere, these techniques can yield useful results in urban areas where sulfur dioxide concentrations are high and the levels of interfering compounds are low. In rural areas this is not the case due to natural production of ammonia and misleading results are obtained. In an alternative wet chemical method, the West-Gaeke, in which SO_2 is collected by reaction with potassium or sodium tetrachloromercurate(II) forming the disulfitomercurate(II) and then determined colorimetrically after addition of pararosaniline methyl sulfonic acid, interferences have been greatly reduced or eliminated by refinements to the method. By using flow systems, continuous monitors based on this reaction have been developed with a response time of several minutes.

Very fast-response analysis of sulfur dioxide is possible using the flame-photometric sulfur analyser in which gaseous sulfur compounds are burned in a reducing hydrogen–air flame and the emission of the S_2 species at 394 nm is measured. The method is very sensitive but unless used with gas chromatographic separation is not specific to sulfur dioxide, but rather gives a measure of the total volatile sulfur content. Alternatively SO_2 may be analysed with a fast response time by excitation at 214 nm and the resultant fluorescence measured.

Some of the considerations that will affect the choice of an analytical method may thus be summarized:

- —sensitivity (depends on detection limit and pollutant levels);
- —specificity (to allow unequivocal determinations);
- —response-time;
- —response-range (particularly linearity of continuous monitors);
- —ease of operation;
- —ease of calibration;
- —cost and reliability;
- —precision and accuracy.

Some of the more commonly used instrumental methods are shown in Table 6 and described comprehensively elsewhere.[58]

Figure 15 highlights the difference between precision and accuracy for an analytical technique. Precision may be defined as the reproducibility of analyses, whereas accuracy is a true measure of the determinant present. In order to assess the precision of an analytical method it is necessary to analyse separate representative sub-samples of the sample under investigation. These sub-samples should be included at intervals during the analysis. In this way the drift in the precision

[58] C. N. Hewitt, (ed.) 'Instrumental Analysis of Pollutants,' Elsevier, London, 1991.

Table 6 *Instrumental analytical methods*

Method	*Sample*[a]	*Specificity*	*Sensitivity*[b]
Gravimetric	SLG	good	>1 μg
Titrimetric	SLG	good	>10^{-7} M in soln.
Visible spectroscopy	SL	fair	>0.005 p.p.m. in soln.
Ultraviolet spectroscopy	SLG	fair	>0.005 p.p.m. in soln.
Flame emission spectroscopy	SL	good	>0.001 p.p.m. in soln.
Atomic absorption spectroscopy	SL	excellent	>0.001 p.p.m. in soln.
Gas chromatography	LG	excellent	>10 p.p.m.
Liquid chromatography	SL	good	>0.001 p.p.m.
Polarography	L	good	>0.1 p.p.m.
Anodic stripping voltammetry	L	good	>0.001 p.p.m.
Spectrofluorimetry	SL	good	>0.001 p.p.m.
Emission spectroscopy	SL	excellent	>0.1 p.p.m.
X-Ray fluorescence	SL	good	>10 p.p.m.
Neutron activation	SL	excellent	>0.001 p.p.m.
Mass spectrometry	SLG	good	>0.003 p.p.m.

[a] S = solid, L = liquid, G = gas
[b] Approximate only, depending upon the particular element being analysed

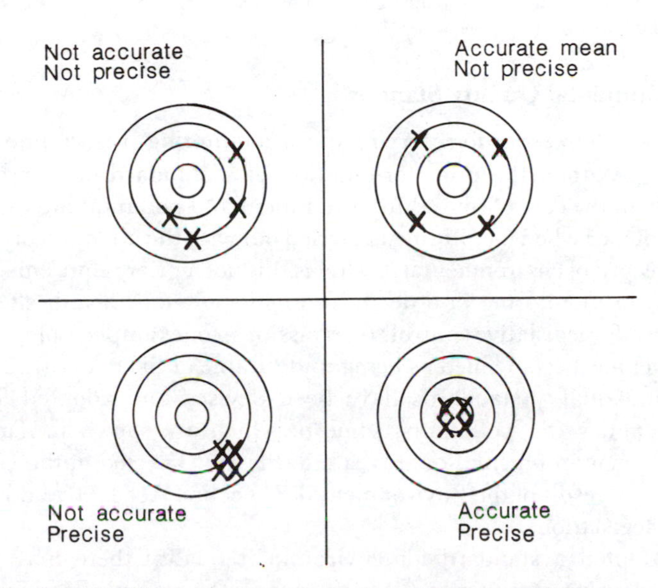

Figure 15 *Schematic illustration of precision and accuracy*

of a technique may also be detected. It is not possible to know how accurate a series of determinations have been, even if they are precise. However, a best estimate of the accuracy of an analytical technique may be found from the regular analysis of national and international standard reference materials. These have undergone inter-laboratory analyses, using different techniques and calibration standards, for a variety of species.

Table 7 *Environmental quality standards for air*

Pollutant	*Regulating body*	*Measurement period*	*Concentration*
Suspended particles[a] (smoke)	EC	Yearly median of daily values	80(68) $\mu g\ m^{-3}$
		Winter median of daily values	130(111) $\mu g\ m^{-3}$
		98% of daily values below	250(213) $\mu g\ m^{-3}$
Sulfur dioxide	EC	Yearly if smoke <40 $\mu g\ m^{-3}$	120 $\mu g\ m^{-3}$
		Yearly if smoke >40 $\mu g\ m^{-3}$	80 $\mu g\ m^{-3}$
		Winter if smoke <60 $\mu g\ m^{-3}$	180 $\mu g\ m^{-3}$
		Winter if smoke >60 $\mu g\ m^{-3}$	130 $\mu g\ m^{-3}$
		98% if smoke <150 $\mu g\ m^{-3}$	350 $\mu g\ m^{-3}$
		98% if smoke >150 $\mu g\ m^{-3}$	250 $\mu g\ m^{-3}$
Particulate matter (PM_{10})	US	Annual geometric mean	65 $\mu g\ m^{-3}$
Sulfur dioxide	US	Annual arithmetic mean	80 $\mu g\ m^{-3}$
Carbon monoxide	US	8 hour max.	10 000 $\mu g\ m^{-3}$
		1 hour max.	40 000 $\mu g\ m^{-3}$
Ozone	US	1 hour max.	240 $\mu g\ m^{-3}$
Nitrogen dioxide	US	Annual arithmetic mean	100 $\mu g\ m^{-3}$
Lead	EC	Annual mean	2 $\mu g\ m^{-3}$

[a] Values in parentheses relate to corresponding measurements of smoke by the UK smoke stain reflectance method

6.4 Environmental Quality Standards

Of the six possible reasons for carrying out a monitoring programme outlined above, four rely upon the prior formulation of a standard of environmental quality. Only in the case of source apportionment and pollutant interaction and pathway studies or when monitoring is carried out with the intention of obtaining a historical record of environmental quality is this not a prior requirement. There is little point in monitoring in order to pinpoint pollutant health effects or to assess the need for legislative controls on emissions, for example, unless a certain pollutant level has been defined as being undesirable or likely to cause damage. Environmental quality standards have been devised and adopted for many atmospheric and water pollutants, some of which are shown in Table 7. A number of new environmental quality standards are expected in the UK in the early 1990's as a result of the Environmental Protection Act 1990 and European Community legislation.

Having adopted a standard for environment quality there may be great difficulty in ensuring compliance.[59] In the case of, for example, lead in drinking water, various reduction strategies are possible culminating in the wholesale removal of lead pipes (although this may not entirely solve the problem when lead-based solder is used with copper pipe) and in the case of primary air pollutants similar 'simple' remedies are possible. The difficulties arise in the case of secondary pollutants (*i.e.* those formed within the atmosphere itself) or for

[59] R. M. Harrison, 'Priorities in Air Pollution Abatement,' Proc. Int. Conf. Environmental Contamination, London, CEP Ltd, Edinburgh, 1984.

pollutants with both primary and secondary origins. In the case of atmospheric suspended particles (smoke) both primary and secondary sources are important. Primary emissions have in recent years been greatly reduced by the use of efficient control techniques on industrial sources and substantial change in domestic fuel usage away from coal towards cleaner fuels. However secondary particles, produced in the atmosphere by formation of the ammonium salts of strong acids from industrial emissions of SO_2, NO_x, and HCl, together comprise a substantial proportion of the atmospheric aerosol. Thus reduction of primary emissions of particles does not necessarily ensure a reduction in atmospheric concentrations or compliance with a standard. One secondary air pollutant that is likely to prove difficult to adequately control in the next decade or so is ozone. This is formed in the lower atmosphere by reactions involving several primary pollutants, and our present understanding of the chemistry of the atmosphere is probably insufficient to accurately predict the effect of control strategies.

7 REMOTE SENSING OF POLLUTANTS

Many highly sophisticated techniques are now available for the remote sensing of atmospheric and water pollutants. However their use is almost exclusively restricted to specialized monitoring exercises due to the very considerable capital cost of the instrumentation. Probably the cheapest and most widely used methods are those of aerial photography, including infra-red sensing, and optical correlation spectrometry. Uses of aerial photography include the monitoring of liquid effluent dispersion using dye tracers and conventional colour film. Infra-red photography has been used for monitoring the condition of crops and forests. Airborne heat-sensing infra-red linescanning equipment has been routinely used to monitor thermal plumes in waters receiving industrial effluents and have also been used for the detection and mapping of oil spills at sea using thermal infra-red data from satellites.

The most common use of the correlation spectrometer in air pollution analysis is for the determination of sulfur dioxide and nitrogen dioxide concentrations in plumes from tall stacks, and provides a good technique for studying the transport and dispersion of a plume. The instrumentation may be ground-based in a mobile laboratory or airborne, in which case the plume is viewed from above.

The use of tunable lasers allows long-path absorption measurements of a range of gaseous pollutants such as NO, NO_2, SO_2, CO, and O_3 and minor reactive species such as HO. Another optical technique used for measuring HO concentrations, laser-induced fluorescence, is subject to a number of possible interferences, the greatest problem being that of self-generation of hydroxyl.[60] Reliable measurements of this particular species are of great importance due to its dominant role in the chemistry of the troposphere. In long-path laser absorption methods a detector is used to monitor absorption of specific wavelengths in the light path. In lidar techniques, however, the back-scattered radiation from a laser is monitored. By using a pulsed system the time taken for

[60] C. N. Hewitt and R. M. Harrison, *Atmos. Environ.*, 1985, **19**, 545–554.

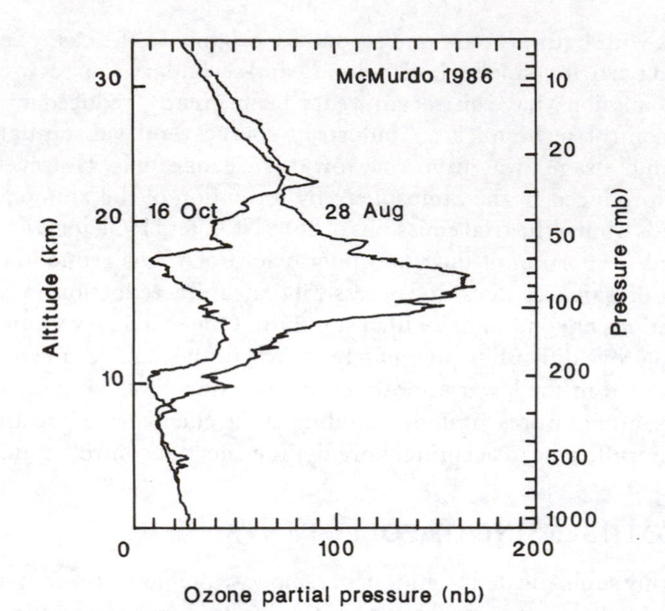

Figure 16 *Ozone profiles from balloon soundings over McMurdo (South Pole) during 1986 (after Hoffman* et al., Nature, *1987,* **326**, *59–62)* (Reproduced with permission from 'Stratospheric Ozone', Department of the Environment, HMSO, London, 1987)

receipt of back-scatter can be related to the distance of travel, allowing spatial resolution of pollutant concentration data within the light path. By monitoring back-scatter intensity at two close wavelengths, one strongly absorbed by the species of interest and one unabsorbed, the species' concentration may be inferred as well as its spatial distribution. Care is required to avoid spectral interferences but this method has been successfully used for measurement of sulfur dioxide up to a range of *ca.* 2 km. Further significant developments of laser methods using the Raman back-scatter, which is highly characteristic of the scattering molecule, are likely.

Case study 7: Measurement of stratospheric ozone.[61] There are several techniques by which stratospheric ozone may be measured. From the ground, the total ozone in a column extending vertically into the atmosphere can be measured passively with a Dobson spectrophotometer. This spectrophotometer is also used to estimate the vertical distribution of ozone, by observing the zenith clear blue sky, while the sun traverses a range of solar zenith angles. Scattering from different altitudes in the atmosphere allows the height profile of ozone to be deduced. Greatly improved vertical resolution of ozone concentrations has been achieved using the differential absorption laser (DIAL) technique. Ozone profiles have been measured from spectrophotometers carried aloft by balloon or rocket. The balloon-sondes provided evidence of a reduction in stratospheric ozone over Antartica during October 1986 (Figure 16).

[61] UK Stratospheric Ozone Review Group, Stratospheric Ozone, HMSO, London, 1987.

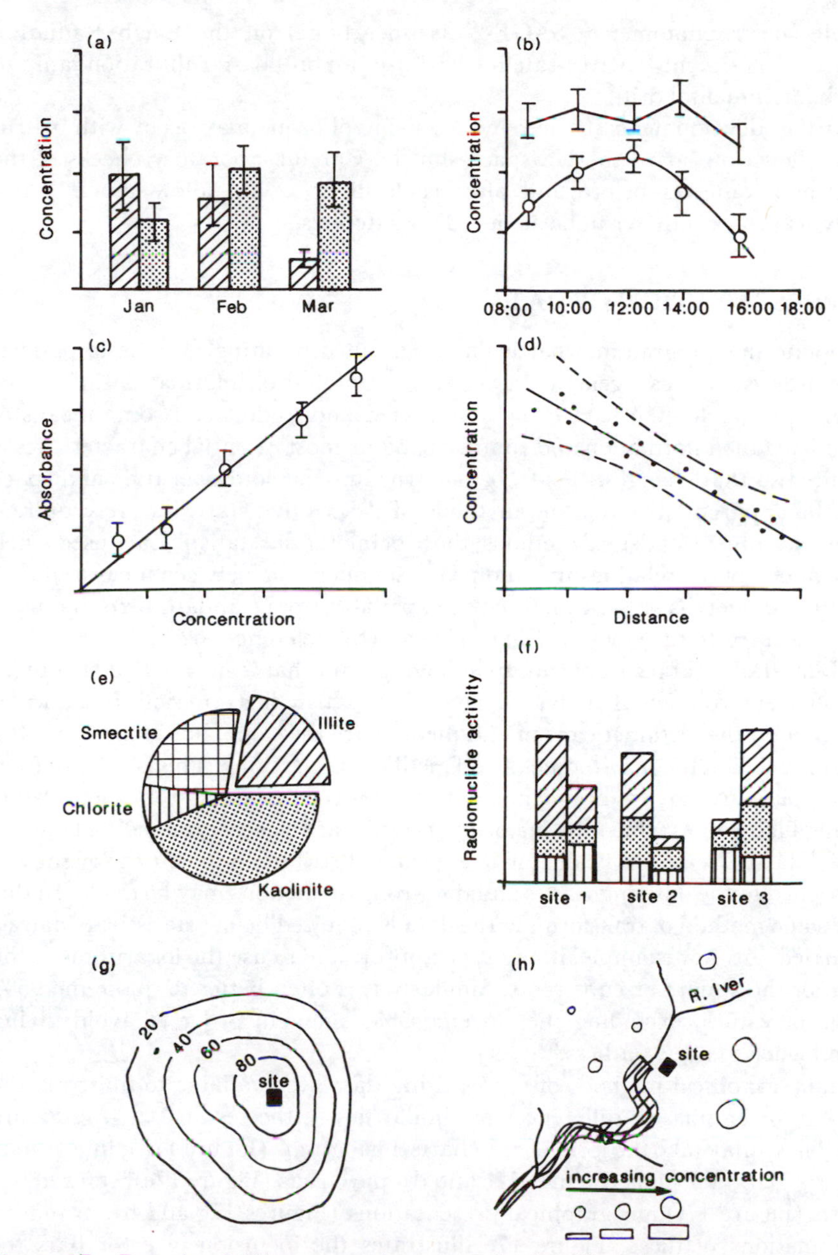

Figure 17 *Examples of different approaches to presenting data (for explanation see text)*

A major achievement in the remote sensing of ozone has been the verified measurements of ozone profiles obtained from satellites. Measurements have been made in the ultra-violet (SBUV/TOMS), visible (SAGE), and infra-red (LIMS) spectral regions on board the Nimbus 7, Applications Explorer II, and Solar Mesospheric Explorer satellites respectively. An improved version of the visible spectrophotometer (SAGE-2) is operational on the Earth Radiation Budget Experiment (ERBE) satellite and uses an on-board calibration lamp to check instrumental drift.

Further developments in the remote sensing of ozone may occur with instruments flown on board the US space shuttle. Careful laboratory checks of the spectrophotometers before and after each flight would allow much better analytical control than can be achieved for satellites.

8 PRESENTATION OF DATA

A monitoring programme, particularly one incorporating automatic or fast-response systems, can generate a very large amount of information in a short time. In order for the data to be assimilated and understood some means of organizing the information and summarizing its most essential characteristics is required so that changes, trends, or patterns in behaviour over time and space may be apparent. For this the methods of descriptive statistics are required. Another group of statistical methods, those of inferential statistics, are used when information of the relationships and processes operating between measurements is required. Details of these methods are available from standard texts and from books devoted to the environmental sciences (*e.g.* references 62 and 63).

Many statistical tests depend upon having data that is normally distributed, but often environmental analytical data do not satisfy this criterion. In a normal distribution the arithmetic mean and median are the same, but in log-normally distributed data the *geometric* mean and median are the same. This is the situation that applies to many environmental data sets and comes about from a few results having high values whilst the majority of results are closely grouped together. If such data are treated as belonging to a normal distribution too much weight will be applied to the outlying values and the wrong deductions may be made. In this case some method of transforming the data is required before statistical analysis is carried out. For example, it might be appropriate to use the logarithms of the data, or the square or cube roots. Similarly it is often better to quote the 95% range of values, excluding the extreme 5%, again in order to avoid giving prominence to a few outliers.[64]

Many graphical methods of representing data are available to illustrate the changes or emphasize differences or similarities in the results[65,66] (*e.g.* Figure 17). These may take the form of bar charts/histograms (Figure 17a), line graphs (Figure 17b), *X Y* plots (Figures 17c and d), pie charts (Figure 17e), stacked bar charts (Figure 17f), geographical presentations (Figures 17g and h), or indeed combinations of these. Figure 17c illustrates the inclusion of error bars (or

[62] C. Chatfield, 'Statistics for Technology,' 3rd Edn, Chapman and Hall, London, 1983, 381 pp.

[63] R. M. Haynes, 'Environmental Science Methods,' ed. R. M. Haynes, Chapman and Hall, London, 1982, 404 pp.

[64] London CEP Ltd, Edinburgh.

[65] T. Schneider, 'Automatic Air Quality Monitoring Systems,' in 'Manual on Urban Air Quality Management,' ed. M. J. Suess and S. R. Craxford, WHO, Copenhagen, 1976.

[66] R. A. Deininger and K. H. Mancy, 'Storage, Retrival, Analysis, and Dissemination of Water Quality Data,' in 'Examination of Water Pollution,' 1982.

standard deviations) in a graphical presentation and also a regression line. Data exhibiting an exponential decay will plot as a straight line or a log-normal graph (Figure 17d) and regression analysis may be performed with the inclusion of 95% confidence limits. Where separate components of a parameter are measured, the relative magnitude of those components may be represented on a pie chart (Figure 17e) or stacked bar chart (Figure 17f). The overall magnitude of the parameter can be proportional to the diameter of the pie chart or height of the bar chart. These types of graphs are useful for displaying speciation data. In cases where the geographical distribution of data is important then contour plots may be appropriate (Figure 17g). If data are limited or specific to certain types of location (*e.g.* road or river) they may be displayed as in Figure 17h where the diameter of a circle, width of bar, or length of line is proportional to the magnitude of the parameter of interest.

It should always be borne in mind that the rigour applied to the design, operation, and execution of a monitoring programme must also be applied to the treatment given to the resultant information and to the deductions made from it.

CHAPTER 8

Ecological and Health Effects of Chemical Pollution

S. SMITH

1 GENERAL INTRODUCTION

Chemical pollution is 'the introduction by man into the environment of substances liable to cause hazards to human health, harm to living resources and ecological systems, damage to structure or amenity, or interference with legitimate use of the environment'.[1,2]

There are, on the one hand, a group of substances which we recognize as pollutants and, on the other, organisms on which pollutants exert their effect. A basic maxim of toxicology is that all substances administered at sufficiently high doses are harmful to biota. Pollutants, by definition, are first introduced into the environment and then dispersed along pathways during which they may undergo transformations to more innocuous forms or to forms that are potentially more hazardous. One important aspect which influences the impact of a pollutant is the amount discharged into the environment. Conceivably, an organism may be able to cope with a small amount of a very toxic substance, whereas an otherwise essential one may reach overwhelming proportions and adversely affect many organisms. Pollutants encompass a broad range of chemical and physical properties and these properties strongly influence both the behaviour of pollutants in the environment and the manner in which they exert their effect on biological systems.

Man is responsible for releasing vast quantities of many different chemical substances into the environment each year; the majority of these anthropogenic substances are waste products generated by industry and society consuming the manufactured goods. Man-made fabrics and fibres, pharmaceuticals, fertilizers, pesticides, paints, and building materials, as well as chemicals for industrial processes, are just some of the products of the chemical industry that are integral to almost every aspect of modern living. Many such substances are natural constituents of the environment, others are man-made synthetic chemicals.

[1] M. W. Holdgate, 'A Perspective of Environmental Pollution', Cambridge University Press, Cambridge, 1979.

[2] Royal Commission on Environmental Pollution, 10th Report, Cmnd, 9149. HMSO, London 1984.

Inevitably wastes generated during manufacture, the substances themselves, and perhaps their degradation products, are released into the environment. Some such as dioxin and methyl cyanide, are released as a result of sudden accidents; others such as sulfur dioxide, carbon dioxide, and oxides of nitrogen from fossil fuel combustion, lead from petrol-driven engines, and mercury from chlor-alkali plants, are continuously discharged, while still others such as pesticides are intentionally released by man. In addition, hazardous chemical wastes are buried either on land or at sea.

Pollutants exert their effect on individuals, but recognition of an effect is only seen when large numbers and even whole communities are affected. The rate of supply or the amount of pollutant reaching a receptor (organism, population, or community) is referred to as the exposure and an effect is a biological change caused by an exposure. Directly toxic substances must gain access to the exterior membranes or interior components of tissues and cells in order to exert their effect. The amount that is taken into an organism is referred to as the dose; essentially it is a function of concentration and the period of exposure, although dose is perhaps more correctly defined as the amount of substance received at the site of effect. In many situations, however, this is impracticable since the site of action is not known.

A pollutant may be supplied in one large package (acute exposure); alternatively, an equivalent amount may be supplied at a lower concentration over an extended period (chronic exposure). Similarly, adverse effects or damage induced by a pollutant are frequently referred to as being acute or chronic. Acute effects are observed almost immediately following exposure; they are rarely reversible and are very often fatal. Chronic effects or damage follow a period of prolonged exposure.[1] Various biochemical and physiological disturbances may become apparent which are usually sub-lethal effects. They may be outwardly manifested as visible or clinical symptoms of damage, *e.g.* chlorotic regions on plant leaves, and inability to maintain homeostatic balance, co-ordinate activities, or breed. These symptoms are often quite diffuse and not specific to a particular pollutant. Following cessation of exposure, chronic damage is frequently reversible, although continued exposure may prove fatal. However, it may be argued that any severe disturbance will impair the efficiency of an individual and so shorten its life.

As well as the direct impact of pollutants on living species, other secondary effects require careful consideration. These relate to community and habitat changes that may follow initial pollution damage. The obvious examples include disturbance of predator–prey relationships following a dramatic decline of one or more species in a foodweb; reduced turnover in biogeochemical cycling of major elements if, say, decomposer organisms are affected. These secondary impacts are a feature of the ecosystem level of organization.

Pollution studies are very wide ranging and so extend over a multitude of situations involving interactions of pollutants with living organisms. Damage to certain groups of organisms arouses more concern than others. Clearly man and his resource species—livestock, crops, and fisheries, for obvious social and economic reasons are two groups which engender most concern. Surveying the range of effects of pollutants, it is evident that these also can be arranged on a scale of increasing concern, with cellular disruption arousing less interest than a situation proving acutely toxic.[1]

2 EXPOSURE

Exposure is the amount of pollutant reaching a receptor, and it is commonly expressed as the concentration of pollutant in the media (*e.g.* air, water, soil) and in food that is available to the receptor (individual, population, community). It is important to take into account the period of exposure to estimate the rate of supply to a receptor and in turn the intake (or dose). In the case of the more persistent pollutants, the amount or concentration in an organism or some part of an organism is frequently a function of exposure. Consequently such measurements can be used to indicate exposure.

Exposure may arise from a single source of supply, as in the transfer of phytotoxic gases from the atmosphere to the leaf surfaces of plants. Alternatively more than one pathway may contribute to the total exposure, for example man's exposure to lead comes from water, air, and food.

Routine measurements of exposure are frequently very involved and are not always representative of the actual amount that reaches a receptor. In the aquatic environment the concentration of pollutants vary with season, time of day, magnitude of freshwater run-off, depth of sampling, intermittent flow of industrial effluent, and hydrological factors such as tides and currents. Man's exposure to atmospheric pollutants such as Pb, SO_2, and smoke can be particularly difficult to assess. Many samples are required to iron out variation due to methods of collection and analysis as well as actual spatial and time differences. Very often assessments are based on data from a small number perhaps only one monitoring station. The size distribution of particles containing lead (Pb) can vary considerably between different locations. It has been shown that, in London (UK) 60% of particles were less than 0.3 μm and only 1% were above 10 μm, whereas in Los Angeles (USA) only 30% were less than 0.3 μm and a greater proportion were above 10 μm. This is of considerable importance for estimating lead exposure to man by inhalation. This issue is further confounded by the fact that man is a very mobile animal. Many adults spend a part of their day at work and children are at school and consequently such groups may experience different exposure regimes during such times. We all spend extended periods indoors removed from the outside air and certain susceptible groups such as the aged would probably spend most of their time inside. Furthermore, such habits as tobacco smoking, which is a very direct form of air pollution, quite obviously increases a person's exposure to an array of substances and it is well known that cigarette smoking provides the major supply of cadmium (Cd) to habitual smokers; smoking 20 cigarettes can result in a daily intake of 2–4 μg Cd.[3,4]

[3] T. W. Clarkson, L. Friberg, G. F. Nordberg, and P. R. Sager (ed), 'Biological Monitoring of Toxic Metals', Plenum, New York and London, 1988.

[4] L. Friberg, G. F. Nordberg, and V. B. Vouk (ed.s), 'Handbook on the Toxicology of Metals', VII, 2nd Edn., Elsevier, 1986.

Much of the research on the effects of gaseous air pollutants on plants has been done in the controlled environment of growth chambers. The concentration of the gas in question in the inlet or outlet of the chambers is frequently equated with plant exposure. These values however, can be quite different from the actual exposure at the leaf surface. Design and size of growth chamber, density of plants, and air velocity are just three of many plant and environmental factors that can influence the concentration gradient between ambient air around the plants and that in contact with the leaf surfaces. The crux of the matter is the effect of these and other factors on the thickness of the boundary layer of air that surrounds the leaves. Gases must pass through this almost laminar flow of air by the relatively slow process of molecular diffusion to gain access to the leaf surfaces. In situations where varied responses had been reported for seemingly equivalent exposures, it is now apparent that at least a part of the discrepancy can be explained by the different ways in which exposure was determined.

Whenever an element such as a metal, as opposed to a molecular species, is considered as a pollutant, the form in which it is exposed to a receptor influences its bioavailability and hence its potency. As a rule the free ion (*e.g.* Cd^{2+}, Cu^{2+}, Pb^{2+}) is the most bioavailable and therefore the most toxic, although organometallic compounds such as methylmercury and tributyl tin are more completely absorbed than their free ion counterparts. In soils and aquatic systems, metals are partitioned between solid and liquid phases and within each, further partitioning or speciation occurs among specific ligands, determined by ligand concentration and the strength of each metal–ligand association. Consequently at any one time, the amount of free-ion available for uptake by organisms is less, and in many instances much less, than the total concentration.

3 ABSORPTION

Absorption is the process whereby a substance traverses the body membranes. By and large, lipid soluble substances are absorbed more efficiently than polar water soluble substances. Lipid soluble substances diffuse across the lipid bilayer of membranes and those that are more lipid soluble (*i.e.* those with a larger octanol–water partition coefficient) tend to be absorbed more efficiently. Membrane proteins serve to transport polar substances such as metals, although certain forms which are more lipid soluble or of reduced polarity traverse membranes by passive diffusion and this is the most likely explanation for the essentially complete absorption of methylmercury.

The characteristics of the interface between the environment and an organism strongly influence absorption of a substance. In mammals the main routes of entry for environmental chemicals are through the lungs, the skin, and the gastrointestinal tract. The physico-chemical properties of a pollutant and the nature of exposure strongly influence the amount presented at each portal of entry. In terrestrial mammals, including man, penetration through the skin is an unimportant route of entry for environmental pollutants. It is a membrane which is relatively impermeable to aqueous solutions and most ions, although many of the man-made organic chemicals, because of their lipid solubility, can penetrate the dermal barrier. Such situations tend to be confined to occupational exposures.

Atmospheric pollutants occur as gases or as particulate matter. The site and extent of absorption of inhaled gases are, for the most part, determined by their water solubility. For instance, sulfur dioxide (SO_2) is a soluble gas, which is mainly absorbed in the upper respiratory tract whereas a less soluble gas such as nitrogen dioxide (NO_2) reaches the lower airways. Inhalation is the most important route of uptake for chemical mercury vapour, and the major site of absorption is alveolar tissue where about 80% of inhaled mercury vapour is absorbed.[3,4,5]

The fraction of inhaled particulate material deposited in the various parts of the respiratory tract is a function of particle size, and the fraction absorbed from the tract is dependent on the chemical nature of the aerosol. Only particles less than 2 μm diameter penetrate as far as the alveolar region. Larger particles are trapped in the upper tracheobronchial and nasopharyngeal regions where they may be either absorbed or transported up the pharynx entrained in mucus propelled by ciliary action. Subsequently, these particles are swallowed and become available for absorption in the gastro-intestinal tract. Particles taken up from the alveoli (the most important site of absorption) may pass directly into the bloodstream or be retained in lung tissue. For respirable particles of lead, at a particle size of 0.05 μm about 40% may be retained but for larger sizes, *e.g.* 0.5 μm only about 20% is deposited and probably only about 50% of the deposited lead is absorbed into the blood, although this will depend on the solubility of lead compounds in the particles. Deposition in the alveolar region of inhaled cadmium (Cd) aerosols up to 2 μm diameter is 20–35% of which less than 50% is absorbed; the rest is exhaled or swallowed.[3,4]

In humans, absorption through the gastro-intestinal tract is an important route of entry for many environmental chemicals. The circulatory system is closely associated with the intestinal tract, and once toxicants have crossed the epithelium, entry into capillaries is rapidly effected. Veinous blood flow from the stomach and intestine introduces absorbed materials to the hepatic portal vein, resulting in transport to the liver, which is the main site of metabolism of foreign compounds. Exposure to metals in the general environment is usually greater via food and drink than via air. However, the absorption of ingested lead and cadmium in adults is normally relatively low, being about 10% for lead and 5% for cadmium. Although absorption is influenced by various dietary factors and increased absorption of lead has been found in cases of low dietary calcium. Absorption of dietary lead is much higher in young children. In the case of inorganic mercury compounds absorption from foods is about 7% of the ingested dose; in contrast, gastro-intestinal absorption of methylmercury is practically complete.[2,4,5]

The gills of aquatic animals present another type of interface between the external medium and an organism and are an important route of entry for water-dispersed pollutants. They present a large surface area for diffusion, and at the

[5] World Health Organization, Geneva, Environmental Health Criteria, 'Mercury,' 1976, **1**.

same time continual circulation of water across the gill filaments ensures maximum exposure.

One other type of external surface membrane worthy of note is that of the leaves of plants. Leaf surfaces are covered by a waxy cuticular layer, interspersed with stomatal pores which may occur on both leaf surfaces or on a single surface (usually the lower). Stomata, as well as regulating uptake of carbon dioxide and water loss, are the major route of entry for gaseous pollutants. Rate of uptake of gaseous pollutants into a leaf is a function of several physical factors; one such factor is the resistance to diffusion caused by the boundary layer which in turn depends on the velocity and turbulence of airflow over the surface, and another factor is the stomatal resistance. Before a pollutant can gain access and cause injury within a plant cell it must first enter into solution in the extracellular water enveloping the cell wall. In solution, SO_2 is active in the form of either HSO_3^- or SO_3^{2-}. Ozone (O_3) is less soluble, but since it is a highly reactive molecule, it is thought that decomposition products such as hydroxyl radicals and other free radicals are important reactive species produced from reactions involving organic compounds.

Changes in stomatal aperture induced by air pollutants has attracted considerable attention. This is not surprising since any change in gaseous flux to and from metabolic sites in mesophyll tissues may eventually affect overall growth and yield of plants. The effect of SO_2 on stomatal aperture is very complex, initial studies found enhanced stomatal opening in *Vicia faba* plants exposed to SO_2. It was also shown that relative humidity strongly influenced the direction of the response; humidities above 40% enhanced stomatal apertures, whereas at humidities below 40% apertures decreased. A host of different species have now been investigated and it would appear that there is no uniformity of response between species; both concentration of pollutant gas and duration of exposure influence the outcome.[6]

4 INTERNAL PATHWAYS

Once inside an organism, a pollutant may follow a number of different pathways. In general terms four main routes can be identified (see Figure 1): some molecules are metabolized (converted) into other compounds which are frequently less toxic to the organism than the parent compound, although some are more toxic. A second pathway involves storage in certain issues, *e.g.* lead in bone, cadmium in kidney, and DDT in tissue with a high fat content. Thirdly, a pollutant and its metabolites may be excreted from an organism. Because metabolites are often more easily excreted than the original pollutant, metabolism is frequently an essential preliminary to excretion. The remaining fraction is available to exert an effect at the site of action. The fundamental effect is a biochemical event, for example lead inhibits the action of ALA-D, an enzyme in

[6] J. Wolfenden and T. A. Mansfield, 'Acid Deposition—Its Nature and Impacts', ed. F. T. Last and B. Watling, Proc. Roy. Soc. Edinburgh, 1991, Section B, **97**, p. 117.

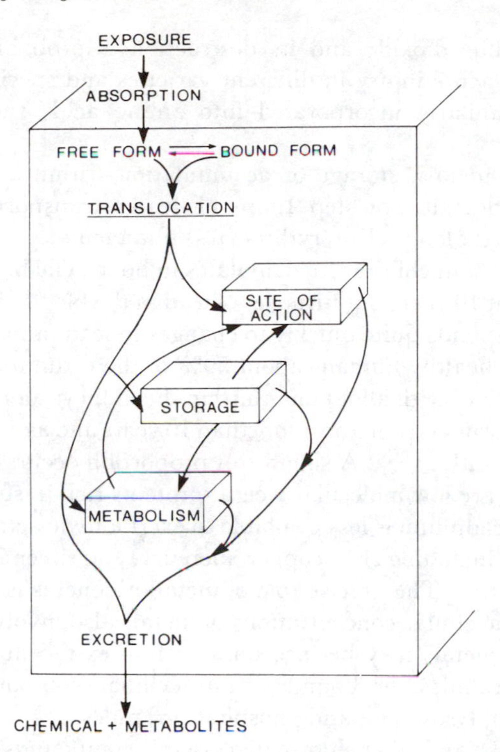

Figure 1 *Diagrammatic representation of the possible internal pathways followed by pollutants* (Adapted with permission from T. A. Loomis, 'Essentials of Toxicology', 2nd Edn., Lea and Febiger, Philadelphia, 1974)

the haem synthesis pathway, although the significance of this to the overall toxicity of lead is not clear. Perhaps surprisingly, the primary toxic lesion of many pollutants is unknown. At low exposures, it is envisaged that most of the pollutant would be stored and/or metabolized and excreted and little if any would be available to have toxic effect. As exposure is increased these controlling processes are progressively overwhelmed, primary biochemical lesions appear which in turn leads to major physiological damage and ultimately death of the organism.

Internal pathways of the major phytotoxic gases are short. They exert their effect near the point of entry. For mesophyll tissues of plant leaves, for example, much of the damage due to sulfur dioxide involves disruption of chloroplasts and depression of photosynthesis and also changes in stomatal aperture. It is widely believed that much of the effect of ozone involves permeability changes in the membranes of palisade cells. Pollutants like ozone are so transitory that it is virtually impossible to detect them within tissues; their presence is detected from the effects induced. While exposure to sulfur dioxide can increase plant sulfur content three- to four-fold, it is rapidly incorporated into organic molecules, notably the amino acids, glutathione, and cysteine. In fact the balance between incoming sulfur dioxide and its destruction is probably a critical factor in resistance to acute injury in different varieties and species of plants. Nitrogen dioxide is similarly incorporated into amino acids such as glutamine and asparagine.

For many metals, storage or accumulation within a tissue or an organ is essentially a detoxication step. Inorganic lead is transported in the bloodstream attached to red blood cells (erythrocytes) but a major proportion, about 90% in adults and 70% in children, accumulates in bone. The biological half-life in this tissue is about 10 years but turnover of lead in the bloodstream and soft tissues is rapid and responds quite quickly to changes in lead intake and exposure.[3,4]

In normal healthy humans about 50% of the cadmium in the body is in the kidneys and liver with about one-third in the kidneys alone. The biological half-life in these tissues is probably more than 10 years and as a consequence cadmium accumulates with age.[3,4] A significant proportion occurs bound to metallothioneins, which are low molecular weight proteins rich in sulfydryl groups. In this bound form cadmium is less available to exert a toxic action. A number of other trace metals, including zinc, copper, mercury, and silver also induce and bind to metallothioneins. The precise role of metallothioneins is unclear and the regulation of intracellular concentrations of metals also involves other mechanisms, for example metals may be incorporated into extracellular structures such as carbonate granules, or bound to intracellular components such as nuclei, mitochondria, lysosomes, and phosphate granules.

DDT, PCBs, and other chlorinated organic compounds are concentrated in fat deposits of a wide range of fauna, including birds and mammals. Tissues with the largest amount of fat accumulate the highest concentrations of these lipid-soluble compounds. In normal circumstances there is a slow turnover of these compounds in the body's fat reserves. However, during periods of stress and starvation, mobilization of the fat deposits can release the stored organochlorines into the bloodstream which in turn may induce toxic effects and ultimately death. Laboratory studies have shown that starved animals will sometimes die of DDT poisoning whereas those adequately fed appear unharmed.[7] The large number of guillemot deaths recorded in September 1969 from the coasts around the Irish Sea may have been due to a combination of stress and mobilization of organochlorine residues, perhaps PCBs, into the blood stream.[1]

A feature of most, if not all, vertebrate and invertebrate animals is a built-in capacity to biotransform a wide range of foreign compounds to less toxic entities that are more easily eliminated from the organism. In higher organisms biotransformation mainly takes place in hepatic tissues and it essentially consists of two phases; phase 1 is a mixed-function oxygenase system (MFO) which oxidizes foreign compounds by such mechanisms as hydroxylation, dealkylation, and epoxidation; phase two involves conjugation reactions by which glucoronide and sulfate derivatives are formed from the oxidized products of phase one. The net effect is to convert lipophilic foreign compounds to water-soluble metabolites

[7] F. Moriarty, 'Ecotoxicology: The Study of Pollutants in Ecosystems', 2nd Edn., Academic Press, London, 1988.

which facilitates their elimination from the animal. Numerous hydrocarbon compounds, including various phenolic and benzene compounds, are known to undergo biotransformation to water-soluble conjugates. The influence of biotransformation in regulating tissue concentrations and hence the toxicity of a substance can be deduced from studies which have used inhibitors of the biotransformation pathways. Salicylamide is a potent inhibitor of the phase II glucuronide conjugation reactions and it has been shown that pretreatment of fish with salicylamide significantly increases the toxicity of phenol as a consequence of inhibition of the formation of phenylglucuronide. In the case of the organophosphorus insecticide parathion, the metabolite paxon is the actively toxic product but pretreatment of fish with sesmex (an inhibitor of MFO reactions) prevents the oxidation of parathion to the active paxon with the result that its toxicity was much reduced. In animal tissues DDT is rapidly converted to the more persistent DDE and this explains the fact that DDE, rather than DDT is the most widespread and abundant form in wildlife.[7]

A fundamental relationship in pollution and ecotoxicology studies is that which relates exposure, body burden, or accumulation, and toxicity of a substance. Residues in an organism (or some part of it) form an important link between exposures and biological response or damage. For the most persistent pollutants, such as metals and organochlorine compounds, enhanced exposure from food or the surrounding media generally results in a greater concentration in an organism. When uptake (absorption) of a substance exceeds elimination (including metabolism) accumulation occurs in the whole organism or some part of it. Eventually a steady state may be attained when uptake and elimination are equal and at this stage the amount in the organism is proportional to the average daily intake and inversely proportional to elimination rate.

For many organic chemicals, the bioconcentration factor (*i.e.* the ratio of the concentration of the chemical in the organism to the exposure) is proportional to the lipid solubility and inversely related to the water solubility of the compounds. It can be estimated from the n-octanol–water partition coefficient.

Many studies have examined the relationship between blood lead concentration and concentration of lead in air, water, and diet. A curvilinear plot is obtained for the total intake *versus* blood lead concentration, indicating that successive increments in intake or exposure result in progressively smaller contributions to blood lead concentrations (Figure 2). This relationship can form the basis of predicting one parameter from the other. The accumulation and elimination rate of methylmercury in man conform to a single exponential first-order function (see Figure 3). The interesting feature of this model is that, assuming a constant hair-to-blood ratio, mercury analysis of segments of hair can be used to predict both the body burden and blood concentration of mercury at the time the hair segment was laid down.

Originally the 'concentration along the food chain' concept of biomagnification derived from the much publicized event in Clear Lake, California. It was established that deaths of western grebes (*Aechmophorus occidentalis*) had resulted from spraying the lake with DDD. It was also suggested that the DDD had reached the grebes by accumulation and concentration along the food chain.

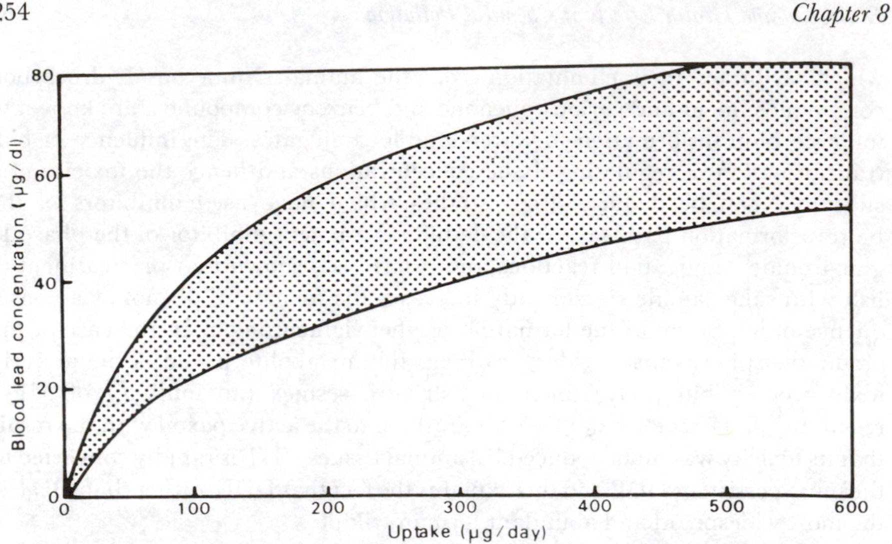

Figure 2 *Relationship between blood lead concentration and total lead uptake for adults, showing the range of interpretation from previously published data considered by Royal Commission on Environmental Pollution* (Reproduced with permission from Royal Commission on Environmental Pollution, HMSO, London, 1983, Ninth Report, Cmnd. 8852)

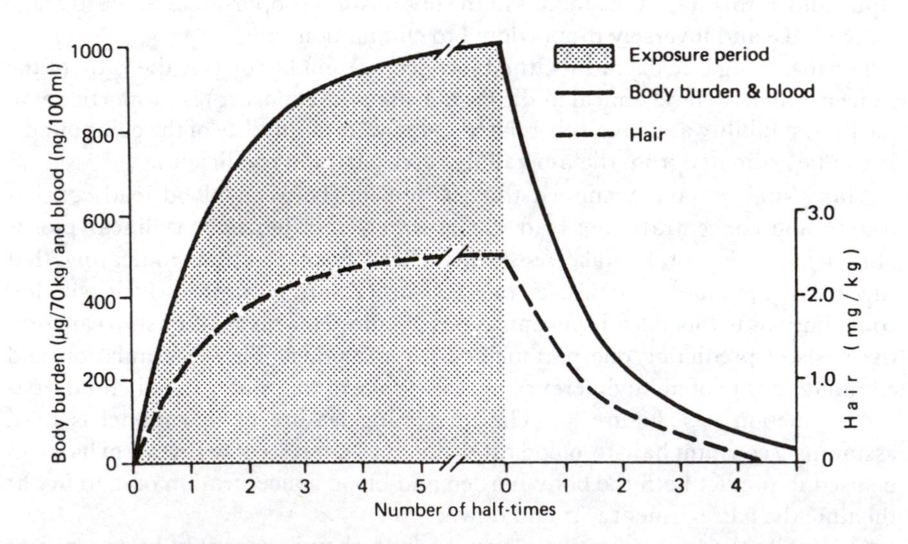

Figure 3 *The changes in the body burden and hair and blood concentration of mercury during constant daily exposure (shaded area) and after exposure. This calculation was based on a daily intake of 10 μg of methylmercury during the exposure period, an elimination half-time of 96 days, and a hair-to-blood concentration of 250* (Reproduced with permission from WHO, Environmental Health Criteria, 'Mercury', 1976, **1**)

Much stress was placed on the fact that carnivorous fish had higher concentrations than herbivorous fish. This view has since been questioned, if only because it is too simplistic.[7]

Organisms in aquatic systems can acquire residues from both food and the surrounding medium. In laboratory based studies comparing equivalent exposures from both routes, uptake directly from the water medium would seem to be more important. However in reality substances which are lipophilic, will strongly partition to the abiotic and biotic phases in the system and the concentration in water would be very small, therefore exposure to such substances would be mainly from ingesting contaminated food.

In terrestrial ecosystems most if not all lipophilic substances including pesticides are transferred via the food chain. Seed corn dressed with dieldrin undoubtedly killed many pigeons in England in the 1950s and 1960s; the poisoned pigeons were responsible for killing predators such as foxes and perhaps peregrine falcons. It is common to find that residues of organochlorine compounds are highest in carnivorous birds, followed by insectivorous, and then herbivorous birds.

5 SOME GENERAL ASPECTS OF POLLUTION STRESS ON INDIVIDUALS, POPULATIONS, AND COMMUNITIES

As a general rule, increasing pollution exposure results in progressively more severe and damaging effects. Holdgate[1] has developed a simple but nevertheless useful model which summarizes the interaction of pollution with living receptors (see Figure 4). The so-called 'cascading effect' suggests that some degree of homeostatic control or carrying capacity operates at each level of organization. As exposure increases in magnitude, the capacity of cells, tissues, whole organisms, populations, and communities are successively overwhelmed and damaged. Disruption at the biochemical level may be quite specific but at higher exposures as physiological systems are affected, more general symptoms of injury become apparent; reproductive capacity and recruitment may be affected and behavioural responses may be evident, especially in cases where neurotoxic substances are involved. At still higher exposures, fatalities become more commonplace and so eventually, as these become as significant as losses due to natural causes, noticeable reductions in population size result.

This then sets the stage to consider another important aspect of pollution studies: the diverse response of organisms to the same exposure. Variation in sensitivity is apparent between:

(1) individuals:
(2) groups or populations differentiated with regard to sex, stage of development, nutritional well-being, *etc*.;
(3) varieties or strains;
(4) species.

In general, differences between individuals are much less evident than those between different species. Differences in sensitivity to pollution may be viewed

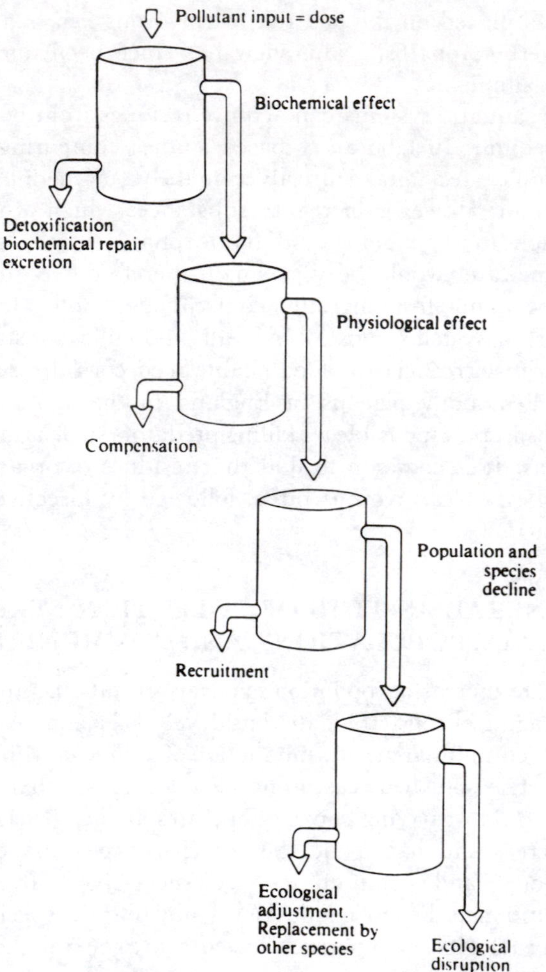

Figure 4 *Crude model of 'cascading' effect of pollution, as a series of reservoirs. Effects do not spill over from one level to another unless inputs exceed capacity of relieving systems depicted on left of each reservoir. Variation between individuals can be thought of in terms of variation in capacity of these systems, which may also be increased by evolution, producing more resistant organisms* (Reproduced with permission from M. W. Holdgate, 'A Perspective of Environmental Pollution', Cambridge University Press, Cambridge, 1979)

broadly in terms of differences in homeostatic capacities between various groups of organisms, *i.e.* differences in rate of absorption, metabolism, storage, and excretion.

(1) In a set of individuals a measured response will reveal a group which is sensitive, another which is moderately sensitive, and one that is relatively resilient; a spread or statistical distribution about a median value is usually apparent. For example the rate of elimination of methylmercury from the

human body is subject to individual variation and a person with a slow elminination rate would accumulate a higher body burden than another with a more rapid rate.

(2) Very often the early stages of life are particularly vulnerable to pollution stress. For example children are known to be more susceptible than adults to lead poisoning, and depletion of fish stocks in acidified lakes is primarily due to high mortality among eggs, larvae, and fry. At the other end of the age scale, the elderly and especially those with a history of heart and lung disease are the most vulnerable during severe air pollution episodes. In humans receiving similar lead exposure, higher blood lead concentrations are generally found in males than females.

(3) Certain varieties of tobacco (*Nicotiana tabacum*) show different degrees of sensitivity to photochemical oxidants. Three in particular, which are described as supersensitive, resistant, and intermediate, have proved useful biological indicators of elevated ozone levels. However, an interesting development of prolonged exposure to certain pollutants is that selection processes may generate resistant or tolerant varieties. This is especially the case of fast breeding forms of life. Examples are as follows:

(*a*) Genotypes tolerant of one or more trace metals (*e.g.* Cu, Zn, Pb, Ni) have been identified in many plant species growing on metal-rich substrates. Such substrates include soils of mineralized areas, metal mine wastes, and soils contaminated by metal-rich fumes originating from smelters, refineries, and automobiles.[8]

(*b*) A genotype of the grass *Lolium perenne*, known as Helmshore (named after the district 15 miles north of Manchester where it was first discovered), has been found to be more resistant to sulfur dioxide than the cultivated variety, S23.[9]

(*c*) An enormous number of insect pests have developed resistance to various insecticides and in the late 1970s resistant varieties had been recorded in over 400 species of insects and mites.[7]

(*d*) In copper- and zinc-rich sediments from old metal mine workings in estuaries of South-west England, tolerant varieties of the polychaete *Nereis diversicolor* can be found.[10]

(*e*) Darker, melanic forms of many species of the larger moths, *e.g. Biston betularia*, occur with increasing frequency in areas within and adjacent to the industrial regions of Britain. The distinctive feature of these darker varieties is the greater preponderance of melanic granules in the cuticle and wing scales. The significance of the adaptation would seem to lie in concealment from bird predation on darker surfaces which are typical of areas affected by air pollution.[7]

[8] M. H. Martin, and P. J. Coughtrey, 'Biological Monitoring of Heavy Metal Pollution—Land and Air', Applied Science, London and New York, 1982.

[9] J. N. B. Bell, and C. H. Mudd, 'Effects of Air Pollutants on Plants', ed. T. A. Mansfield, Cambridge University Press, Cambridge, 1976, p. 87.

[10] G. W. Bryan, 'Pollution and Physiology of Marine Organisms', eds. W. B. Vernberg and F. J. Vernberg, Academic Press, 1974, p. 123.

(4) Species, even closely related ones can differ quite markedly in their responses to one form of pollution or another. This is an important consideration in the assessment of the ecological effects of pollution since it is common to find that communities which are subject to pollution stress undergo a reduction of species diversity as the most susceptible ones disappear first, to leave behind fewer and more resilient types. Some notable examples include forest ecosystems grossly affected by air pollution and freshwater systems subject to organic pollution, artificial eutrophication, or acidification. In certain situations changes in community structure can be used as a sentinel of the degree of pollution stress, *e.g.* biotic indices that are used to indicate water quality. The selective sensitivity of lichen species to gaseous air pollutants and, in particular sulfur dioxide, is another noteworthy example. Corticolous lichens (*i.e.* those growing on tree trunks) along a gradient of sulfur dioxide as say from a rural area to an industrial centre, reveals that the lichens progressively become impoverished as higher sulfur dioxide concentrations are encountered and commonly within urban-industrial centres there is a complete absence of lichens. Hawksworth and Rose[11] devised a scale comprising of 11 zones based on presence and abundance of corticolous lichens and a map of England and Wales depicting these lichen zones shows them to be closely related to the winter mean sulfur dioxide concentration for that particular year of the survey (1972–73).

In the actual environment, differences in species sensitivity may be more apparent than real because certain organisms may experience greater exposure as a consequence of pollutants becoming more concentrated in specific parts of an ecosystem such as the upper soil layers or the sediments at the bottom of lakes and estuaries. Those organisms which inhabit these areas and in particular those that by virtue of their feeding habits, *e.g.* earthworms or filter feeders such as mussels, will experience greater exposure from ingesting contaminated particles.

Fatalities due to pollution do not necessarily affect the overall size of a population. Organisms die from a variety of natural causes, and unless pollution is on a par with these, its influence on population size would be unlikely to be significant. In broad terms a population is controlled by birth rate, death rate, and the balance between immigration and emigration. For many wildlife populations, recruitment (*i.e.* the number of 'recruits' produced annually) has a strong influence on population size. This is certainly the case for most commercially important fish species in which fluctuations of the early life stages account for much of the fluctuation in overall population size and environmental factors, rather than spawning stock, control the number of early larval and juvenile stages.

Natural populations of animals and plants are subject not only to short-term changes but also long-term fluctuations of population size which are often large-scale and irregular events that may take place rapidly. These changes are often due to changes in weather patterns such as periods of lower

[11] D. L. Hawksworth and F. Rose, *Nature (London)*, 1970, **227**, 145.

or higher temperatures. Clearly for disturbances caused by pollution to be apparent in a population, they must be on a scale over and above fluctuations due to natural causes. In addition, other activities of man, *e.g.* over-exploitation of a natural resource can cause significant perturbations.

Whether or not pollution exerts a permanent effect on the population size will depend on the persistence of the polluting agent. Once the stress has been withdrawn, populations have a remarkable ability to recover from large-scale fluctuations, due to whatever cause. The rapid recolonization of the tidal river Thames is a striking example of recovery of an ecosystem from pollution. Although, to conclude that recovery is complete assumes a precise knowledge of the situation prior to pollution onslaught and this is rarely known.

In general terms, the ecological effects of pollution are reductions in species diversity, productivity, and biomass. These trends have been described as a retrogression to an earlier successional stage of ecological development. However, perception and evaluation of such changes at a community level is difficult, since even the simplest of ecosystems contain a considerable number of species, populations interactions are complicated, and measurement of pollutant exposure in the various niches of the system can be very problematic.

6 METHODOLOGY

In practice, three general methods of approach are employed in assessments of ecological and health effects of pollutants:

(1) standardized laboratory-based experiments;
(2) field surveys and epidemiological studies;
(3) field experiments and microcosm or mesocosm studies.

(1) Experimental studies may be concerned with uptake, accumulation, biotransformation, and the toxicity of a substance or mixtures of substances. Acute toxicity tests are widely used and they usually form the basis of assessing the potential hazard of chemicals. The LD_{50} (or LC_{50})—the dose (or concentration) which is lethal to 50% of a test population over a set period of time (24, 48, or more usually 96 hours)—is the parameter most commonly derived in such tests. These simple tests which are carried out under standardized conditions provide a useful means of screening chemicals and ascertaining their relative toxicity. However alteration of the experimental conditions can substantially modify toxicity values. All too frequently acute toxicity tests are based on a few well-known easily reared test species and, as a result marked differences in sensitivity of wildlife species are overlooked. Many incidents of poisoning in birds by carbamates and organophosphorus compounds could have been avoided had LD_{50} data extended to more than one species. In one incident a number of geese (*Anser anser* and *Anser fabalis brachyrhynchus*) died from ingestion of lethal doses of carbophenothion, which is one of the organophosphorus insecticides that has

replaced dieldrin as a seed dressing protecting winter wheat from wheat bulb fly. Original registration data, based on toxicity to chickens indicated a low avian oral LD_{50}; however it is now known that the two species of geese are very much more sensitive to carbophenothion than other bird species particularly chickens.[7]

Chronic toxicity tests by their nature are undertaken at lower exposures for more prolonged periods and therefore they provide a more realistic indication of toxicity. Response criteria that have been used to indicate chronic stress include growth, feeding rate, respiration, reproductive activity and success behaviour, and bioaccumulation. An integrated physiological response, that has been used in chronic toxicity studies in the laboratory and the field, is scope for growth using the common mussel, *Mytilus edulis*. A detailed description of the methodology of measuring scope for growth may be found in Widdows and Johnson (1988).[12] Basically, the index measures the energy available for activity, growth, and reproduction from assimilation of food, after respiratory and excretory requirements have been satisfied. In the presence of a pollutant this energy may decrease in response to a decreased feeding rate and expending metabolic energy to deal with stress. With regard to the ecological effects of offshore oil production in the North Sea, there is good agreement between scope for growth response in mussels exposed to elevated polycyclic aromatic hydrocarbons and community responses such as changes in species diversity.[12]

(2) Field surveys and epidemiological studies, in basic terms, relate pollutant exposure to a biological response. Much of the remainder of this chapter examines various aspects of this relationship. However some general points are worth noting. Field evidence relating a pollutant to an adverse response does not constitute a cause–effect relationship, but merely demonstrates a statistical correlation. There are many practical problems concerned with estimating the true exposure of pollution to a group of organisms. Equally it is often difficult to obtain a true or representative indication of size of population, because of uneven distributions and fluctuations over time. The impact of pollution on wildlife depends as much, perhaps more, on the duration of damage as on immediate losses.

(3) Field experiments, in many ways bridge the important gap between purely experimental methods and field surveys. This type of approach has proved particularly valuable in the study of the effects of gaseous air pollutants on plants. Basically plant growth chambers are set up in the field, one set receives unfiltered air containing air pollutants at ambient concentrations and the other set receives air from which the pollutants have been removed by filtration. A number of studies, some of which are referred to in subsequent paragraphs, have used this approach to good effect. It is important to realize that this is still fundamentally a correlative response,

[12] J. Widdows and D. Johnson, 'Biological Effects of Pollutants', ed. B. L. Bayne, K. R. Clarke, and J. S. Gray, Marine Ecology Progress Series Special Issue, Amelinghausen: Inter Research, 1988, **46**, 113.

since effects are related to the pollutants which the operator chooses to measure.

In Ontario, Canada, some 46 small lakes in an area containing several hundred lakes have been set aside for experimental research. In this Experimental Lakes Area, whole lakes have been fertilized with nutrients to study eutrophication or acidified with inputs of sulfuric acid to monitor community changes.[13]

Increasingly, microcosms and mesocosms are used to examine the fate and effect of chemicals in the environment. Such systems include artificial ponds and streams which are colonized with natural assemblages of plants and animals and then subjected to different degrees of chemical stress. In other situations a part of the lake, estuary, or the sea may be segregated using enclosures such as large rubber tubes and the entrapped communities exposed to various combinations of pollutants.

7 IMPACT OF MAN ON WATER QUALITY

7.1 Organic Pollution

Much of the early concern regarding water quality focused on what is now commonly referred to as organic pollution. This is the discharge of untreated or partially treated human waste, and biodegradable industrial waste into water courses. In the 200 years leading up to the middle of the present century, population, urbanization, and industrialization increased apace which in turn generated ever increasing quantities of waste. Organic pollution has proved to be a major factor in the deterioration of water quality in many stretches of rivers in the United Kingdom and elsewhere in the developed world. Conditions generally reached their worst in the 1950s and early 1960s. As a consequence of more efficient and widespread sewage treatment and greater control over industrial discharges, there has been a marked overall improvement in many rivers.

A very useful and simple model of the chemical, physical, and biological changes that take place in a river following a discharge of organic biodegradable waste is reproduced in Figure 5 from Hynes.[14] Suspended and dissolved solids increase markedly below an outfall. The effluent introduces a significant organic load into the river and this is indicated by the increase in BOD or biological oxygen demand.

In breaking down organic matter micro-organisms consume the dissolved oxygen in the water, with the result that very low or near-zero levels of oxygen can occur. In such situations nitrogen remains in the reduced form and ammonia concentrations build up to toxic levels. Micro-organisms are the main feature in these regions and a benthic community, known as sewage fungus, frequently

[13] D. W. Schindler, K. H. Mills, D. F. Malley, D. L. Findley, J. A. Shearer, I. J. Davies, M. A. Turner, G. A. Linsey, and D. R. Cruikshank, *Science*, 1985, **228**, 1395.

[14] H. B. N. Hynes, 'The Biology of Polluted Waters', Liverpool University Press, Liverpool, 1960.

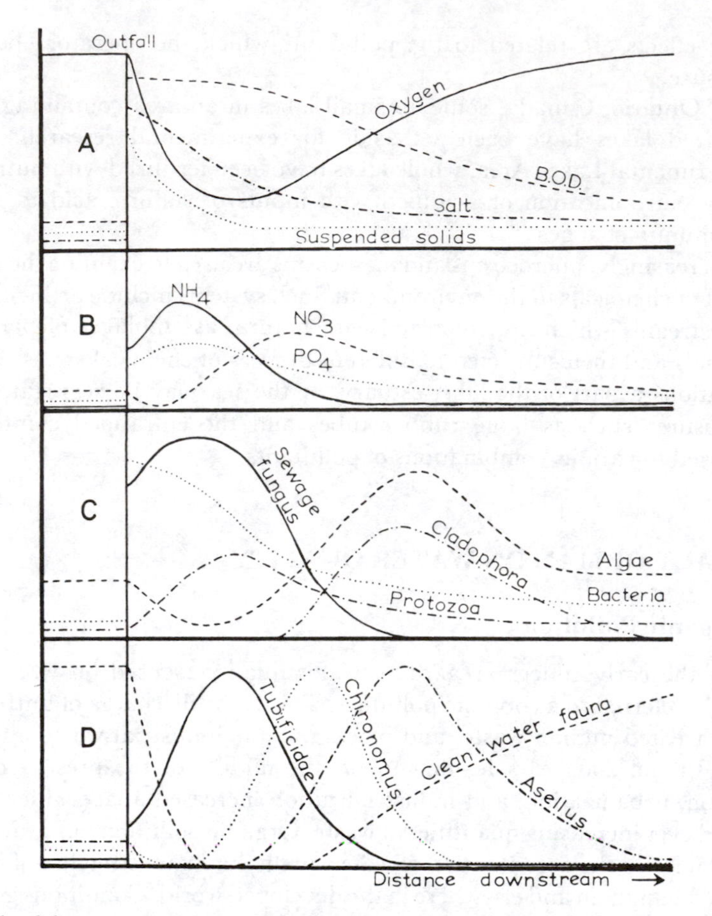

Figure 5 *Schematic representation of the changes in water quality and the populations of organisms in a river below a discharge of an organic effluent:* A, *physical changes;* B, *chemical changes;* C, *changes in micro-organisms;* D, *changes in macro-invertebrates* (Reproduced with permission from H. B. N. Hynes, 'The Biology of Polluted Waters', Liverpool University Press, Liverpool, 1960)

develops. A typical sewage fungus community, which is often a white or light brown slime in appearance, is dominated by bacteria but includes fungi, protozoa, and algae (see Mason[15] for a more detailed account).

In areas where oxygen levels are much reduced and light penetration is limited because of the suspended solids, plant growth is adversely affected. As conditions improve downstream and nutrients are released from breakdown of organic matter, primary productivity may be quite excessive.

The macro-invertebrate fauna differ markedly in their susceptibility to organic pollution. Some are remarkably tolerant, *e.g.* tubificid worms and chironomus larvae, and once relieved from the constraints of competition as the more

[15] C. F. Mason, 'Biology of Freshwater Pollution', Longman Scientific and Technical, 2nd Edn., 1991.

sensitive species die-off, their population size can increase enormously. For example, even during the early stages of recovery of the tidal Thames, tubificid densities in the region of Crossness and Beckton sewage outfalls on occasion exceeded 600×10^3 m^{-2}.

As a general rule, fauna which inhabit areas of sediment accumulation tend to be more tolerant of organic pollution, whilst the most sensitive ones are more commonly found in the eroding and more swiftly flowing waters. Stoneflies (Plecoptera) and mayflies (Ephemeroptera) are some of the more sensitive groups. The reappearance of *Asellus* sp. and *Gammarus* sp. indicate improved conditions.

Many fish species are particularly sensitive to reduced dissolved oxygen concentrations. However, since fish are considerably more mobile than other members of the ecosystem they may avoid heavily polluted waters. Nevertheless, migratory fish such as salmon (*Salmo salar*) and sea trout (*Salmo trutta*), which require well oxygenated waters, may fail to survive passage through grossly polluted sections of the lower reaches of rivers and estuaries, *e.g.* the tidal river Thames.

Clearly the amount of organic waste discharged into a water body strongly influences the response of the system, as does the number of discharges along the length of a river. At one time, extensive stretches of a number of rivers in the UK became grossly polluted and largely devoid of life, especially where one polluted region merged with another downstream, *e.g.* River Trent and tributaries. A further aspect is that conditions often vary from season to season. For example, periods of low flow in late summer and autumn coincide with times of high biological activity, as a consequence the third quarter of the year, commonly experiences the lowest dissolved oxygen concentrations, as exemplified by the Tidal River Thames. Estuaries and tidal rivers are in many respects special cases, they are complicated systems and the ebb and flood tides can carry organic wastes upstream and downstream of the point of discharge. The pollutant load tends to have a longer residence in these systems and the overall dilution and mixing is more difficult to predict.

The physical, chemical, and biological parameters discussed above provide the basis for assessing water quality. Recent statistics show that the length of grossly polluted and non-tidal rivers and canals has decreased steadily from the late 1950s through the 1980s.[16] These improvements have been achieved as a result of the provision of new and extended sewerage systems and sewage works and treatment plants for industrial effluents that were built mainly during the 1960s and 1970s. At present over 80% of all sewage produced in England and Wales is subject to secondary sewage treatment and much of the remainder is directly discharged to the sea. Substantial improvement in the quality of many rivers, *e.g.* the River Trent, has occurred over the past 30 years. However, significant parts of the catchment of the River Mersey in North-west England are still grossly polluted. In many Third World countries untreated sewage is

[16] Department of the Environment, Digest of Environmental Protection and Water Statistics, HMSO, 1989, No. 12.

discharged directly into rivers and it is becoming apparent that severe water pollution problems exist in many of the Eastern European countries.[15]

One example involving other pollutants in addition to organic waste is that of the river Ebbw in South Wales. For many years, this river has received industrial wastes from the steel works and coal mining activities in the area, and for most of the past 100 years the river has been devoid of life for much of its length. Although sewage discharges contributed to the poor water quality of the river, a major cause of deoxygenation was the considerable input of ferrous salts from the steel works. In addition other toxins were present in high concentrations, including ammonia, phenols, and heavy metals (Cr, Ni, Cu, Zn, Cd, Pb). Following installation of waste treatment facilities at the steel works, and improvements to the coal washeries and sewage treatment facilities in the early 1970s, there was a marked improvement in the quality of the river and recolonization by several species of plants, macro-invertebrates, and fish. For example, the period of survival of trout retained in cages only 5 km downstream of the steel works increased from about 0.5 hours in 1970 to several days after 1974. Further downstream fish generally survived a test period of 2–3 months.[17]

The tidal Thames has suffered two periods of gross deterioration, once in the 1850s and then again in the 1950s.[18,19] During both these periods the river was compared to an open sewer. However, in the 15 years between the early 1960s and the late 1970s conditions markedly improved. The underlying reason for the periods of deterioration was the increase in the population of London. By the early part of the nineteenth century many of the tributaries flowing through the city received all kinds of waste. Despite these inputs the Thames itself must have been satisfactory as records indicate that salmon were still caught regularly. A significant turning point was the introduction of water closets from about 1810 onwards, which emptied directly into the sewers that drained into the Thames. By the 1850s the smell of the river was notorious and 1858 became known as the 'Year of the Great Stink'. Many people, especially the poor, relied on the Thames for their water supplies; not surprisingly a number of serious Cholera epidemics occurred and in 1849 and 1854 about 20 000 deaths due to this disease were recorded.

Some improvements in the condition of the river flowing through London were achieved between 1850 and 1900 by transporting the entire sewage load of the metropolis in a system of sewers eastwards and discharging into the river further downstream at Barking. However recovery was short lived as the population of the capital continued to rise and so the quantity of polluting matter of all kinds discharged into the tideway also increased. Numerous inefficient sewage works contributed to the deterioration of the river and the newly introduced synthetic detergents containing alkyl arylsulfonates, which resisted biodegradation, interfered with the transfer of oxygen from the atmosphere across the air–water

[17] R. W. Edwards, 'Ecological Assessment of Environmental Degradation, Pollution, and Recovery', ed. O. Ravera, Elsevier, Amsterdam, 1989, p. 159.

[18] L. B. Wood, 'The Restoration of the Tidal Thames', Adam Hilger Ltd., Bristol, 1982.

[19] M. J. Andrews, 'Effects of Pollutants at the Ecosystem Level', eds. P. J. Sheehan, D. R. Miller, G. C. Butler, and Ph. Bourdeau, J. Wiley, 1984, SCOPE 22, p.195.

interface. As a result the river's capacity for self-purification was much reduced. By the 1950s up to 40 km of the river below London was regularly deoxygenated in the summer and autumn periods.

Fish stocks in the tidal Thames, particularly migratory fish such as salmon and trout, began to decline in the second and third decades of the nineteenth century. Although a slight revival towards the end of the century is noted from records of lampern and smelt in the reaches upstream of London, this improvement was short-lived since, with the exception of eels, no fish were recorded from about 1920 to about 1960 in the river between the Lower Thames near Richmond (20 km above London Bridge) and Gravesend (40 km below London Bridge).

The most significant influence on the improvement of water quality in the Thames has been the modernization of the region's major sewage treatment facilities at Crossness and Beckton. Between 1964 and 1980 the number of species of fish recorded in the metropolitan reaches increased from 3 to 98. Previous to this date the macro-invertebrate fauna was solely based on a super-abundance of one species of tubificid. Since 1976 a complex community has developed with ragworm; *Nereis* sp., sand hopper; *Corophium* sp., and the shrimp group; *Crangon* sp., *Gammarus* sp., and mysids; *Praunus* sp. and *Neomysis* sp. now occupying central positions in the food web whereas formerly it was solely oligochaete worms (see Figure 6).

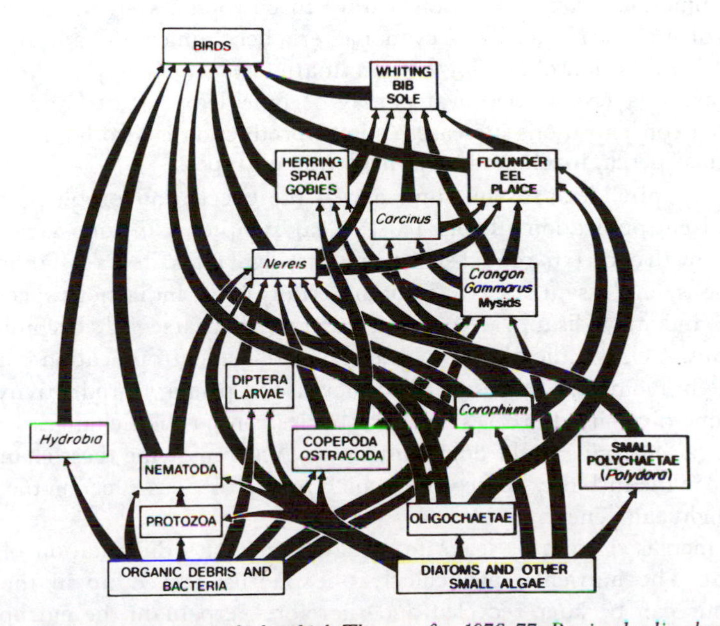

Figure 6 *Food web occurring in the brackish Thames after 1976–77. Previously oligochaete worms occupied a central position, providing the major diet item for both waterfowl and fishes* (Reproduced with permission from 'Effects of Pollutants and the Ecosystem Level', eds. P. J. Sheehan, D. R. Miller, G. C. Butler, and Ph. Bourdeau, J. Wiley, 1984, SCOPE 22, p. 195)

7.2 Artificial Eutrophication

Primary productivity of aquatic systems is essentially controlled by the availability of the major plant nutrients, nitrates and phosphates. Enrichment of waters with nitrates and phosphates may result from natural sources such as a forest fire and from various human activities, *e.g.* domestic sewage and drainage from farmland.

Lakes may be classified according to their nutrient status, those with a nutrient-poor status are oligotrophic and those which are enriched with nutrients are eutrophic. Lakes generally become more eutrophic with age, however early in the 1960s it was noticed that numerous lakes and reservoirs were undergoing eutrophication in a matter of decades or even years as a consequence of nutrient enrichment from human activities. Accelerated or artificial eutrophication is also increasingly found in estuaries and coastal waters.

Typically, oligotrophic lakes are deep water bodies which remain fully oxygenated throughout the year. Their low nutrient status ensures that primary productivity is low, although the diversity of the algae is high and often dominated by Chlorophyceae. Similarly animal production is low with salmonids (*e.g.* trout, char) and coregonids (whitefish) being the main types present. In contrast, in response to the larger nutrient pool in eutrophic waters, high rates of photosynthesis in the growing season result in the upper epilimnion layer becoming super-saturated with oxygen. Blooms of algal growth which are rare in oligotrophic lakes are a common feature in eutrophic systems and generally species diversity is reduced with cyanobacteria becoming dominant. At the same time, decomposition of sinking organic matter consumes oxygen in the lower hypolimnion layer and these waters may become deoxygenated and as a result ammonia concentrations increase. Animal production is also high with coarse fish such as perch, roach, and carp the dominant types.

Mason[15] provides a useful summary of the effects and problems posed by artificial eutrophication (Table 1). Typical symptoms include algal blooms, heavy growth of certain rooted aquatic plants, algal mats, deoxygenation, and, in some cases, unpleasant tastes and odours. Many of the indigenous species of fish decline and may disappear, with less attractive coarse fish becoming more abundant. A clear indication of the effect of artificial eutrophication is a marked increase in biomass, as a result of an increase in primary productivity. At the same time diversity decreases and the species composition changes with blue-green algae emerging as the dominant group. Accompanying these changes is an increase in the turbidity of the water which has an obvious effect on the depth to which light can penetrate.

The increased biomass leads to a higher rate of sedimentation of organic material. The nutrients, particularly phosphorus, locked up in the bottom sediments can be later recycled and therefore perpetuate the eutrophication process in the water body. This is often an underestimated influence on the eutrophication of the overlying waters. Lake Trummen in Sweden is a shallow, formerly oligotrophic lake, that received considerable inputs of wastewater between the 1930s and 1950s. Following diversion of the polluting waters, no

Table 1 *The Effects of Eutrophication on the Receiving Ecosystem and the Problems to Man Associated With These Effects*

Effects
(1) Species diversity decreases and the dominant biota change
(2) Plant and animal biomass increases
(3) Turbidity increases
(4) Rate of sedimentation increases, shortening the life span of the lake
(5) Anoxic conditions may develop (particularly in the hypolimnion of stratified lakes)

Problems
(1) Treatment of potable water may be difficult and the supply may have an unacceptable taste or odour
(2) The water may be injurious to health
(3) The amenity value of the water may decrease
(4) Increased vegetation may impede water flow and navigation
(5) Commercially important species (such as salmonids and coregonids) may disappear

(Reproduced with permission from C. F. Mason, 'Biology of Freshwater Pollution', 2nd Edn., Longman Scientific and Technical, 1991)

significant recovery took place and it was revealed that the black sediment laid down during the period of pollution was releasing nutrients and causing extensive blooms of blue-green algae; improvement only occurred when the offending sediment was removed by dredging.

An OECD study[20] investigated the relationship between nutrient load in waters and the degree of eutrophication. Statistical analysis of data from some 150 lakes, ranging from 'pond size' to the North American Great Lakes showed that phosphorus was the main limiting factor controlling the development of artificial eutrophication. Phosphorus cycling between sediments and water is therefore one of the key issues in the overall process.

Substantial inputs of sewage into Lake Washington (Northwestern USA) in the 1960s increased the nutrient load in the lake. For example, winter phosphate levels increased from about 8 $\mu g\ l^{-1}$ in 1950 to around 50 $\mu g\ l^{-1}$ by the early 1960s, over the same period summer chlorophyll levels (used an index of phytoplanktonic production) increased from 1 $\mu g\ l^{-1}$ to 30–40 $\mu g\ l^{-1}$ and transparency of the lake decreased four-fold. In 1950 only about 7% of the July–August phytoplanktonic crop consisted of blue-green algae but by 1962 this figure had increased to 95%. Despite the high productivity levels, severe oxygen depletion never occurred. In the five years following improvements to the sewage disposal system, conditions in the lake improved considerably.[21]

In the UK, examples of water bodies which have attained a highly eutrophic state include Lough Neagh in Northern Ireland and the Norfolk Broads. Analysis of sediment core material from Lough Neagh has shown that over the past 50

[20] OECD, 'Eutrophication of Waters—Monitoring Assessment and Control', OECD, Paris, 1982.
[21] R. D. Gulati, 'Ecological Assessment of Environmental Degradation, Pollution, and Recovery', ed. O. Ravera, Elsevier, Amsterdam, 1989, p. 81.

years, the productivity of diatoms has increased markedly while their diversity has decreased and the dominant forms have changed. In the Lough blue-green algae have increased in abundance.[15]

Many urban coastal areas worldwide are affected and/or threatened by eutrophication, *e.g.* Baltic and North Seas. A recent estimate of the external supplies of N and P to the Baltic gave figures of 980×10^3 tonnes of N and 50×10^3 tonnes of P, which are higher by factors of 4 and 8 respectively than inputs at the turn of century. Over extensive coastal areas of the Baltic Sea many eutrophic-related changes have been reported. These include increased nutrient levels, increased chlorophyll-a concentrations and primary productivity. The incidence of algal blooms, particularly of the toxic variety has increased. In many years, water transparency has decreased with the result that the depth distribution of macroalgae such as *Fucus vesiculosus* has diminished. Sedimentation rates of organic matter to the sea floor are increasing as are the frequency and severity of oxygen deficiency in the bottom waters which in turn has depleted bottom fauna populations. Over the last 15 years the SE Kattegat on the Swedish west coast has shown signs of increased eutrophication, and since 1980 in particular there has been an increased incidence of mortalities to many benthic organisms including suspension-feeding bivalves, Norway lobster, and fish as a consequence of hypoxic conditions. In recent years oxygen deficiencies have been detected in German and Danish coastal bottom waters of the North Sea and there has been an increased occurrence of algal blooms, particularly the toxin producing varieties over extensive areas of the coastal waters of Holland, Denmark, and Norway.[22]

8 ECOLOGICAL AND HEALTH EFFECTS OF THE MAJOR AIR POLLUTANTS

8.1 Health Effects

Health effects of air pollution in the general population are associated with an increase in mortality and worsening of heart and lung conditions during episodes of dense smogs within which certain air pollutants accumulate to high concentrations. A number of incidents have been reported in different parts of the industrialized world, for example in the Meuse Valley in 1930 and in Donora in 1948. The most notorious incident occurred in London (UK) in early December 1952 when 3500–4000 deaths above the norm were recorded during an exceptional smog episode that lasted, unremittingly, for 5 days.[23,24] At the time, much of the air pollution was due to coal combustion, with numerous near-ground level sources generating considerable quantities of smoke and sulfur dioxide. The fog

[22] *Ambio*, Special Issue, 1990, **19**.
[23] Ministry of Health, United Kingdom, Report on Public Health and Medical Subjects, No. 95, HMSO, London, 1954.
[24] World Health Organization, Geneva, Environmental Health Criteria, 'Sulfur Oxides and Suspended Particulate Matter', 1979, **8**.

developed across the capital as a consequence of a very stable high pressure zone and an inversion layer. This prevented the dispersal of pollutants and concentrations built up to very high levels, the particles of smoke acted as condensation nuclei which added further to the density of the fog. The epidemic went largely unnoticed by the population of London and it was only when the death certificates for the whole of the London area were later examined that the sudden upsurge in the number of deaths became apparent. Deaths were mainly confined to the elderly and those people with a history of heart and lung diseases. The central areas of London, where the fog was at its densest and most persistent, showed the greatest increases in mortality; some 200% more than the average for that time of year. Estimates indicate that sulfur dioxide and smoke concentrations (48 h mean) during the London episode attained very high values and were in the region of 3.7 mg m^{-3} (1.3 p.p.m.) for SO_2 and above 4.5 mg m^{-3} for smoke. The London fog episode was also at the time of the Smithfield Agricultural Show at Earls Court and it is interesting to note that a number of cows also died during the course of the show.[23]

It is generally accepted that the combined effects of sulfur dioxide and smoke particles on the respiratory tract were responsible for the excess deaths and for exacerbating heart and lung disease in susceptible people. To the healthy majority of the populace of London the pollutant ladened fog was merely an inconvenience. Epidemiological studies established that the spatial and temporal trends of sulfur dioxide and smoke correlated closely with the distribution of the enhanced mortalities, but other pollutants would also have been elevated, *e.g.* carbon monoxide, sulfates, and sulfites. However there are insufficient data on such substances to allow exposure–response relationships to be established.[24]

Follow-up studies in other large cities concentrated on more moderate day-to-day variations of mortality and morbidity in relation to pollution levels. A WHO Task Group[24] provided a summary of the salient features of many of these studies and the collated data have formed the basis for developing short- and long-term exposure–effect relationships. It is important to emphasize that concentrations of sulfur dioxide and suspended particulate matter vary considerably from place to place and from one time period to the next. Secondly most monitoring data relate to levels prevailing in the outdoor environment and take little account of indoor exposure which is usually lower but it is where most people spend much of their time. The elderly and the chronically sick spend most of their time indoors. Also there is no consideration of occupational exposure which can be significant and smoking habits which is the most direct form of air pollution. Notwithstanding, short-term exposures of 500 μg m^{-3} (24 h mean) for both sulfur dioxide and smoke can be expected to result in excess mortality among the elderly and the chronically sick. Exposures of 250 μg m^{-3} (24 h mean) to both pollutants are likely to lead to worsening of the condition of patients with existing respiratory disease. With regard to long-term exposure, increased prevalence of respiratory symptoms among both adults and children and increased frequency of acute respiratory illnesses in children are more likely to occur when annual mean concentrations of sulfur dioxide and smoke exceed 100 μg m^{-3}. Based on these relationships, and incorporating a margin of safety, WHO advocate, as a guideline, that 24 h mean values of sulfur dioxide and smoke should remain below 100–150 $\mu g\ m^{-3}$, and with an annual mean below 40–60 $\mu g\ m^{-3}$.[24]

The relationship between photochemical smog episodes and human health are less clear cut. Early epidemiological studies, based mainly in Los Angeles, failed to establish direct relationships between increased mortality rates and the frequency of smog episodes. Although eye irritations (perhaps due to peroxyacyl nitrate compounds) and upper respiratory tract discomforture are common complaints during smog episodes. Increased breathing difficulties in heavy smokers and asthmatics and reduced performance in people indulging in physical activity are associated with periods of high oxidant concentrations.[25] Laboratory and field studies of adults who exercise heavily for short periods of time have provided evidence for the existence of short-term reversible decrements in pulmonary function to ozone concentrations at or near the USA National Ambient Air Quality Standard of 0.12 p.p.m. In recent years there are indications, in several parts of the world, that high ozone levels are associated with an increased risk of asthmatic attack. There is evidence that the prevalence of asthma may be increasing in a number of countries, including the United States and it is also possible that asthma is becoming more severe. Recent population-based studies have shown that respiratory function may be reduced proportionately to increasing concentrations of ambient O_3. Children attending a summer camp showed an increase in respiratory symptoms with increases in ozone concentrations above 120 p.p.b.[26,27]

8.2 Effects on Wildlife

In the decades following the industrial revolution in Europe and North America various sensitive plant species showed poor growth within the adjacent to urban areas. Lichens, in particular, and also certain conifer species are now known to be especially sensitive to sulfur dioxide and other pollutant gases. The effect of air pollution on terrestrial ecosystems such as forests is dependent on the nature of the pollutants and the magnitude of exposure. At relatively low exposures ecosystems absorb or act as a sink to the influx of pollutants and little or no harm can be detected, in certain circumstances such as inputs of nitrogen compounds into a nitrogen deficient forest increased growth may even be observed. At intermediate or moderate exposure, physiological damage occurs, growth may be reduced, and sensitivity to disease may increase. In grossly polluted situations, widespread destruction of vegetation and denudation of soils occurs.

Immediately surrounding heavily industralized areas and, most notably, in the vicinity of large point sources such as metal smelting operations, fumes rich in

[25] World Health Organization, Geneva, Environmental Health Criteria, 'Photochemical Oxidants', 1979, **9**.
[26] M. Berry, P. J. Lioy, K. Gelperin, G. Buckler, and J. Klotz, *Environ. Res.*, 1991, **54**, 135.
[27] D. M. Spektor, G. D. Thurston, J. Mao, D. He, C. Hayes, and M. Lippmann, *Environ. Res.*, 1991, **55**, 107.

sulfur dioxide, and metals have caused extensive damage to vegetation. Well known examples are found around large smelters like those at Sudbury and Wawa, Ontario, and at Copper Hill, Tennessee, where hundreds to thousands of square kilometres of landscape have been denuded. Smaller examples are found in the Nangatuck Valley in Connecticut, around a zinc smelter at Palmerton, Pennsylvania, and the Lower Swansea Valley in the UK. Many of these activities have either closed down or emission controls have been installed and so the gross disturbances observed previously around such sources are increasingly uncommon. However severe damage to forests still occurs at very polluted sites in Europe, notably in Czechoslovakia, East Germany, and Poland and the nature and extent of this damage is only recently coming to light.

Extensive areas of Europe and North America are subject to moderately elevated levels of gaseous air pollution, for example ozone has caused widespread damage to forests in California. A point worth emphasizing is that a polluted air mass rarely contains a single phytotoxic agent. In recent years it has been increasingly realized that regional scale air pollution is concerned with complex gas–aerosol mixtures that may include varying proportions of SO_2, nitrogen oxides (NO_x), ammonia (NH_3), ozone (O_3), and acid aerosols.

For a number of years, attempts to establish conifer plantations in southern parts of the Pennine Hills in the North of England generally proved unsuccessful. A survey, recording frequency of Scots Pine (*Pinus sylvestris*) in the area, revealed a corridor about 50 km wide, downwind of the major conurbations of Greater Manchester and Merseyside, where the species was either absent or very sparse (Figure 7). A significant negative correlation was noted between frequency of Scots Pine and mean winter sulfur dioxide concentrations in the area, and no

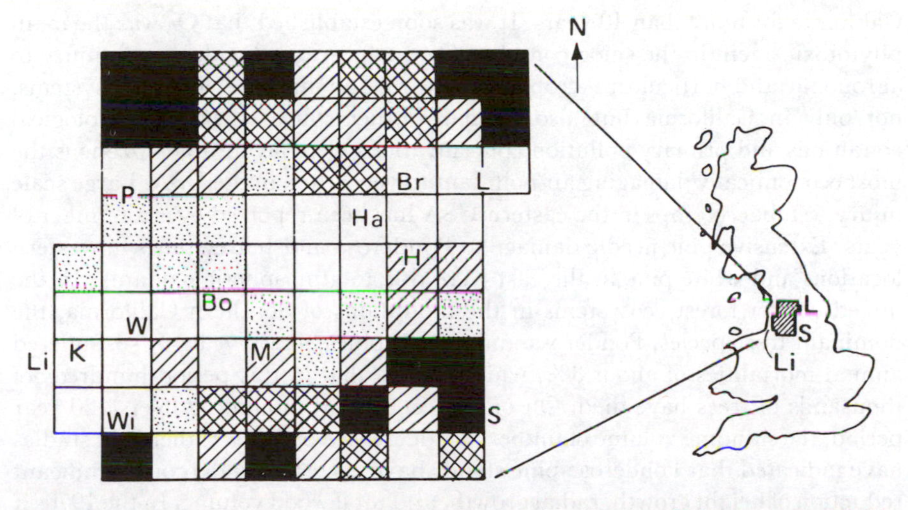

Figure 7 *Distribution of Scots pine in the Southern Pennines. Number of 2 km squares occupied by at least one specimen per 10 km traverse of the 10 km squares shown:* □ 0–0.9; ▒ 1–1.9; ▨ 2–2.9; ▩ 3–3.9; ■ 4–4.9. *The towns marked are: Bo, Bolton; Br, Bradford; H, Huddersfield; Ha, Halifax; K, Kirby; L, Leeds; Li, Liverpool; M, Manchester; P, Preston; S, Sheffield; W, Wigan; Wi, Widnes* (Reproduced with permission from *Environ. Pollut.*, 1977, **14**, 63)

trees occurred wherever SO_2 concentrations exceeded 0.076 p.p.m. Field observations have also established that indigenous grassland in the polluted Rossendale area of Lancashire grew better than imported cultivars. In recent years, a general decline of sulfur dioxide concentrations in the area has coincided with a marked improvement in conifer trials.[28] General improvement in urban air quality can be seen in a number of areas of the UK from the gradual recolonization by lichens on the bark of trees, including parts of London.[29]

Growth chamber experiments which subject plants to unfiltered (ambient) air and filtered (control) air provide strong supportive evidence of air pollution damage to vegetation. One of the earliest studies (1950–51), examined the effect of pollutants in the air of a Manchester suburb on the growth of perennial ryegrass, *Lolium perenne*. Significant reductions in growth were found in the grass exposed to the ambient air which had a mean SO_2 concentration of 0.07 p.p.m. This study occurred prior to any improvement in urban air quality. More recent investigations in the Sheffield area found reduced yields in plants exposed to much lower mean SO_2 concentrations. In other studies using a more realistic open-top design of chamber, growth reductions only occurred at higher SO_2 exposures. The differences may be due to the different design of the chambers, although other factors may be involved. For instance, chronic SO_2 injury can be enhanced if plants are exposed to SO_2 during winter, when they are growing more slowly. Furthermore in these studies, only SO_2 was measured and at least in the more recent cases, it is now evident that other phytotoxic pollutants would also have been present, *e.g.* NO_x and O_3. The Sheffield studies coincided with some of the highest O_3 concentrations yet recorded in Britain.[28]

Vegetation injury caused by photochemical smog was first reported in the Los Angeles basin in 1944 and it has continued to be a chronic problem in southern California for more than 40 years. It was soon established that O_3 was the main phytotoxic agent in the smog complex. Ozone has caused widespread injury to agronomic and horticultural crops and natural and managed forest ecosystems, not only in California but also in many other states where meteorological conditions and primary pollution concentrations were favourable. Ozone is the most economically damaging air pollutant to vegetation in the USA. Large scale injury to tobacco crops in the eastern USA has been reported over a number of years. Extensive pine needle damage to Ponderosa and Jeffrey pines in western locations and white pine in the east were due to atmospheric oxidants. In the mixed conifer forest ecosystems in the mountains of Southern California, the dominant tree species, Ponderosa and Jeffrey pines, for 30 years or so, suffered annual mortalities of about 3%, which means that in this period hundreds of thousands of trees have died. On one selected plot, monitored over a 20 year period, the standing volume of timber had decreased by 28%. Other field studies have indicated that Ponderosa pine stands have generally undergone significant reduction of height growth, radial growth, and total wood volume. In the 1970s it

[28] J. N. B. Bell, 'Gaseous Air Pollutants and Plant Metabolism', eds. M. J. Koziol and F. R. Whatley, Butterworth, 1984, Chpt. 1, p. 3.
[29] C. I. Rose and D. L. Hawksworth, *Nature (London)*, 1981, **289**, 289.

was estimated that in the San Bernardino forest 46 230 acres had suffered severe ozone-type injury, 53 920 acres moderate injury, and 60 800 acres light or no injury.[30]

Evidence that ambient ozone concentrations in Britain during summer can inflict deleterious effects on vegetation comes from a series of experiments, using open-top chambers, at a rural site near Ascot in Berkshire, some 32 km west of London. At this location SO_2 and NO_x concentrations are generally low, but episodes of high ozone concentrations have been regularly recorded. During three such incidents between 1978 and 1983, *Pisum sativum*, *Trifolium repens*, and *T. pratense*, grown in open-top chambers and receiving unfiltered air, developed visible leaf necrosis typical of O_3 damage. Concentrations of O_3 during the course of these experiments exceeded 0.1 p.p.m., which is considered to be the threshold above which visible leaf necrosis appear in these sensitive plants. Field observations in the surrounding areas revealed symptoms typical of O_3 injury in a variety of crop plants. It has since been realized that *Pisum sativum* crops grown in the UK for a number of years, frequently developed necrotic lesions typical of ozone injury without the cause being known.[31]

In recent years there has been considerable concern over the decline in forest health in western Europe. The problem first came to light in the late 1970s and early 1980s in West Germany when Norway spruce (*Picea abies*), the dominant silviculture tree in mid-Europe, began to show increasing signs of defoliation and needle discolouration. Extensive death of the fine root system has also been reported. These symptoms have since been observed in the UK, France, the Benelux countries, Scandinavia, the Alps (Switzerland and Austria), Czechoslovakia, and Poland. The decline in forest health is not confined to Norway spruce, other conifer and broad leaved species have been affected. It has been argued that whilst regional declines of individual species can occur, for example due to disease epidemics, the synchronous decline of several species implicates a common cause such as air pollution. In 1987 and 1988 a survey of the severity of damage across the European Community (EC) showed that the forests of southern Europe are in better health than those of central and northern Europe. In the major coniferous forests of northern Europe, 15%, 22%, and 28% of *Pices abies*, *Pices sitchensis*, and *Abies alba* trees, respectively, are unhealthy. Among the hardwoods of northern and central Europe, 16%, 15%, and 12% of *Quercus robur*, *Quercus petrea*, and *Fagus sylvatica*, respectively, show symptoms of damage.[32]

Over the past decade, thousands of papers and reports have been published throughout Europe on the subject of forest decline. However, definitive cause–

[30] P. R. Miller, and J. R. McBride, 'Air Pollution and Forest Decline', eds. J. B. Bucher and I. Bucher-Wallin, 1989, p. 61.

[31] M. R. Ashmore, Proceedings of an International Workshop on 'The Evaluation and Assessments of the Effects of Photochemical Oxidants on Human Health, Agricultural Crops, Forestry, Materials, and Visibility', Swedish Environmental Research Institute (IVL), Goteborg, 1984, p. 92.

[32] M. R. Ashmore, J. N. B. Bell, and I. J. Brown (eds.), 'Air Pollution and Forest Ecosystems in the European Community', Air Pollution Research Report 29, Commission of the European Communities, 1990.

effect relationships of forest decline have yet to be established. In the early stages, forest damage was attributed to acid deposition, through its effect on leaching nutrients from foliage, loss of nutrients from soils and increasing bioavailable forms of toxic metals such as aluminium in soils. It is now believed that the actual explanation probably involves a combination of factors which include natural and pollution stresses.

It is difficult to reconcile pollution patterns with the distribution of forest decline on a regional scale such as Europe. The distribution of air pollution which is typically a complex mixture of gases (SO_2, NO_x, O_3), acidic aerosols, and ammonium sulfate varies markedly from region to region and over short distances within regions. Equally forest condition varies enormously across the region and between the major forest species. Recent field observations on Norway Spruce have shown that damage is associated with areas of Europe where air pollution is greatest, in particular deposition patterns of S and N pollutants and summer concentrations of O_3.[33]

Forest decline is frequently associated with foliar deficiency of certain nutrients, and in particular magnesium deficiency, *e.g.* at elevated sites in West Germany, although in others areas potassium deficiency has been found. However, the cause of the nutrient deficiency is unknown. There are reports of increased foliar leaching of mineral nutrients due to acidic deposition or other pollutants but this is thought to be of secondary importance and the supply of mineral nutrients from the soil is thought to be more critical. There is growing evidence that the deposition of sulfur and nitrogen pollutants has significantly modified soil chemistry and plant nutrition. Increased nitrates and sulfates in the soil solution can increase the leaching of cations such as calcium and magnesium from soils and enhance soil acidification. Soil solution chemistry, notably decreases in calcium and/or magnesium to aluminium ratio, is known to affect root development and hence water and nutrient uptake.[32,33]

In addition to effects on photosynthesis and stomatal conductance, there is also evidence that certain gaseous air pollutants acting singly and in combination (SO_2, SO_2/NO_2, and O_3) have an effect on the relative distribution of growth above and below ground, with root growth being more severely affected than shoot growth. This may be a consequence of interference with phloem translocation and assimilation in the plant. Other effects of air pollutants include damage to epicuticular wax layer of leaves and needles which can lead to increased water loss and also there are reports that air pollutants can delay the onset of winter hardening. It is well known that plants in polluted areas are more susceptible to attack by insect pests and it has been shown in experiments that exposure of plants to small or medium doses of SO_2 and/or NO_2 results in increases of the population growth of aphids feeding on the plants. This in turn can result in significant increases in pest damage to plants.[6,32]

Therefore the cause of forest decline is likely to involve a number of factors,

[33] E.-D. Schulze, and P. H. Freer-Smith, 'Acid Deposition—Its Nature and Impacts', eds. F. T. Last and B. Watling, Proc. Roy. Soc. Edinburgh, 1991, Section B, **97**, p. 155.

including air pollution and environmental factors interacting in some way, with one set predisposing forest stands to attack and damage from another set.

9 ECOLOGICAL EFFECTS OF ACID DEPOSITION

In recent decades, numerous streams, rivers, and lakes in various regions of western Europe and north-eastern North America have become progressively more acidic. It is now generally accepted that acid deposition with major acidic, or acidifying ions is the cause of freshwater acidification in geologically sensitive areas throughout western Europe and north-eastern North America. Geologically sensitive areas are those with slowly weathering granites and gneiss rocks that support acidic and weakly buffered soils. Freshwater acidification is typified by a loss of acid neutralizing capacity (essentially the carbonate buffering system), decrease in pH by as much as 1–2 pH units, increases in sulfate, nitrates, ammonium ions, aqueous aluminium, and metals such as manganese and zinc. Acidification is responsible for the loss and depletion of fish populations from numerous freshwater ecosystems in parts of Norway, Sweden, UK, Canada, and USA.

Freshwater organisms generally maintain their internal salt concentration by active uptake of ions (in particular Na^+ and Cl^-) from water against a concentration gradient. Substantial experimental evidence and field data indicate that fish mortality at low environmental pHs is primarily caused by a failure to regulate internal salt concentrations at gill surfaces. A characteristic symptom of acid-stressed fish is therefore a reduction in the salt or electrolytic ion concentration of blood plasma.[34]

At an early stage it was realized that toxicity is not simply a function of acidity. Field data indicated that fish mortality occurred in waters of pH 5, whereas laboratory experiments with purely acid solutions failed to reproduce these results and significant mortalities only occurred at pH 4. It has since been shown that an important influence on the toxicity of acidified rivers and lakes is the dissolved concentration and chemical species of aluminium. Aluminium is more soluble in acidic environments and various surveys of lakes and rivers have revealed that concentrations of aqueous aluminium increases in acidified waters. Because of reactions with hydroxide radicals, aluminium toxicity to fish is pH dependent and the most bioavailable forms of aluminium occur at pH 5. Several reports have shown the effects of the interaction of acidity and soluble aluminium on fish survival. The survival of brown trout fry is reduced in concentrations of aluminium of 250 $\mu g\ l^{-1}$ and greater, especially if calcium is low. More significantly, sub-lethal effects have been found at much lower concentrations, for example Figure 8a shows that the growth rate of brown trout is reduced at a concentration greater than 20 $\mu g\ l^{-1}$, a concentration more than an order of magnitude lower than the total aluminium typical of any acid waters. Moreover the most toxic combinations were found with elevated aluminium concentrations at pH 5.

[34] H. Leivestad and I. P. Muniz, *Nature (London)*, 1976, **259**, 391.

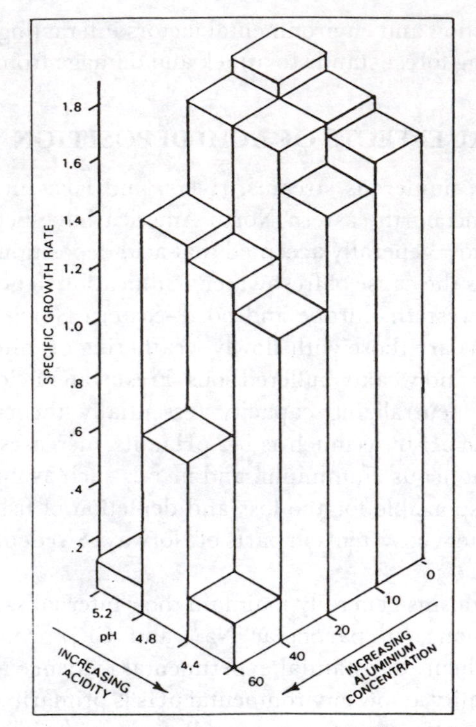

Figure 8a *Growth rate of yearling brown trout in a range of pHs and aluminium concentrations*

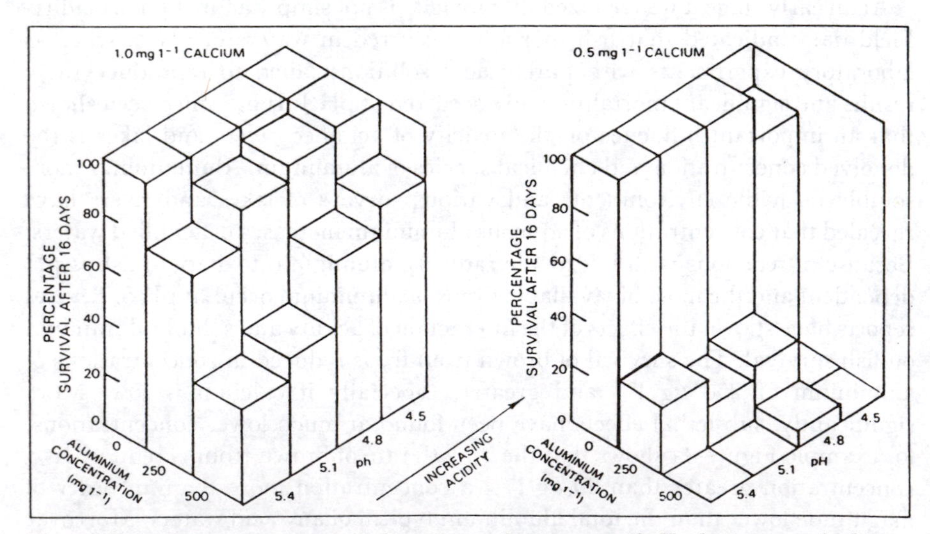

Figure 8b *The percentage survival of brown trout fry in a range of pH, aluminium, and calcium concentrations*

(Reproduced with permission from D. J. A. Brown, 'Acid Deposition, Sources, Effects, and Controls,' ed. J. W. S. Longhurst, British Library Technical Communication, 1989, p. 107.)

Figure 8b demonstrates that calcium has a strong influence on the toxicity of acid waters; increasing the calcium concentration from 0.5 to 1.0 mg l^{-1} offset the effect of increasing aluminium and acidity on the survival of brown trout. Sulfates, fluorides, and soluble organic matter in acidified waters which form complexes with aluminium can have a significant ameliorating influence on aluminium toxicity.[35]

In summary, loss of fish populations can be expected in acidified clear water lake ecosystems, low in organic matter and low in calcium. Such environments tend to be characteristic of mountain areas and in southern Norway in particular these are the regions where the loss of fish populations was first recorded.

The effects of acidification are not restricted to fish, all facets of freshwater ecosystems are affected. The assemblages of all the major groups, micro-organisms, plants, and animals, alter considerably and there is a general tendency for a reduction in species diversity at all trophic levels. For many groups this is accompanied by a reduction in productivity and a loss or decline of populations, *e.g.* several species of crustacea (snails and crayfish), amphibians, and fish.

The following summary of the effects of acidification on the major trophic levels in freshwater systems is based on a recent review of the subject by Muniz:[36]

Decomposers. Accumulation of undecomposed and partly decomposed organic matter is taken to indicate reduced activity of bacteria, fungi, and protozoans in acidified waters. Some field and experimental studies have shown decreased rates of decomposition, but others have found no evidence of any changes.

Primary Producers. Reductions in the numbers of species of several phytoplanktonic algae, especially the green algae (Chlorophyceae) have been reported. Dinoflagellates dominate the plankton of many acid lakes. Fewer species of periphyton occur on submerged surfaces and these habitats are commonly superseded by mass encroachments of filamentous algae (*e.g. Mougeotia*, *Zygogonium*, and *Spirogyra*). In many lakes in Sweden macrophyte populations of *Lobelia* have declined and have been invaded by *Sphagnum* species. Synoptic surveys generally indicate that biomass and primary productivity are less in acidified lakes, although experimentally acidified lakes have shown no change and even increases in biomass.

Zooplankton. Species diversity decreases with increasing acidity, particularly below pH 5.5–5.0 and this is usually followed by a reduction in biomass. Acid sensitive species of *Daphnia* and *Cyclops* frequently disappear below pH 5.

Benthic macro-invertebrates. Many species of mayflies, amphipods (freshwater shrimps), crayfish, snails, and clams are very sensitive to acid conditions. In Norway, snails were found to be absent in waters below pH 5.2 while mussels disappeared at pH 4.7. Recruitment failure and reduced growth are important causes of the elimination of species. Crayfish decline has been

[35] D. J. A. Brown, 'Acid Deposition, Sources, Effects, and Controls', ed. J. W. S. Longhurst, British Library Technical Communication, 1989, p. 107.

[36] I. P. Muniz, 'Acidic Deposition—Its Nature and Impacts', eds. F. T. Last and R. Watling, The Royal Society of Edinburgh, 1991, Section B, **97**, p. 228.

due to the effects of acidity on the eggs and juvenile stages, and the moulting stages are particularly sensitive. A similar situation has been found for freshwater shrimps, mayflies, and snails.

Fish. There are numerous examples of the loss of individual species and changes in the composition of fish communities in acidified waters. Minnows seem to be some of the most sensitive fish and population declines often start below pH 6. An approximate order of sensitivity to acid stress is as follows (beginning with most sensitive): roach, minnow, Arctic char, trout, European cisco, perch, and pike, and eels are consistently the most tolerant species.

The younger life-stages, particularly newly hatched or 'swim-up' fry just starting to feed are the most sensitive stages to acid stress. As a consequence many acid-stressed populations are dominated by older fish due to recruitment failure.

Amphibians. Reproductive failure is the main effect of acid-stress on amphibians. In Sweden populations of the frog, *Rana temporaria* have been lost and reproductive failure of the toad, *Bufo bufo* has occurred.

Birds. Studies have shown that the breeding density of Dippers in Wales (songbirds that feed almost exclusively on aquatic invertebrates) has decreased as acidity has increased. This has been related to the decrease in abundance of mayflies and caddis fly larvae in acidified streams. In general, some species of song-birds, that breed alongside acidified lakes and streams produce smaller clutches, and thinner eggshells, breed later with fewer second clutches. Also evidence indicates that their young grow less rapidly and suffer more mortality. Although the loss or decline of populations of animals and plants is largely attributable to changes in water chemistry, indirect effects may also cause additional stress. These essentially arise out of changes in predator/prey relationships caused by a decline in food resources. For example, instances of increases in phytoplankton biomass may be due to a decrease in the performance of herbivores. The decline of the dipper is strongly associated with the disappearance of its prey from acidified streams.[36]

Experimental acidification of entire lakes has yielded very valuable information and increased understanding of the effects of acidification on communities and ecosystem function. Lake 223, in the Experimental Lake Area of Canada, was acidified over an 8 year period during which time the pH was gradually reduced from 6.8 to 5.0. One of the surprising features of the study was that primary production and decomposition showed no overall reduction. However, the composition of the phytoplankton and zooplankton communities showed distinct changes. At pH 5.9 several key organisms in the lake's food web were severely affected; the opossum shrimps (*Mysis relicta*) declined from almost 7 billion to only a few animals and fathead minnows (*Pimephales promelas*) failed to reproduce. At pH 5.6 the exoskeleton of crayfish (*Oronectes virilis*) after moulting hardened more slowly and remained softer. The animals later became infested with a microsporozoan parasite.

When the acidification of Lake 223 was reversed, several biotic components recovered quickly. Fish resumed reproduction at pHs similar to those at which it failed on acidification. The condition of lake trout improved as the small fish on which they depended for food returned. Many species of insects and crustaceans returned as the pH was raised.[13,37]

10 ECOLOGICAL AND HEALTH EFFECTS OF METAL POLLUTION

10.1 Episodes of Poisoning in Humans

Three forms of mercury are important in the environmental cycles of the element, they are elemental mercury (Hg^0), divalent mercury (Hg^{2+}), and methylmercury (CH_3Hg^+). Methylmercury is a particularly toxic species of mercury and it is easily absorbed across the external membranes of animals.

Several major episodes of mercury poisoning in the general population have been caused by consumption of methyl- and ethyl-mercury compounds. Methylmercury is a neurotoxin and the main clinical symptoms of poisoning reflect damage to the nervous system. The sensory, visual, and auditory functions, together with those of the brain areas, especially the cerebellum, concerned with co-ordination, are the most common functions to be affected. Symptoms of poisoning progressively increase in severity in line with increased exposure, as follows: (1) initial effects are non-specific symptoms which include paraesthesia, malaise, and blurred vision; (2) in more severe cases, concentric constriction of the visual field, ataxia, dysarthria, and deafness appear more frequently; and (3) in the worst affected cases, patients may go into a coma and die. The effects in severe cases are irreversible due to destruction of neuronal cells. There is a latent period, usually of several months between the onset of exposure and the development of symptoms.[5,38]

10.1.1 The Iraqi outbreak. This epidemic of methylmercury poisoning occurred in agricultural communities in Iraq in the winter of 1971–72. Over 6000 people were admitted to hospitals in provinces throughout the country and over 400 people died in hospital with methylmercury poisoning. The poisonings stemmed from mis-use of imported high-grade seed grain treated with alkylmercury fungicide. The imported seed was intended for sowing but in many areas the grain was ground directly into flour and used in the daily baking of homemade bread. Depending on the number of loaves consumed, individual exposure ranged from a low non-toxic intake to a prolonged intake over 1–2 months. Because of the brief exposure period, epidemiological investigations began 2–3 months after it had ceased and in most cases after the onset of poisoning. This made calculation of ingested dose, and body burden of mercury at the time of exposure to methylmercury, difficult to estimate.

[37] D. W. Schindler, T. M. Frost, K. H. Mills, P. S. S. Chang, I. J. Davis, L. Findley, D. F. Malley, J. A. Shearer, M. A. Turner, P. J. Garrison, C. J. Watras, K. Webster, J. M. Gunn, P. L. Brezonik, and W. A. Swenson, in 'Acid Deposition—Its Nature and Impacts', eds. F. T. Last and B. Watling, Proc. Roy. Soc. Edinburgh, 1991, Section B, **97**, p. 193.

[38] World Health Organization, Geneva, Environmental Health Criteria, 'Methylmercury', 1990, **101**.

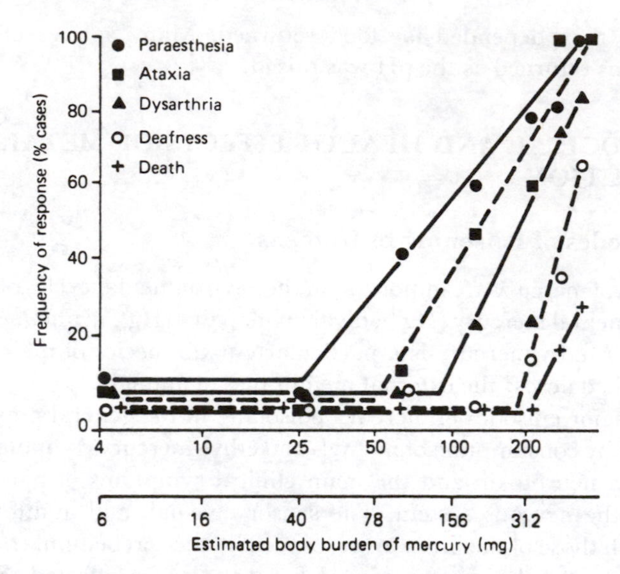

Figure 9 *The relationship between frequency of signs and symptoms of methylmercury poisoning and the estimated body burden of methylmercury (the two scales of the abscissa result from different methods of calculating body burden of methylmercury)* (Reproduced with permission from *Science*, 1973, **181**, 230)

Figure 9 shows both dose–effect and dose–response relationships that have been established between symptoms of poisoning and the estimated body burden at the time of cessation of ingestion of methylmercury in bread. An increased body burden of mercury is associated with an increase in the severity of symptoms experienced by patients. The dose–response curve for each sign or symptom shows the same characteristic shape, a horizontal and a sloped line which is referred to as a 'hockey-stick' line. The horizontal line represents the background or general frequency of each symptom and the sloped line shows that an increased body burden is associated with an increasing frequency of each symptom. The intersection of the two lines has been taken as the 'practical threshold' of the mercury related response and this increases with increasing severity of the effects; for paraesthesia it is at a body burden of about 25 mg Hg, for ataxia it is at 50 mg, for dysarthria it is at about 90 mg, for hearing loss it is at about 180 mg, and for death it is over 200 mg. These relationships do not demonstrate cause and effect and the only proof that methylmercury produced the above effects is that the effects followed a known high exposure to methylmercury, the frequency and severity of these effects increased with increasing exposure to methylmercury, the effects are similar to those seen in other outbreaks of methylmercury poisoning, and the major signs have been reproduced in animal models.

10.1.2 The Minamata and Niigata Outbreaks. In Japan, two major epidemics of methylmercury poisoning have occurred, one in the Minamata Bay area and the other in Niigata. In each case the problem arose as a result of the local population consuming seafood contaminated with methylmercury. Mercury compounds, including methylmercury were released from industrial sources into the aquatic environment and subsequently this resulted in the accumulation of methylmercury in seafood. These outbreaks of poisoning were first discovered during the 1950s and the early 1960s and, by the mid 1970s, about 1000 cases (with 3000 suspected cases) in the Minamata area and over 600 in the Niigata area had been recorded. By and large symptoms of poisoning followed a very similar pattern to the Iraqi epidemic, although an important difference was in the nature of the exposure which was lower but more prolonged. Moreover, the latent period was longer and in some isolated cases it was as long as 10 years between initial exposure and the onset of symptoms. In other cases it was observed that clinical symptoms worsened with time, despite reduced or discontinued exposure. All this would appear to be related to long-term accumulation of mercury in the brain.[5,38]

Because mercury even from a natural sources in fish is predominantly in the methylmercury form, there is concern that certain groups of people who depend largely on a fish diet will be exposed to slightly elevated intakes throughout their lives and hence possibly accumulate mercury to toxic levels. One such group is the Canadian Indians but because of several confounding factors, not least of which are high incidences of malnutrition and alcoholism, the assessment of health risk due to methylmercury is difficult to ascertain. One study, involving 35 000 samples obtained from 350 communities, found that over two-thirds had mercury blood concentrations within normal limits ($<20\ \mu g\ l^{-1}$) but 2.5% (over 900 individuals) had levels in excess of $100\ \mu g\ l^{-1}$ and which therefore could be considered as a group 'at risk' and therefore in need of close surveillance.[5]

Mercury analysis of segments of hair can provide a meaningful index of past exposures and hence body burdens (see Figure 3). As a guideline, blood mercury concentrations of 200–500 $\mu g\ l^{-1}$, hair concentrations of 50–125 $\mu g\ g^{-1}$, and a long-term intake of 3–7 $\mu g\ kg^{-1}$ body weight are likely to be associated with the onset of the initial symptoms of methylmercury poisoning, such as paraesthesia.

Clinical and epidemiological evidence indicates that prenatal life is more sensitive than adult life to the toxic effects of methylmercury. The first indications came from Minamata in the early stages of the outbreak, where it was found that mothers who were only slightly poisoned gave birth to infants with severe cerebral palsy. A similar situation has been reported in the Iraqi outbreak of 1971–72 with infants which had been prenatally exposed showing severe damage to the central nervous system. Using the maximum maternal hair concentration during pregnancy as an index, the lowest level at which severe effects have been observed was $404\ \mu g\ g^{-1}$. More recent follow-up studies with infants in Iraq have found evidence of psychomotor retardation (delayed achievement of development milestones, a history of seizures, abnormal reflexes) at maternal hair levels well below those associated with severe effects.[38]

Lead is a neurotoxin and the overt toxic effects of lead have been known for many centuries. Probably the first reported cases of lead poisoning due to environmental sources was a group of children diagnosed as having lead palsy by clinicians at the Brisbane Hospital in Queensland, Australia at the turn of the century. A total of ten cases of lead poisoning were found by Health Officials and it was later shown that the source of the lead was lead-based paint which was turning to powder on the walls of homes and railings.

Lead poisoning due to exposure to lead-based paint has affected a large number of children over the years. Most of the epidemiological studies have been centred on the United States of America. The disease is confined to children, especially those living in inner city areas in dilapidated buildings with surfaces of flaking and peeling lead-based paint. Children playing in the vicinity can take in particles and flakes of paint by inhalation and hand-to-mouth activities. Certain children have a craving for eating non-food items such as flaking paint. This habit called 'pica' and the more normal hand-to-mouth activities of children are capable of introducing excessive amounts of lead into the body, *e.g.* a square centimetre of paint may contain over one milligram of lead.

The disease in the USA was neglected for a long time and the full extent of the number of children who have suffered from lead poisoning and the number at risk from the disease emerged over several decades. The probable reasons for this time lag include the poor socioeconomic status of the children in the high risk groups and the difficulties in recognizing the disease. In 1971, the disease became fully recognized when the Government of the USA introduced 'The Lead-based Paint Poisoning Prevention Act'.

It is important to emphasize that 40 years or so ago, a child was diagnosed as suffering from lead poisoning only when acute encephalopathy was evident. Typically this was characterized by a progression to intellectual dullness and reduced consciousness and eventually to seizures, coma, and in very severe cases death. Although unknown at the time, acute encephalopathy would only develop when lead concentrations had exceeded 80–100 $\mu g\ dl^{-1}$ of blood. At about 80 $\mu g\ dl^{-1}$ severe but not life-threatening effects on the CNS can be expected. Lead encephalopathy in children is usually accompanied by peripheral neuropathy, especially foot drop, and general weakness. Acute renal damage is common in severe cases. Less severe symptoms associated with lead poisoning which have been recognized in recent years are deficits in neurobehavioural development and effects on the synthesis of haem in the blood. These effects can be expected to occur at blood lead concentrations of 20–30 $\mu g\ dl^{-1}$ and even less (see Table 2).[39,40]

The whole issue of childhood lead poisoning is further complicated by the fact that there are other significant sources of lead in the urban environment, most notable of which are lead from car exhausts and from dusts and soils contaminated with lead from industry, paint, and petrol. Lead exposure and blood lead levels among the general population in Europe and in particular in the USA have declined substantially in recent years. A major part of this reduction has been due to the switch from leaded to unleaded petrol.[41]

[39] H. L. Needleman (ed.), 'Low Level Lead Exposure: The Clinical Implications of Current Research', Raven, New York, 1980.
[40] Royal Commission on Environmental Pollution, 9th Report, Cmnd. 8852, HMSO, London 1983.
[41] EPA, 'Air Quality Criteria for Lead', EPA-600/8-83-028 US EPA, Environmental Criteria and Assessment Office, Research Triangle Park, North Carolina, June 1986.

Table 2 *Lowest Blood Lead Concentrations Association With the Onset of Observable Effects in Children*

Lowest effect blood lead level (μg dl^{-1})	*Neurological effects*	*Haem synthesis effects*	*Other effects*
10–15 (prenatal and postnatal)	Deficits in neurobehavioural development; electrophysiological changes	ALA-D[a] inhibition	Reduced gestational age and at birth; reduced size up to age 7–8 years
15–20		FEB[b] elevation	Impaired vitamin D metabolism; Py-5-N[c] inhibition
<25	Lower IQ, slower reaction time		
30	Slowed nerve conduction velocity		
40		Reduced haemoglobin; elevated CP[d] and ALA-U[e]	
70	Peripheral neuropathies	Overt anaemia	
80–100	Encephalopathy		Colic, other GI[f] effects: kidney effects

[a] Δ-aminolevulinic acid dehydratase; [b] Free erythrocyte prophyrins; [c] Pyridine-5-nucleotide; [d] Coproporphyrin; [e] Δ-aminolevulinic acid excreted in urine; [f] Gastro-intestinal;
Reproduced with permission from J. S. Lin-Fu, 'Low Level Lead Exposure: The Clinical Implications of Current Research,' ed. H. L. Needleman, Raven Press, New York, 1980.

The main focus of attention in recent years has been the effects associated with blood lead concentrations (PbB) less than 40 μg dl^{-1}, these include effects of prenatal and early childhood exposures on physical and neurobehavioural development of children. Behavioural and attentional deficits as rated by teachers (*e.g.* disordered classroom activity, restlessness, easily distracted, not persistent, inability to follow directions, low overall functioning) have been significantly associated with children's tooth and PbB levels. On the basis of these studies, it has been concluded that neurobehavioural deficits may occur at PbB levels at and below 30 μg dl^{-1}.

There is also evidence that low but elevated levels of lead may be associated with effects on some complex cognitive functions including learning, visual-perception skills, and IQ scores. However attempts to attribute subtle deficits in child development to lead to exposure are controversial as many other factors (genetic, nutritional, medical, educational, and parental and social influences) can strongly influence the development of a child. From an assessment of studies in which confounding factors were adequately taken into account, it has been suggested that PbB levels in the range of 50–70 μg dl^{-1} tend to be associated with about a 5 point reduction in IQ. There remains considerable uncertainty about the influence of lead on IQ scores of children with PbB levels below 40 μg dl^{-1}.[41]

Prenatal exposure to lead has been associated with a reduction in the mental development of infants. The effect of environmental exposure to lead on children's abilities at the age of 4 years was studied in a cohort of 537 children born during 1979 and 1982 to women living in the vicinity of a lead smelter at Port Pirie in Australia. The study indicated that elevated PbB concentrations in early childhood had deleterious effect on mental development up to the age of 4 years.[42]

10.2 Effects on Wildlife

In various parts of the world, non-ferrous mining and smelting operations have had a marked impact on surrounding environments. Typically elevated metal concentrations are found in vegetation and soils in areas adjacent to such metal sources, with the highest concentrations occurring nearest the source but decreasing quite sharply away from it. Contamination arises in the main from atmospheric deposition of metal-rich particulates on to vegetation surfaces rather than uptake from soil. One consequence of this is that grazing animals are particularly liable to ingest toxic doses of metals and over the years there have been many reports of such incidents; for example livestock fatalities caused by lead occurred in a lead-mining area of the west of Ireland during the 1960s. In the 200 years leading up to the early 1970s, smelting of various metals took place in the Lower Swansea Valley, Wales, UK. Horse fatalities in the area, at various times in the past have been attributed to metal poisoning. In a more recent incident (1968–71) 112 horses died in the vicinity of the one remaining lead–zinc smelter (which closed in 1971) and lead poisoning was confirmed in about half of these and strongly suspected in the remainder. Although one victim analysed in a separate study, also had high levels of cadmium in its kidneys.[43]

In terrestrial ecosystems affected by elevated metal deposition, the highest accumulations of metals are found in the surface layers and in the overlying litter layers of soils. The metals remain in these upper layers of the soil compartment for some considerable time and are therefore a continued source of metals to the soil communities. The upper soil horizons are an important focus for a number of

[42] A. J. McMichael, P. A. Baghurst, N. R. Wigg, G. V. Vimpani, E. F. Robertson, and R. J. Roberts, *New Eng. J. Med.*, 1988, **319**, 468.

[43] Welsh Office, 'Report of a Collaborative Study on Certain Elements in Air, Soil, Plants, Animals, and Humans in the Swansea/Neath/Port Talbot Area, Together With a Report on a Moss Bag Study of Atmospheric Pollution Across South Wales', Welsh Office, 1975.

soil biological processes concerned with nutrient cycling and ecosystem productivity. A number of studies have shown that in metal contaminated soils in the vicinity of major metal smelters, the thickness of the leaf litter layer has actually increased as a result of a reduced rate of litter decomposition. The densities of soil and litter arthropods, particularly mites, and microbial activity have been shown to be much reduced in metal contaminated soils. A brassworks in a small town in south-east Sweden has been the subject of a long-term study of the effects of metals on the surrounding coniferous forests. The industry emits almost exclusively zinc and copper and because there is no primary smelter emissions of acidic compounds like sulfur dioxide are insignificant. At metal contaminated locations near to the works the worst effects were observed in the organic topsoil layers; undecomposed litter had accumulated in thick layers, soil micro-organisms and animals were reduced in both numbers and species composition, and soil biochemical processes were much reduced.[44]

Feral pigeons (*Columba livia*) occur in large numbers in many urban environments. They are a gregarious type of bird which occupy fairly specific territories. Populations in a city environment may be expected to experience a range of different lead exposures in common with changes in traffic density which tend to be highest in central areas. Tissue lead concentrations in three populations from London were found to follow closely the magnitude of environmental lead contamination, with the highest levels in the population nearest the city centre. At the same time, several symptoms of lead intoxication were detected, especially in individuals with the largest lead burdens. These included increased kidney weight, presence of renal intranuclear inclusion bodies, altered mitochondrial structure and function, and suppression of ALA-D activity in blood, liver, and kidney.

Incidents of lead poisoning have been recorded for many years in waterfowl which have ingested spent lead gunshot pellets. These birds need to take in coarse grit and sand to aid break-up of ingested plant material in their gizzards and as a result lead shot may also be ingested. Estimates indicate that in the USA 3000 tonnes of spent gunshot enter the environment each year, which accounts for about two million duck deaths. Fatalities due to this cause have been recorded in many countries and in the UK 500 mallards were fatally poisoned each year, although this is small compared with the over-wintering mallard population and the direct losses from wildfowling.

In the UK ingestion of lead fishing weights has been the cause of fatalities in populations of the mute swan (*Cygnus olor*) on stretches of the rivers Thames, Avon, and Trent. Post-mortem examination of swans collected from these rivers in the 1970s and 1980s revealed that over 50% had died from lead poisoning. It was not unusual to find an average of about 7–9 lead weights in a swan's gizzard and a high proportion of all birds examined had at least one lead weight in their gizzard and intestinal tract.

Background blood lead concentrations in swans are generally in the order of less that 40 $\mu g\ dl^{-1}$, whereas in a survey of 300 blood samples from the Thames

[44] G. Tyler, *Ambio*, 1984, **13**, 18.

63% had PbB levels greater than 100 μg dl^{-1}. In a survey of 37% cygnets from the Thames in 1980, 25% had PbB levels over 100 μg dl^{-1} and by the end of that particular summer 51% of the cygnets had died; this contrasts with a 2% rate of mortality on rivers in the Cotswolds which are free from this form of lead contamination. Swans affected by lead poisoning show a progressive loss of weight and a drooping of the lower part of the neck due to muscle paralysis; this extends to the rhythmic muscular contractions in the oesophagus and so passage of food through the gut is prevented and the oesophagus becomes packed with food. These symptoms are also accompanied by severe liver and kidney damage.[40]

During the mid-1970s at certain locations along the British east coast and the French west coast, the newly introduced Pacific oyster, *Crassostrea gigas*, started to show poor growth and unusual thickening of their shells. In France it was noted that oysters with these symptoms were invariably in the vicinity of yachting marinas and it was later shown that the tissues had a high tin content. In the 1980s it was firmly established that antifouling paints containing tributyl tin (TBT) were responsible for these effects. This was confirmed by laboratory tests which reproduced shell-thickening at the same concentrations of TBT found near to yachting marinas and it was also shown that the tissue concentrations of TBT associated with these effects were the same in field surveys and laboratory experiments.[45]

11 HYDROCARBONS IN THE MARINE ENVIRONMENT

Crude oil is a complex mixture of a great variety of substances with many different physical and chemical properties. The physical and chemical composition of oils of different origins also vary widely. In March 1978, the supertanker *Amoco Cadiz* released most of its 223 000 tonnes cargo of Iranian and Arabian crude oil into the coastal waters of Brittany. The oil was light, it had a low viscosity, and it contained 30–35% of highly toxic aromatic hydrocarbons, 39% saturated hydrocarbons, 24% polar material, and 3% residuals and there were more than 300 compounds in the fresh crude oil. The fate or the weathering of the oil is important, for immediately following a spillage, processes such as evaporation, dissolution, microbial, and photochemical degradation, *etc.* play major roles in altering the nature and composition of the oil. In general terms, these processes primarily affect the lower molecular mass fractions and an ageing or weathered oil is more viscous as it is made up of a greater proportion of higher molecular weight compounds. The more persistent and larger polynuclear aromatic hydrocarbons (PAHs) and their transformation products at this stage assume greater ecological significance. Despite sensitive and modern sophisticated instrumentation only a small spectrum of the compounds in crude oil are routinely determined. Consequently, it is not easy to identify the specific toxic components in oil. The lighter aromatic fractions tend to be the most toxic

[45] M. A. Champ and P. F. Seligman (ed.), 'Tributyl Tin: Environmental Fate and Effects', Elsevier Applied Science, 1990.

components and the high carbon-number aromatics, PAHs, although persistent, are among the least acutely toxic, although many are known to be carcinogenic.[46,47]

A major oil spillage in the marine environment has immediate catastrophic results. However, such events constitute a relatively small percentage of the total quantity of petroleum compounds discharged each year. Chronic inputs from shipping, offshore wells, industry, rivers, and the atmosphere together make a very significant contribution. The fate and effects of these discharges are particularly difficult to monitor and assess, as the more toxic fractions which tend to be more volatile, soluble, and easily degraded in the environment are continually being removed from the oil. In consequence, much of our knowledge and understanding comes from investigations undertaken following a major accident.

Accidents of this nature in the marine environment have had considerable impact on marine birds and the adjoining coastal ecosystems. Bearing in mind that changes in the composition of the oil occur fairly rapidly, and much of the acutely toxic components disappear in a matter of days. Deleterious effects are evident at sea, but because of the added influence of dispersal, they tend to be short-lived covering a time course of days or at most weeks. The effect of oil reaching the coastal environment depends on the distance of travel and therefore the time taken for the oil to reach the coast. In this respect it is interesting to compare the *Torrey Canyon* incident (a spillage of 100 000 tonnes some 200 hundred miles off the coast of Cornwall) with that of the *Amoco Cadiz*. By the time that oil was stranded on the Cornish beaches from the *Torrey Canyon* it was almost biologically inert and the damage that ensued was largely caused by dispersants used to clean the beaches. In the case of the *Amoco Cadiz* the spillage was only 1.5 nautical miles from the shore and a major part of the more toxic aromatic fraction was still present when the oil reached the coast.

Oil washed onshore interacts with a variety of coastal features, extending from high-energy eroding rocky promontories to low-energy accumulating environments. The processes that degrade oil in the open water operate in the coastal environment but to differing degrees. In high-energy and medium-energy coastal systems the hydraulic action of breaking waves mechanically disperse and erodes the oil rapidly. Consequently, oil persistence is less of a problem for the intertidal communities in these environments. By contrast, oil may persist for many years in the low-energy environments that include lagoons, estuaries, and marshes, since in these areas wave energy is low and degradation is dependent on the slower processes of microbial degradation and dissolution. With time, attention shifts from the immediate short-term acute effects of oil and focuses on the more persistent long-term effects of PAHs in the low-energy coastal systems.

About 360 km of the Brittany coastal environment was polluted by oil following the wrecking of the *Amoco Cadiz*. This included rocky and sandy shores, salt marshes, and estuaries. During the first weeks after the disaster a very

[46] Royal Commission on Environmental Pollution, 8th Report, Cmnd. 8358, HMSO, London 1981.
[47] R. B. Clark, *Philos. Trans. R. Soc London, Ser. B*, 1982, **297**.

heavy mortality or 'acute mortality crisis' affected the intertidal and subtidal fauna. Populations of bivalves, periwinkles, limpets, peracarid crustaceans, heart urchins, and seabirds were most severely affected. Populations of polychaete worms, large crustaceans, and coastal fishes were less affected. Highest mortalities were found in a 5 km radius around the wreck and in locations further afield where the oil was blown and accumulated ashore by the wind. At one of these locations (St. Efflam) for example, mortalities included 10^6 heart urchins (*Echinocardium cordatum*), 7.5×10^6 cockles (*Cardium edule*), and 7×10^6 other bivalves (Solenidae, Mactridae, Veneridae). Within a radius of 10 km of the wreck about 10^4 fish were found dead, mainly wrasses (Labridae), sand eels (*Ammodytes* sp.), and pipe fishes (*Syngnathidae*), as well as large crustaceans (*Cancer crangon, Leander seratus*). There was no evidence of a severe impact of oil pollution on any intertidal or subtidal species of algae. Delayed effects on mortality, growth, and recruitment were still observed up to three years after the spill. For example, estuarine flat fish and mullets exhibited reduced growth, fecundity, and recruitment, and many were affected by varying degrees of fish-rot disease. Populations of clams and nematodes in the meiofauna declined one year after the spill. Weathered oil remained for a number of years in low-energy areas although the biological consequences of this were unknown.

Initially, recolonization favours those species with a high fecundity, a short life cycle, and/or planktonic stage in their life cycle. A fauna of cirratulid and capitellid polychaete worms became dominant in sandy to muddy areas in the years following the spill. Recruitment of clam populations were very unstable and on the basis that full recovery takes several generations, species of such organisms as clams and fish with life expectancies of 5–10 years, would not be expected to attain stable populations for up to 30 years, and even longer for seabirds.

Seabird mortality due to oil pollution generates considerable public outcry and such incidents are mainly associated with major or acute oil pollution episodes. In addition, various surveys indicate that large numbers of seabirds are killed by oil in the North West Atlantic throughout the winter, peaking in January–March. These winter deaths are associated more with the ongoing and diffuse chronic oil pollution problem, although clearly various stresses due to the weather are likely to influence mortalities. There has been considerable concern that local populations of certain species are particularly at risk from oil pollution. However, of the many species of seabirds relatively few, notably the auks (Razorbills, Guillemots, Puffins, *etc.*) and the diving sea ducks (Eiders, Scoters, *etc.*) have suffered severe mortalities. These species are particularly susceptible to oil pollution since they spend almost their entire lives at sea, collect their food by diving, and, in most cases, have a low breeding rate. Furthermore, these birds are highly gregarious, particularly in their breeding and wintering areas, so a localized spillage can inflict large numbers of casualties.

Several reasons have been suggested for the high mortality rates among oiled birds. Oiling of feathers causes dissolution of the natural oils which can lead to loss of waterproofing and, in turn, loss of both control over body temperature and buoyancy and death by drowning is a common outcome. Preening with subsequent ingestion of oil may produce a number of pathological conditions, *e.g.*

fatty degeneration of the liver, toxic nephrosis, enlargement of the spleen. There is also some evidence that ingestion of crude oil can disrupt the intestinal absorptive mechanisms, impairing water, and ion absorption.

About 10 000 dead oiled birds were picked up after the *Torrey Canyon* incident in 1967 and about 4500 after the *Amoco Cadiz* ran aground. It is generally accepted that these figures represent only a portion of the total killed; for example, it has been suggested that mortalities due to the *Amoco Cadiz* incident were probably in the region of 15 000–20 000. For western European beaches as a whole in the winter of 1980/81, which at the time was one of the worst for seabird mortality, about 60 000 dead oiled birds were recorded. During the period 1971–79, it was estimated that about 3000–16 000 birds died each year from oil pollution in British waters. However, the number of birds killed annually by oil (tens of thousands per annum) is small compared with losses due to natural causes (hundreds of thousands). Furthermore, monitoring of breeding seabird colonies indicates that, in general, populations of all species are increasing although, not surprisingly, some local areas may exhibit decreases.[46,47]

A summary of the major biological effects in the aftermath of several major spills in the 1960s and 1970s reveals a broad range of longer-term effects which extend from abnormalities in development and recruitment disorders to large-scale community perturbations. Although it is generally believed that most populations and communities do recover following the disappearance of the oil.[47]

The impact of two contrasting refinery–effluent discharges on a rocky-shore (Milford Haven, Wales) and a saltmarsh/seabed community (Southampton Water, England) have been followed for a number of years. Monitoring has continued at both locations more or less since the commissioning of the refineries. In 1974, in the vicinity of the effluent of the refinery at Milford Haven, the density of barnacle and limpet populations (*Petella vulgata*) was considerably reduced. Laboratory and field experiments showed that this was due to inhibition of larval settlement. When the refinery at Milford Haven closed in 1983, the barnacle and limpet (*Patella vulgata*) populations over the next few years showed a steady recovery. The saltmarsh communities in Southampton water which were extensively damaged during the 1950s and 1960s have shown a gradual recovery over a number of years, in response to marked improvement in the quality of the effluent.[48]

The Royal Commission on Environmental Pollution in 1981,[46] further endorsed in the following year by the Royal Society,[47] reached the conclusion that oil spills are unlikely to cause long-lasting damage to the marine environment and that oil pollution generally does not constitute a chronic threat to either marine ecosystems or indirectly to man. Nevertheless, short-term effects of oil spills, though local in character, can be serious in terms of damage to beaches and amenity, damage to fisheries, and killing of seabirds. That being said, it is still the case that we know surprisingly little about the effects of oil on natural populations

[48] B. Dicks (ed.), 'Ecological Impacts of the Oil Industry', Institute of Petroleum, J. Wiley and Sons, 1989.

and communities.[49] Of particular concern are potentially vulnerable ecosystems of the polar and tropical regions. Jackson *et al.*[50] investigated the effects of spillage of over 8 million litres of crude oil into a complex region of mangroves, seagrasses, and coral reefs just east of the Caribbean entrance to the Panama Canal. At the time it was the largest recorded spillage into coastal habitats in the tropical Americas. The study is significant because many of the habitats damaged by the oil had been studied since 1968, following an earlier oil spill in the region and also because observations of the effects of the spill began as the oil was coming ashore. This is shown in Figure 10 where the per cent cover of the dominant epibiota on the roots of mangroves in three different habitats are compared before and after they were affected by oil and also for habitats which remained free of oil. In each of the oiled habitats the cover of the major groups was greatly reduced. In the oiled habitats most of the roots sampled for epibiota were dead, broken, or rotting and so the habitat will not be restored until new trees grow. Seagrass, intertidal reef flat, and subtidal reef habitats were similarly damaged. Of particular note was the extensive mortality of subtidal corals and infauna of seagrasses as this contradicts the view that these habitats are not affected by oil spills. Also, sublethal effects of the corals including bleaching or swelling of tissues, conspicuous production of mucus, and dead areas devoid of coral tissue may affect the long-term well-being of the habitat.

12 ECOLOGICAL IMPACT OF ORGANOCHLORINE COMPOUNDS

A wealth of literature exists on the relationship between DDT (and its metabolite DDE) and the phenomenon of eggshell thinning in various bird populations.[7,51] It was first discovered in the peregrine falcon (*Falco peregrinus*) in the UK. This particular falcon is widespread throughout Eurasia and North America, feeding almost entirely on live birds caught in flight. From about 1955 onwards, for no obvious reason, numbers of falcons rapidly declined in southern England and subsequently this decline spread northwards to parts of the Scottish highlands. By 1962, in the UK, some 51% of all known pre-war territories had been deserted and this figure was as high as 93% for southern England. At the time biologists in several countries were also investigating populations of birds that had slowly declined to critical levels. It emerged that the species all had residues of DDT, its metabolites, other chlorinated hydrocarbon insecticides, polychlorinated biphenyls (PCBs), and other chemicals in their tissues.

In the quest to understand the relationships involved in population declines of peregrine falcons and sparrow hawks in the UK, Ratcliffe[52] began to examine

[49] National Research Council, 'Oil in the Sea: Inputs, Fates and Effects', National Academy Press, Washington, DC, 1985.

[50] J. B. C. Jackson, J. D. Cubit, B. D. Keller, V. Batista, K. Burns, H. M. Caffey, R. L. Caldwell, S. D. Garrity, C. D. Getter, C. Conzalez, H. M. Guzman, K. W. Kaufman, A. H. Knap, S. C. Levings, M. J. Marshall, R. Steger, R. C. Thompson, and E. Weil, *Science*, 1989, **243**, 37.

[51] T. J. Peterle, 'Wildlife Toxicology', Van Nostrand Reinhold, New York, 1991.

[52] D. A. Ratcliffe, *J. Appl. Ecol.*, 1970, **7**, 67.

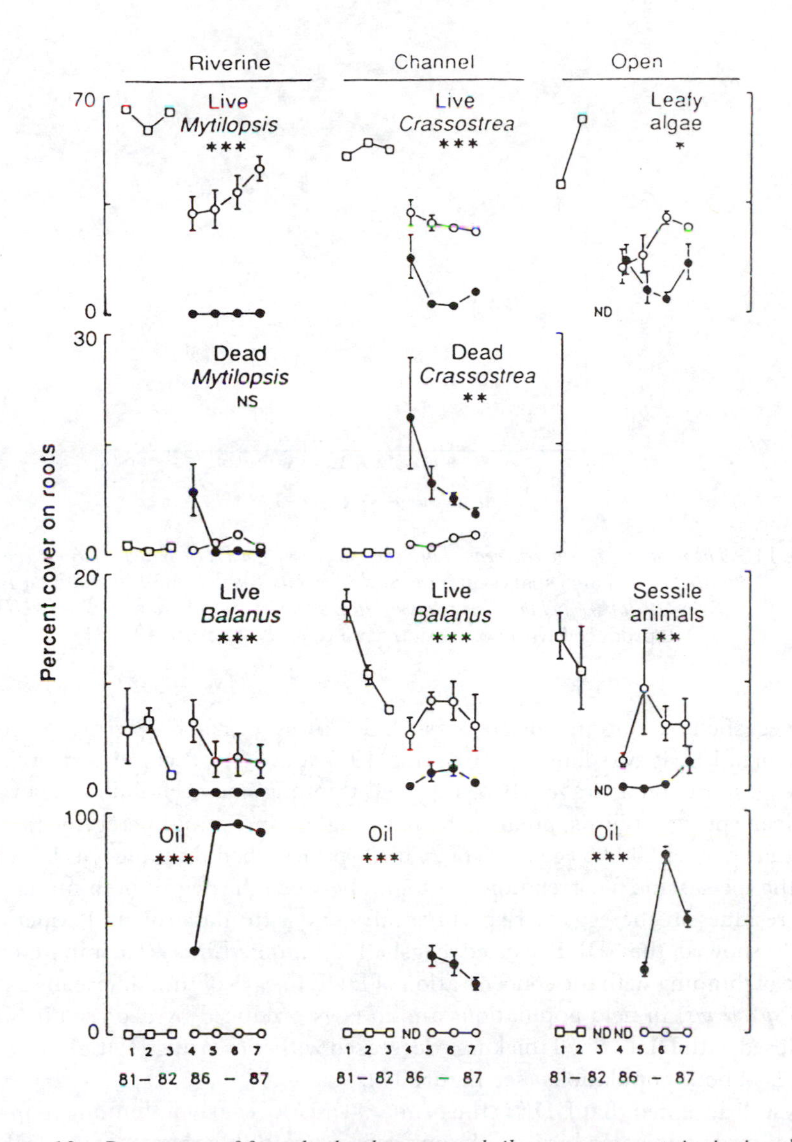

Figure 10 *Percent cover of formerly abundant taxa and oil on mangrove roots in riverine, channel, and open coast habitats before and after the oil spill.* □, *Prespill data:* ○, *unoiled sites;* and ●, *oiled sites. Means are plotted* ±1 SE, *converted from arcsine transformations. Some error bars lie within the plotting symbols, except for sampling time* 3, *when* n = 1. *ND, no data. 'Leafy algae' include* Polysiphonia, Acanthophora, *and* Ceramium *as common genera; the most abundant 'sessile animals' include hydroids and sponges. Sampling dates:* 1 = *September to October 1981*; 2 = *January 1982*; 3 = *June 1982*; 4 = *July to August 1986*; 5 = *October to November 1986*; 6 = *February 1987*; and 7 = *May 1987. Results of repeated measures analysis of variance for oiled and unoiled sites after the spill are shown on each graph, NS,* $P > 0.05$; *, $P < 0.05$; **, $P < 0.01$; ***, $P < 0.001$ Reproduced with permission from *Science,* 1989, **243**, 37.

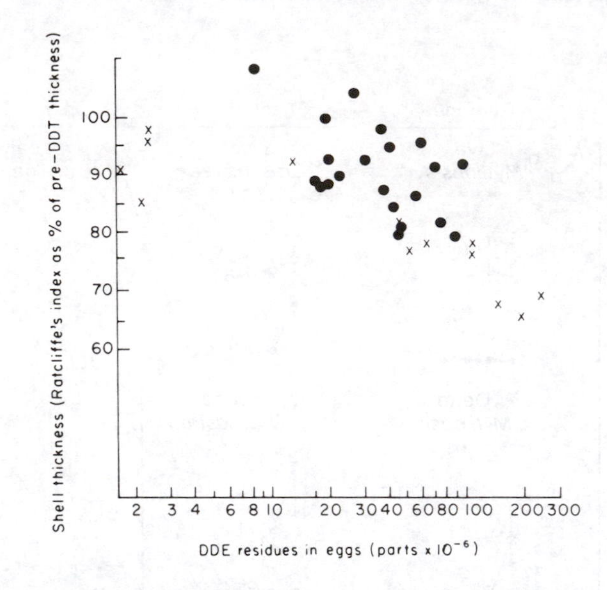

Figure 11 *The relationship between mean clutch shell thickness and DDE residue of American kestrel eggs* (Falco sparverius) *collected from wild populations in Ithaca, New York, during 1970* (●) *and the same relationship experimentally induced with dietary DDE* (x) (Reproduced with permission from *Appl. Ecol.*, 1975, **12**, 781)

raptor eggshells. Using an index of eggshell thickness (eggshell weight/egg length × egg breadth) it was found that during 1947 and 1948 this index decreased significantly by, on average, 19%. Eggshell thinning has been found in several American species of birds, notably the bald eagle, osprey, and peregrine falcon. In a study of over 23 000 eggshells of 25 bird species, shell thinning was found in 22 of the species and a correlation was found between the degree of thinning and DDE residues in the eggs.[53] Experimental work with mallards and American kestrels showed that DDE caused eggshell thinning. Lincer[54] compared the degree of thinning with the concentration of DDE in eggs of the American kestrel (*Falco sparverius*) in field populations and in eggs produced by captive birds fed food dosed with DDE. Shell thickness decreased with increasing DDE residues in the eggs of both populations (see Figure 11).

It is well accepted that DDE is the primary cause of eggshell thinning in many kinds of birds and some bird species are more susceptible than others to DDE-induced shell thinning. Eggshell thinning was definitely a major cause of low reproductive success and population decline in some species, but chlorinated hydrocarbons probably contributed to declines in other ways. In general terms, reproductive trouble tends to increase as shells become thinner and thinning of over 20% is likely to result in reproductive failure and population decline. In the case of peregrine falcon populations in the UK, the incidence of broken eggs

[53] D. W. Anderson, and J. J. Hickey, Proceedings of the XVth International Ornithological Congress.
[54] J. L. Lincer, *J. Appl. Ecol.*, 1975, **12**, 781.

within clutches rose from the normal 4% to 39% during the period 1951–66. Although thinner eggshells resulted in an increased breakage of eggs and, in turn, a reduction in breeding success of a particular pair of falcons, it was realized that the overall size of the UK population was unaffected by this at the time. The actual population decline took place at least five years after the onset of eggshell thinning. This coincided more closely with the upsurge in use of dieldrin as an anti-fungicidal seed dressing. Although it has never been substantiated, it is widely believed that the abrupt decline was due to a combination of factors, involving eggshell thinning and the later ingestion of toxic doses of dieldrin derived from a diet of contaminated pigeons.[7,51]

There are examples of incidents concerning bird populations in which chlorinated hydrocarbons have been implicated but the exact cause has never been fully resolved. In the autumn of 1969, 15 000 dead and dying seabirds were washed up on the coasts around the Irish sea. Nearly all the birds were guillemots (*Uria aalge*) and almost all the victims were adults just after the annual moult. The bulk of the birds were washed up after storms in late September, but even during the previous fine calm weather some dead and dying birds were spotted. Chemical analyses of dead guillemots for a variety of pollutants revealed the presence of high concentrations of DDE and in particular very high levels of PCBs in their livers. It has been suggested that moulting, followed by the stress of withstanding stormy conditions, resulted in mobilization of fat reserves and in turn release of toxic doses of PCBs into the blood stream.[1]

During the 1970s extensive surveys on the Canadian side of the Great Lakes revealed reproduction problems in fish-eating birds and initially it was thought that organochlorine compounds were involved. In one study eight colonies of herring gull (*Larus argentatus*) were investigated. Two colonies from Lake Ontario showed very poor reproduction and in another two colonies from Lake Erie hatching was reduced, but in the remaining colonies from around Lake Huron and Lake Superior gulls reproduced normally. Both altered behaviour of the adults (such as decreased nest attentiveness) and some agent(s) present within the eggs may have been involved. From 1976 onwards, populations tended to reproduce normally, indicating that the causative agents had diminished or disappeared. PCBs have been implicated to explain behavioural changes in the adults. Also TCDD residues were detected in herring-gull eggs collected at the time but egg injection experiments with TCCD have failed to reproduce the observed embryonic mortality.[51]

The toxicity of PCBs to wildlife is quite variable, exposures as low as 0.1 $\mu g\ l^{-1}$ are toxic to certain fish species, while others can survive concentrations of over 1000 $\mu g\ l^{-1}$. Experimental studies with mink and ferret have established that PCBs are highly disruptive of reproductive processes. Mason[15] reports the results of an experiment in which American mink (*Mustela vison*) were fed diets of PCBs. Females with intakes of 3.3 mg PCB per kg food for 66 days, produced fewer young and of those that were born the survival rate was very low. At higher intakes no young were produced at all. Tissue concentration of 50 mg kg^{-1} PCB was associated with the onset of reproductive failure. The otter which is a closely related species to the mink has declined markedly over wide areas of Europe since the 1950s. In particular animals from those populations showing the greatest decline tend to have PCB concentrations in their tissues of over 50 mg kg^{-1}.

In recent decades there has been a rapid decrease of seal populations in the Baltic and in the Dutch Wadden sea. Between 1950 and 1970 the population in the Wadden sea dropped from more than 3000 to less than 500 animals. The reduction in numbers is associated with low breeding success. The populations from both the Baltic and Wadden were found to contain high levels of organochlorine compounds and in particular PCBs.

Reijnders[55] carried out a field experiment with two groups of 12 harbour seals; one group was fed a diet of fish from the Wadden sea that provided a daily intake of 1.5 mg PCBs for two years, the second group received fish from the Atlantic that gave an intake of 0.22 mg PCBs per day. The reproductive success of the group receiving the higher dose of PCBs was significantly lower than the other group with the lower dose of PCBs.

Effects related to contamination of PCBs have been found in the seal populations, they include hormonal imbalances and immunosuppression. In the light of recent large scale seal deaths, which have been found to be due to a virus related to the canine distemper virus, there is some concern that such outbreaks may be related to a breakdown in their immune systems.

13 CONCLUSION

Pollutants are not of a distinct type and all manner of physical and chemical properties are represented in this broad group of substances. Cause–effect or more specifically exposure–effect relationships form an important basis for the perception and evaluation of risk or damage from pollution. For many of the important pollutants, it is perhaps surprising to learn that these basic relationships have yet to be fully established.

The main problem stems from the complexity of the interaction of pollution with the environment. Cause–effect relationships cannot be derived from a single approach, but they are developed from field and experimental investigations and observations. In this regard it is important to establish a consistent relationship between the measured effect and the suspected cause. The observed association should have a reasonable biological explanation. It should be possible to isolate the causal agent and to reproduce the effect under controlled conditions. The cause should normally precede the effect.

In the past, pollution studies were generally initiated in response to an obvious problem. At the present time there is an increasing obligation to forestall such events by attempting to predict ecological and health effects of pollutants. As preventive pollution control strategies are implemented particularly for new chemicals it is important that the hazard evolution procedures take account of all possible risk. The concluding remarks of the Tenth Report of Royal Commission

[55] P. J. H. Reijnders, *Nature (London)*, 1986, **324**, 456.

on Environmental Pollution are still relevant today, 'an important feature of the type of long-term environmental protection policy .... is that it should guard against creating situations which, though they may initially appear innocuous, have the potential for erupting disastrously—in other words, it should ensure that no new "time bombs" are set'.[2]

CHAPTER 9

Regulation and the Economics of Pollution Control

D. J. HUGHES

Humankind possesses vast abilities to modify its environment—simple acts of cutting down trees for fuel or quarrying stone for building illustrate this. Such activities have a long history; but only in the past two hundred years, or so, have advances in industrial and technical processes and increased resource demands, due to massive population growth, led to the impact of human activity on the environment to be so marked as to lead to its perception as dangerous to humanity and, more recently, to ecosystems. It is not, however, always easy to draw distinctions between impacts considered desirable or attractive and those which are deleterious. Building a reservoir may create what is, to some, an attractive aquatic landscape and recreational amenity as well as a utility; to others it may represent destruction of a centuries old landscape with its associated habitats and natural interest. Thus we encounter a fundamental issue in environmental protection; regulation is an ethical matter.

Concerning deleterious activities it is also necessary to make a distinction between those which deplete the environment, such as over fishing, and those which pollute it; *i.e.* to adopt the definition given by Section 1 of the Environmental Protection Act 1990, which pollute 'by the release . . . from any process of substances which are capable of causing harm to man or any other living organisms supported by the environment'. 'Environment' in this context means the mediums of air, water, and land. Many activities may both deplete and pollute; for example mining which uses up natural resources and pollutes the environment by dust, noise, and vibration.

It is important to realize that law does not operate in a vacuum. The form and content of regulation exists within social, economic, political, and ethical contexts: law is a phenomenon within society, being acted on by social forces and acting upon them in return. Law is not a simple list of rules learnt by heart to be parrotted out by law students so that they may enter a lucrative profession. Law is, rather, one of the means whereby society seeks to ensure its continued harmonious operation and the eradication of problems between individuals, between individuals and society as a whole, and between sections of society. Of course individual rules are of great importance, but there are also underlying principles, and modes, techniques, and styles of regulation.

Furthermore because law is a product of society it reflects forces, tensions, pressures, and disagreements within society—very often as they have existed over long periods of time. This explains why law is sometimes apparently quirky, or vague. Projected on a world-wide basis it also helps to explain the problem faced by International Law in securing the compliance of states with its requirements.

The idea of '*Law*' is an expression of humanity's desire to live in decent civilized order, but '*the law*' in any given state will be the product of current and historical societal forces within that state. '*A law*', *i.e.* a particular rule of law will certainly encapsulate values in relation to its subject matter—no matter how purely technical it may appear.

That law operates in an ethical context has been stated; more extensive examination of the interaction between law and economics must now take place.

1 POLLUTION AS AN ECONOMIC PROBLEM

A particularly close relationship between law and economics exists with regard to pollution control. In very basic terms legal systems frequently use economic means to impose sanctions or punishments on wrongdoers, as fines or awards of damages. However, the relationship is much more subtle than that. Economic analysis may be used to investigate and define problems which may then be considered suitable topics for legal regulation. The impact of particular types of legal regulation may be subjected to economic analysis to determine whether a law is working efficiently to produce maximum benefits at minimum cost. Economic analysis is a major determinant of state policy on environmental issues while law is a major means of implementing policy.

Certain basic principles of environmental economics are well established. Economists have studied the adverse consequences of pollution for resources available to society, consequences such as reduced productivity (*e.g.* polluted waters may not support fish life). However, they have also pointed out that reducing pollution itself can have adverse consequences: benefits flowing from polluting activities may be reduced. What emerges is a theory of the 'efficient level of pollution control'. This is a cost benefit exercise whereby the quantity of damage or physical impact of any given pollutant, and its 'damage cost' (*e.g.* the commercial value of trout killed by a polluting release into a river) are compared with control or abatement costs of preventing pollution. These latter costs include resources devoted to abatement technology, and the time and effort that have to be put into, and inconvenience that may result from, avoiding polluting activities. This analysis has usually resulted in findings that an efficient level of pollution control is one that does not eradicate all pollution. Furthermore it may be found that different levels of pollution abatement will result in widely differing costs. Quite moderate costs may be incurred in abating pollution to a given level; to reduce it further may involve considerable increases in costs. Marginal improvements have to be subjected to cost benefit analysis if the maximum benefit is to be efficiently obtained at the lowest cost.

It must not be assumed, however, that one is dealing with an exact science. Though economists utilize statistical methods and established methodologies for obtaining information, information has to be *interpreted*, and there are *assumptions* and *presuppositions* in data interpretation. This can be illustrated by referring to modes of determining what the cost of pollution damage might be with regard to effects on human health. Four basic approaches can be identified: the 'human capital' technique concentrating on earnings lost and health costs flowing from pollution induced illness; the 'willingness to pay' approach, concerned with willingness to pay for reductions in pollution related risks; the 'public expenditure' technique deriving valuations implicitly from public expenditure; and the 'labour market' technique which argues that wage differentials reflect, at least partly, occupational risks. Once value is established, if it is desirable to calculate how pollution abatement might reduce risk of premature death, and thus to estimate total health benefits, the overall size of the polluted population would have to be determined, then predicted changes in mortality, perhaps on a per 100 000 of the population basis, consequent on abatement would have to be multiplied by the 'life value' figure and the size of the exposed population. This leads to an estimate of the value of damage. It should, however, be noted that studies of, for example, the health benefits derivable from abating SO_x emissions from power stations indicated a *range* of benefits (reductions in health care costs and lost earnings) of between \$580–\$14 400 M per annum. Cost benefit analysis here was useful in that it identified an area of uncertainty for policy makers.[1]

The mention of reducing sulfur oxide emissions, particularly SO_2, is a convenient point for a case study which illustrates legal and economic approaches to a particular environmental problem; that of acidification of soils and water. The issues were considered by the Royal Commission on Environmental Pollution in 1984.[2] As part of their investigation they considered the contribution of SO_2 emissions from power stations to total acidification problems; by the mid 1980s some 63% of UK SO_2 emissions came from power stations, 2.4 M tonnes p.a.—the figure was 70% by 1990. There was strong evidence from the National Academy of Sciences in the USA of a proportional relationship between emissions of SO_2 (and NO_x another acidifying 'culprit') and acid deposition.[3]

The Central Electricity Generating Board (CEGB), the authority then responsible for power stations, submitted evidence to the Royal Commission of the costs of various options available for reducing SO_2 emissions (see Table 1).

This economic evidence was principally directed *against* the introduction of Flue Gas Desulfurization (FGD) technology, considered by the CEGB expensive

[1] Organization for Economic Co-operation and Development (OECD), 'The Costs and Benefits of Sulfur Oxide Control, A Methological Study', Paris 1981. See also P. Burrows, 'The Economic Theory of Pollution Control', Martin Robertson, Oxford, 1980, and G. H. Mooney, 'The Valuation of Human Life', Macmillan, London, 1977.

[2] Royal Commission on Environmental Pollution, 10th Report 'Tackling Pollution—Experience and Prospects', Cmnd 9149, HMSO, London, 1984.

[3] National Research Council, 'Acid Deposition: Atmospheric Processes in Eastern North America', National Academy Press, Washington DC, 1983.

Table 1 *Estimates Presented by the Central Electricity Generating Board of the Costs of the Options Available for Reducing Emissions of Sulfur Dioxide from Power Stations*

Option	*Sulfur Reduction*				*Estimated Costs*		*Specific Costs Per Tonne of SO_2 Abated*		
	Annual coal tonnage affected M tonnes	*Sulfur removed* K tonnes annum^{-1}	*SO_2 abated* K tonnes annum^{-1}	*% of CEGB emission*	*Capital cost* £M	*Gross operating cost*** £M annum^{-1}	*Capital investment* £ tonne^{-1} annum^{-1}	*Gross operating cost*** £ tonne^{-1}	*Net operating cost* £ tonne*$^{-1}$
That might be specified now									
1. Exchange higher sulfur coal for lower sulfur UK coal	5	40	80	2.9		10.15			125–188
2. Exchange higher sulfur coal for low sulfur imported coal	5	81	162	5.8		(No additional cost anticipated)			
3. Reduction in sulfur content of fuel oil (from 3% to 1%)	8 mtce	90	180	6.4		(Costs not estimated by CEGB)			
4. Extension of current coal preparation techniques	11½	22–40	45–80	1.6–2.9	45–50	10–20	560–1110	125–440	0–190
5. Provision of flue-gas desulfurization plant:									
(a) Wellman–Lord process on one 3 × 660 MW station	~4	68	136	4.9	117	23.6	860	173	147
(b) Wellman–Lord running at lower removal efficiency	~4	43	86	3.1	86	13.9	1000	161	119
(c) Wellman–Lord retrofitted to one 4 × 500 MW station	~4	72	144	5.1	152	24.0	1055	167	128
(d) Limestone/gypsum process on one 3 × 660 MW station	~4	68	136	4.9	117	17.9	860	132	N/A

That might be specified 5 years hence									
6. Retrofit of advanced techniques to existing coal preparation plants	11½	35–70	70–140	2.5–5.0	55–80	20–30	390–1140	140–430	60–260
7. Extended use of advanced coal preparation techniques	11½	45–110	90–220	3.2–7.9	70–85	20–30	320–940	90–330	40–210
8. Seawater FGD process on one 3 × 660 MW station	~4	68	136	4.9	55	16.1	404	118	118
9. Direct injection of limestone to furnaces		(Not quantifiable at present—current view is that it could be more expensive and less effective than FGD processes)							
That might be specified 10 year hence									
10. Pressurized fluidized bed/combined cycle systems with limestone addition		(Not quantifiable at present—US estimates suggest 80% sulfur removal may be attained at specific costs similar to those for FGD processes)							
11. Gasification/combined cycle systems		(Not quantifiable at present—but high (>90%) sulfur removal attainable)							
12. Alternative advanced coal preparation techniques, *e.g.* magnetic separation of pyrites from coal		(Not quantifiable at present—possibly similar performance to 7 above but could be less costly)							

* Assumes an incidental benefit of £1 $tonne^{-1}$ of coal burned in the case of the coal preparation options and assumes the sale of by-products in appropriate FGD options.
** Annual cost excluding capital charges.
Reproduced from Cmnd 9149 'Tackling Pollution—Experiences and Prospects', HMSO, London, 1984.

and wasteful in comparison to benefits conferred. That, however, was not the only side of the story. As the Royal Commission pointed out, the beneficiaries of reduced SO_2 emissions would not necessarily be those living in the UK. The Organization for Economic Co-operation and Development had, in 1982, concluded, by aggregating data on the economic effects on health, metal corrosion, damage to crops, and acidification of waters, that in Europe as a whole there would be benefits ranging from £80 to £800 for every tonne of SO_2 removed from the atmosphere. The Royal Commission noted there were major practical and methodological problems involved in producing such estimates, with a considerable range of uncertainty, nevertheless concluding that they indicated *where* costs were falling. This illustrates the usefulness of economic analysis as a means of *identifying* problem areas.

The need to avoid treating economics as an exact science was further reinforced by a study of a particular aspect of the acidification issue, namely effects on buildings.[4] The Building Effects Review Group (BERG) argued that it was far from easy to determine the exact damage costs of acidification, pointing out it would be simplistic merely to concentrate on determining only the quantities of material subject to deposition and their new replacement costs. That would, at least, leave out of any calculations maintenance costs of materials and the fact that they wear out naturally and have to be replaced. They argued that uncertainties exist at every stage in determining acid deposition costs on buildings. These stages are, first, determining which buildings are exposed, secondly, what their exposure is and what level of damage may be due to acid deposition, and, thirdly, damage costs or the benefits of its eradication. They also considered that the aesthetic and historical associations of buildings damaged by acid deposition were not quantifiable for the purpose of economic analysis—this, as will become clear later, is open to challenge.

In terms of policy development, the government responded to the 1984 evidence by committing itself to reducing, by the year 2000, annual emissions of SO_2 from all sources by 30% on 1980 levels, though it seems to have been pressure from other states, particularly those of the European Community (EC), which led to current UK policy. In 1986 the CEGB performed something of a 'U'-turn proposing to fit FGD equipment to three existing power stations. Subsequent legal action taken by the EC should lead to major reductions in SO_2 from large combustion plants in three stages, 1993, 1998, and 2003. The economic costs of abatement action are high: 20% of initial capital investment costs in relation to older coal fired power stations, or, overall, 10% of the costs of producing energy generated (or some £6 billion over a ten year period for the UK electricity supply industry), but it has proved difficult to place a precise figure on the benefits of abatement. It is, however, accepted that damage costs outweigh the costs of abatement action.[5]

[4] Building Effects Review Group, 'The Effects of Acid Deposition on Buildings and Building Materials in the UK', HMSO, London 1989, Chpt. 8.

[5] S. P. Johnson and G. Corcelle, 'The Environmental Policies of the European Communities', Graham and Trotman, London, 1989, pp. 139–147.

The case study is not intended to denigrate the value of economic analysis; it intends to place the issue in context. Economic analysis provides no 'magic' answer to the problems of pollution. The same is, of course, true of law: pollution does not disappear merely because it is made illegal. Equally the natural sciences provide no categorically imperative evidence that action on pollution *must always* be taken. The fact that a particular chemical produces a particular instance of pollution in particular circumstances is no reason for precipitate regulatory action. Many types of analysis have to be applied before a policy on the problem can be formulated and an appropriate legal response developed. Economic analysis is of great importance in the policy formulation stage: recent developments in the understanding of how such analysis can be applied promise an even greater role for environmental economics in the future.

The general principles of environmental economics are undergoing rapid development. A leading school of thought favours the notion of 'sustainable development'.[6] This stresses that environmental resources are not freely 'there for the taking', and argues for intra-generational and inter-generational equity—those least advantaged in society should have provision made for their needs, while future generations should receive compensation for any depletion of resources brought about by the activities of the current generation. Put another way the stock of environmental assets, or 'critical environmental capital or assets', should not, overall, diminish as a result of the activities of a given generation.

Such 'capital' clearly includes the mediums of land, air, and water. Sustainable development theorists argue these mediums cannot be treated as free assets, *i.e.* they cannot be treated as 'externalities' to decision making processes. Their use, which includes affectation by pollution, is a cost which the user/polluter must pay rather than the mere unplanned consequence of a decision. Pollution is not a discrete environmental issue separate from resource depletion; all economic activity affects the environment, and vice versa.

Such thinking stresses the need for anticipatory policies (in legal terms the 'precautionary principle'), *i.e.* to reduce pollution levels now, rather than postpone action to some future date when irreversible damage may have eventuated, as appears the case with thinning of the ozone layer and releases of global warming gases. Anticipatory action also reflects the sustainable development principle of inter-generational equity. It places on the shoulders of the current generation an obligation that might prove impossibly burdensome for those to come. This further requires that society's tendency to discount (*i.e.* to reduce the value of) future costs of benefits and burdens, especially in comparison to short-term gains, must be modified. Long-term benefits and burdens must be adequately considered and reflected in decisions if inter-generational equity is to be achieved.

[6] D. Pearce, A. Markandya, and E. B. Barbier, 'Blueprint for a Green Economy,' Earthscan Publications Ltd., London 1989. See also I. Bateman, 'Social Discounting, Monetary Evaluation, and Practical Sustainability', *Town and Country Planning, 1991*, **60**, (6) 174.

Sustainable development thinking stresses its concept of 'development' is wider than simple economic growth. It embraces notions of advancing the well being of individuals, and does not set its face against increases in per capita incomes. It also places value on environmental quality, preservation, and enhancement of freedom from ignorance, poverty, and squalor, promotion of human skills, capabilities, knowledge, and choice, together with the enhancement of self respect and esteem. Central to understanding these ideas is the concept that proper value must be placed on the natural environment and the economic functions it serves. It is a fundamental misconception to treat them as having no value simply because there is no market in which they are traded. The contributions of the natural and human created environment in both direct and indirect terms have an economic value. This must be appreciated if the sustainable development concepts of equity outlined above are to operate. For example, in order to hand on to future generations stocks of environmental assets no less than those we have inherited every attempt must be made to avoid irreversible changes such as species loss or destruction of natural habitats, or, at least, to provide compensation for any such loss by creating other assets. Indeed some argue the application of sustainability principles precludes progress on any activity which permanently destroys critical natural capital such as the atmosphere.

These principles demand means of valuing the hitherto unvalued. Here economic thinking is not fully formed, though monetary terms have to be used as measuring devices. Established economic thinking, as outlined above, shows how pollution abatement can be costed in terms of enhanced crop yields or reduced health expenditure. Sustainable development thinking takes this further by arguing that 'monetization' of environmental values is a measure of society's willingness to devote resources to environmental protection/improvement, and also serves to allow comparisons with the costs of other policies. Traditional cost benefit analysis is thus developed. For an example consider the present system of road planning in the UK. The Department of Transport currently uses a cost benefit system (COBA) in which the costs of road building, the savings in journey times, *etc.* are expressed in monetary terms, while environmental issues are subject only to non-monetized impact statements. Like cannot therefore be compared with like. The extension of cost benefit technique currently taking place enables this omission to be supplied.

But how to decide what the 'value' of the environment is? Some techniques are well established, others are more recent. There is no time here to go at length through these techniques; readers are referred to the work of Professor Pearce[6] and the studies further referred to there. The techniques involve considerable complexity of application, and data collection can be a demanding exercise. Briefly, however, the following techniques are identified by Professor Pearce (*et al.*), to whose work the present author acknowledges indebtedness. It is, first, possible to identify a '*use value*' for the environment made up of '*actual use value*' to those who utilize it to their advantage, for example as anglers or hunters, *plus* an '*option value*' which is, in essence, a 'willingness to pay for the environment', either because a person may have future use wishes, or vicariously because of pleasure derived from knowing others can use the environment, or because there is a desire to preserve the environment for future generations. Use of this technique provides a measure of '*total user value*'. A further technique concentrates on the '*intrinsic value*' of the environment. This is a less well defined concept than 'total user value' but it comprehends human preference *to conserve and preserve features of the environment for their own sake*, not simply because they are, or may be, of use to humankind. Combining 'total user value' with 'intrinsic value' produces '*total economic value*'.

But how are such values measured? There are numerous techniques. 'Direct measurement' techniques concentrate on measuring the monetary values of environmental gains, such as an improvement in air quality. They include the 'hedonic price approach' which relates the value of land to benefits derivable from it, such as output, shelter, access to other land, commercial and recreational facilities, and neighbourhood quality. Statistical methods can be used to identify that part of a difference in price between pieces of land due to given differences in environment, and to infer what a person could be prepared to pay to bring about improved environmental quality. Such techniques, of course, depend on using many variables in calculation of property values. Though it appears they can provide acceptably reliable estimates, they must also be checked against results obtained by other estimation techniques, *actual* figures derived from any 'real market' dealings with land, *and* their own consistency with hedonic prices estimated in similar situations. An alternative technique is 'contingent valuation' which asks individuals 'what would you pay for environmental benefits', or 'given this disbenefit, what compensation would you wish to receive?' To introduce greater accuracy, questions are put as a series of bids, for example: 'would you be prepared to pay £*X*?', 'would you then be prepared to pay £*Y*?' where in each case the 'price' demanded for the benefit rises. There comes a 'sticking point' where the person questioned will refuse to pay more. That indicates his/her highest level of 'willingness to pay'. Obviously the technique has to be applied across a meaningful sample of the population to obtain significant measures of valuation. It is, however, apparently applicable to most environmental situations, and produces useful results, though these must also be checked against any available 'real market' figures and estimates produced by other techniques, *e.g.* those based on land prices as outlined above. The essential point is that various techniques can be used empirically to estimate the 'option' and 'existence' values referred to above. Though these will not be mathematically exact, they do indicate that the environment cannot be treated as free. It has a positive economic value which must be allowed a role in policy formulation and implementation. It is no longer enough to weigh the costs of a development against its benefits, the benefits of *not* going ahead and thus preserving the environment as it is must also be considered. Other valuation techniques include: 'dose response', *i.e.* using statistical methods to create a model of physical damage which relates 'doses' (or levels of pollution) to 'responses' (or levels of damage) to produce estimates of impacts on health which can be monetized using value-of-life estimating techniques (see above), and 'travel cost', *i.e.* placing a value on an amenity by assigning it a 'price' not less than the total costs of travel to the amenity for those who use it.

There are implications for environmental policy formulation and implementation at all levels—international, national, local, and individual. Only the merest outline of these can be given here. Those seeking further enlightenment must turn to Professor Pearce's work.[7] As one instance it is arguable that project evaluation should involve cost benefit analysis where notions of sustainability are integrated into the techniques of analysis. Investment programmes should be checked to assess environmental impact, consequent adjustments being made to ensure that, overall, there is as little depreciation of natural environmental assets as possible—zero depreciation where achievable. Furthermore programmes should include projects which generate positive environmental benefits. The internal balance of investment programmes would thus alter. This would be applicable at the level of individual companies and governments allocating public finance. There are, of course, some projects that are destructive of the environment, for example highway construction or mining. These may take place internationally, as witness the flow of minerals from the Third World to developed nations. The application of sustainability notions will enable a value to be placed on the environmental disruption or depletion of resources, and will concentrate attention on minimizing that figure, and, *inter alia*, compensating those whose loss, in the form of depleted resources, leads to the gain of others.

Further applications of sustainability concepts relate to prices for, and taxes on, goods and services being adjusted to take account of the hitherto unpriced use of the environment, for example as a dump into which waste products are placed. Such adjustments could, for example, be effected by a central governmental decision to incorporate valuations of services provided by the environment for the manufacture of goods in the market price of such goods. Where production entails pollution the price of a unit of production should reflect not only marginal costs of production, but also the cost of pollution damage production entailed. This lies at the heart of the concept of 'the polluter pays' which means that costs of goods and services resulting in pollution in production or consumption (or both) should reflect costs of measures of pollution prevention and control.[8] There are a number of mechanisms whereby this principle can be applied. An environmental standard may be set in relation to a polluting activity with the cost burden of meeting that standard placed on the polluter. The polluting product may be made subject to a levy or tax of some sort. If the producer then passes some of these 'pollution costs' on to consumers there is not a derogation from the 'polluter pays' principle. Consumers participate in pollution by providing demand for the polluting product and must therefore realize that the cost of the product reflects its environmental effects and the use of environmental services. Pollution levies/ taxes in particular are prime examples of market based incentives for sustainable

[7] D. Pearce, A. Markandya, and E. B. Barbier, *Op. Cit.*, Chpt. 4–7.
[8] OECD, 'The Polluter Pays Principle: Definition Analysis, Implementation', Paris, 1975.

development. Where charges on a product bear a real relationship to costs of using environmental services in the course of production, the price of production, and individual products will rise. It then makes sense for producers to consider whether they can eliminate extra costs by appropriate abating measures. A carbon tax with graduated levels of taxation rising in step with any given fuel's carbon content, would be an example of such a tax. Users of fuels containing carbon for energy production would have the choice to spend on introducing technology to clean up emissions, to switch to other less polluting fuels, or to find means of conservation so that, overall, less fuel is used with consequent reductions in emissions. A broadly similar pattern of choice would be available to energy consumers. Such a tax would, however, be unlikely to make a major contribution to reducing outflows of greenhouse gases if it were to be adopted in one or just a few states only.

One final application of the 'polluter pays' principle is the notion of a 'pollution permit' whereby, once an acceptable overall environmental quality standard has been set by the appropriate authority, permits are issued to those who carry on polluting activities. They are then allowed to trade these permits. Polluters who could easily and cheaply abate their harmful activities would, having implemented the necessary measures, be free to sell their permits to those facing higher abatement costs. The total amount of permitted pollution would not increase, and could be at any time further reduced by the standard setting authority 'buying in' permits themselves.

Economics has an increasingly important part to play in the formulation of environmental policy. It has a role alongside that of, and interacting with, law as means of environmental regulation and protection. Indeed economic modes of regulations utilizing market forces as outlined above may be more effective than legal modes based simply on the command and control technique of a legal rule which prohibits a particular activity and imposes a penalty or sanction for breach.

2 LEGAL RESPONSES TO ISSUES OF POLLUTION

Though economic analysis plays an important part in policy formulation, policy and its implementation by law has to take account of other factors. There are historical constraints such as established working and investment patterns to consider, and political pressures from those wishing to promote their own interests. Almost inevitably compromises have to be made. Thus to *eradicate* a particular type of pollution requires the satisfaction of many conditions:

(a) agreed perception of hazards;
(b) appropriate available techniques to deal with the problem;
(c) convergence of interests between those creating the hazard and those who suffer it to make eradication acceptable in economic, political, or social terms;
(d) perception of the issue 'in the round' to ensure co-ordinated and integrated means of dealing with it so that its effects are not merely switched from one part of the environment to another;

(e) strong and effective agencies of enforcement.

If these conditions are not met some element of compromise in the form of legal regulation will result. It will then amount to a control over or a palliative of the problem rather than an eradication of it. Such inbuilt compromise often affects the effectiveness of laws. Other issues also affect effectiveness, such as what is the source of the law in question, who has the task of enforcing it, what is the object of the existence of a particular law, what are the 'ingredients' of this particular form of legal liability and whence were they derived?

3 SOURCES OF LAW

A prime distinction is between International and National (or Domestic) Law. International Law, or the Law of Nations, increasingly concerns itself with environmental issues, but is hindered by long standing adherence to the principle of state or territorial sovereignty whereunder States have the rights to carry on their activities and use their resources within their boundaries for their own benefit. This can lead to deleterious effects for others, as for example by release into the atmosphere from combustion processes of greenhouse or acidifying gases. To counter this, states are beginning to make increasing use of multi-party Conventions—and associated subsidiary agreements known as Protocols, the collective terms for both is 'Treaties'—whereby they agree to refrain from particular activities or to achieve declared ends. Since the Stockholm Conference of June 1972 on the Human Environment, which produced a non legally binding declaration of environmental ethics for states, there have been increasing numbers of treaties dealing with pollution on an international or 'transboundary' basis, for example the Geneva Convention on Long Range Transboundary Pollution of 1979 and the Vienna Convention on Protection of the Ozone Layer of 1985. Each has protocols extending its operation; the Geneva Convention is supplemented by The Helsinki Protocol on SO_2 (1985) and the Sofia Protocol on NO_x (1988), while the Vienna Convention has the Montreal Protocol of 1987, further adjusted in London in 1990. However, here a legal problem is encountered—the fact that a state is a party to a convention does not mean it will become party to its subsequent protocols. The UK and USA, for example, refused to become party to the Helsinki Protocol, thus failing to join the '30% Club' of nations who promised a 30% reduction in SO_2 emissions on 1980 levels by 1993. The UK's argument was it had already reduced emissions by 42% on 1970 levels by the mid 1980s, and would make a further 30% reduction on 1980 levels by the late 1990s.

Herein lies the weakness of treaties: states cannot be forced to become parties and will seek to negotiate them so as to secure their own advantages. This can lead to vagueness in the terminology of treaties and the burdens they impose. The element of compromise is clearly apparent here. Further weakness is exhibited by the lack of effective enforcement mechanisms for treaties. The last resort may be to an action such as international isolation, even armed intervention, or economic sanctions, but these are hardly likely to be regularly used. For most of the time treaties rely for their efficacy on the good will of states. A further considerable weakness is that many states, the UK and USA included, *generally* require legislation to be passed by their national legislatures before the terms of a treaty can be implemented within the state. All in all the International Environmental Law must be regarded as a weak system of regulation.

This focuses attention on national law which in nations of the English speaking tradition subdivides into public law and private law. So far as the latter—the law between individuals—is concerned, the law of torts, (actionable civil wrongs) is most important from an environmental point of view, in particular the Tort of Nuisance. This body of law, the product of judicial decisions in individual cases, has been developing for centuries. It has become an exceedingly technical and complex system of rules, with different types of situation being treated differently. Rivers and streams receive considerable protection with any pollution being viewed as actionable. In other cases a remedy will only be allowed if the judge considers the action of the defendant to be 'unreasonable'—a vague word which enables judges to take into account a whole range of factors, including matters such as the character of the neighbourhood, the social utility of the defendant's operations, the nature of the damage suffered, and whether the defendant was activated by malice. A successful claim in nuisance can result in awards of substantial damages, and orders from the Court known as Injunctions which can require the cessation of the deleterious activity. Nuisance actions are comparatively rare, partly because of the technicality of the law which renders the outcome of litigation uncertain. There is often also an imbalance of economic power between plaintiff and defendant—the former may be a householder of modest means whose property is invaded by emissions, the latter a major company. In the absence of state funding in the form of legal aid, something for which increasingly fewer people qualify in the UK, plaintiffs will not have the same access to legal services and expert assistance as will defendants. The Nuisance may also not affect any one individual sufficiently badly for it to be said that there has been an 'unreasonable' interference with his/her rights.

The shortcomings of private law lead to most environmental law being derived from public law, *i.e.* the law as between the state (*e.g.* central and local government) and the citizen. Public law is principally to be found in Acts or 'Statutes' passed by Parliament at the behest of government ministers, supplemented by regulations, usually known nowadays as Statutory Instruments—SI—made by Ministers, assisted by the Civil Service, acting under powers delegated by Parliament in Acts. Acts and Statutory Instruments are subject to judicial interpretation, giving judges the role of applying the meaning of statutory phrases in individual cases.

One further source of public law for member states, the UK included, is the Law of the European Community. The EC is a unique organization; it is not a nation state in its own right such as, for example, Austria, nor is it a sovereign federation such as the USA, but it is clearly more than a trading bloc or customs union, having increasingly common policies designed to secure the economic advancement of member states by the harmonization of their regulatory legal measures. EC Law rests on the notion that it is a paramount system of law; where there is any clash between EC law and the laws of a member state the former must prevail and be given effect even by the courts of the state involved. Environmental protection has been a declared EC policy objective since the amendments made to the Treaty of Rome 1957 by the Single European Act 1986, though the Community has had an extensive environmental programme since 1972. Many measures have been pursued on the basis that the harmonization of member states' environmental standards would contribute to the advancement of the single or 'common' market which lies at the heart of the idea of the EC. For the future, EC law is likely to be *the* major source for new environmental legislation for the member states, and may well serve as a model for states in other parts of the world. EC Law enshrines the principles that the 'polluter pays', with those that damage should be rectified at source, and preventive action should be taken in respect of environmental damage—this is the legal counterpart of the 'non-irreversibility' principle of sustainable development economics.

The principal means whereby the EC communicates its will on environmental matters to member states are Directives issued under Article 189 of the Treaty of Rome. A Directive sets an objective to be achieved by member states, generally leaving quite a broad discretion to each state as to how that is to be achieved within its boundaries. This does not mean they have discretion to ignore the objective, or that they can claim to comply merely by making administrative arrangements only. States who fail to implement Directives by the set date can be made subject to proceedings before the Court of Justice of the EC pursuant to Article 169 of the Treaty. More rarely in relation to environmental matters the EC will issue Regulations. These are 'directly. applicable' and are law within member states without need for the further legislation on their part. A good recent example is Council Regulation 594/91/EEC which contains the rules for the phasing out of the use, manufacture, import, and export of ozone-depleting substances to implement the Montreal Protocol (see above) within the EC. The detailed influence of EC law on UK law and practice, and its implementation in the UK will be returned to below.

4 THE PERSONNEL OF ENVIRONMENTAL REGULATION

Within the UK a wide range of persons and institutions have responsibilities for environmental regulations, some as regulatory bodies, others in policy formulation. Overall policy is the responsibility of central government, but there is no one ministry which has exclusive jurisdiction over environmental issues. The Ministry of Agriculture, Fisheries, and Food—MAFF—and the Departments of Energy and Transport have portfolios with environmental implications, though the principal ministry is the Department of the Environment (DoE). The DoE is both a policy formulation and implementation body. Headed by a Secretary of State who, as a Cabinet Minister, is responsible to Parliament, the DoE formulates environmental policies and has important supervisory functions in relation to inferior agencies such as local authorities. Implementation is a DoE task via Her Majesty's Inspectorate of Pollution—HMIP—who have responsibilities for regulating the most complex and potentially polluting processes which make emissions to land, air, or water (see further below on the Integrated

Pollution Control—IPC—system). HMIP is appointed by the Secretary of State for the Environment under the Environmental Protection Act 1990, a power which replaces older legislation dating back to 1906.

Certain environmental functions are entrusted by statute to specific central agencies—'quangoes'. The best example is the National Rivers Authority—NRA—which, currently under the Water Resources Act 1991 exists to have oversight over, *inter alia*, national water resources and water pollution issues. That the NRA has teeth, and intends to use them, was shown by its first major prosecution, that of Shell in 1990 for permitting oil to pollute the River Mersey through a fractured pipe: a £1 M fine was imposed at Liverpool Crown Court for the offence. HMIP have responsibilities under the Environmental Protection Act 1990 which supersede those of the NRA though the NRA has powers under the 1990 Act to protect its functions and a memorandum of agreement signed by both bodies is designed to promote their harmonious co-operation.

Local authorities have numerous environmental responsibilities. Waste disposal and control over mineral extraction are primarily the concern of County Councils, *e.g.* The County of Leicestershire, while the subsidiary level of District Councils, *e.g.* The City of Leicester, have responsibilities for environmental health, including regulating less complex processes which pollute the atmosphere. To this end they may appoint inspectors under the 1990 Act. Both Counties and Districts have functions with regard to planning control over the use of land. Counties are largely confined to certain strategic functions such as the drawing up of structure plans which are broad outline documents setting down a general framework for land use within their areas, and the control of mineral extraction.

What the UK currently lacks is a national environmental protection agency—a point of contrast with the USA. In the USA overall responsibility for environmental regulation must be a federal responsibility for otherwise the various states of the Union would go their own ways. The US National Environmental Protection Agency—NEPA—is therefore a body of considerable importance. Neither the UK nor the EC has a single agency with an across the board remit for environmental protection issues, though the setting up of an environmental protection agency has been promised for the UK. It is a further major source of weakness if environmental agencies do not act in a co-ordinated fashion, or are under-resourced in terms of trained personnel and support facilities—a criticism regularly levelled against HMIP and the waste regulatory departments of many county councils. At the international level it should also be noted there is no all embracing environmental enforcement agency, though notions of a 'Trustee for the Environment' or an 'Environmental Ombudsman' with powers to deal with deleterious activities have been mooted. Overall, enforcement may be regarded as the Cinderella of environmental regulation.

5 THE STYLES OF ENVIRONMENTAL REGULATION

There are many diverse styles of environmental regulation; the character of each depends on how various components of control are fitted together. Thus:

(i) there may be an initial authorization—planning and hazardous substance consents—or that plus some continuing form of oversight—IPC authorization;
(ii) conditions may be attached to an authorization, either generally according to statutory requirements—IPC authorizations—or wholly/partly at the discretion of the authorizing agency—planning consents;
(iii) refusal or conditional grant of authorization may allow a disappointed applicant to appeal—planning and IPC issues;
(iv) third party rights, *e.g.* rights for the public to make representations about grants of authorizations, may be guaranteed by law to a greater or lesser degree;
(v) there may be requirements to inform some statutory agency of the proposed carrying on of some activity; this may also give that agency abilities to make representations, or give directions to another agency having the power to authorize the activity—NIHSS situations, see below;
(vi) some activities may be subject to more than one type of control, to satisfy the requirements of one set of controls will not guarantee satisfying the requirements of others—*e.g.* a new installation handling hazardous materials in specified processes, see further below;
(vii) control may be imposed by requiring a person to meet some preset standard such as:

(a) not to create a nuisance or conditions prejudicial to health—reserved nowadays for relatively minor issues,
(b) to conform to an emission limit—*i.e.* not to exceed a maximum concentration or permitted quantity of specified material in a release from a process;
(c) to meet a specification standard, *i.e.* requirements as to how plant and machinery are to be constructed, used and operated;
(d) to meet some overall environmental quality objective, *e.g.* a standard is set for the quality of air with regard to the presence in it of particular pollutants, and polluters are not allowed to reduce that quality—the 'standstill' principle—such standards are variously classified as receptor or ambient standards—in the EC there are 'Air Quality Limit Values'—AQLVs—and in the USA National Ambient Air Quality Standards—NAAQS;

(viii) where a breach of authorization/continuing control occurs enforcement action against guilty parties may be more or less at the discretion of the relevant authority, *e.g.* planning control, or may depend on a present administrative environmental 'tariff' such as 'the more serious the offence, the more likely some sanction will be imposed,' *e.g.* prosecutions by the NRA, while in other cases *any* discharge may generally constitute an offence, *e.g.* those of radioactive material;
(ix) control may involve recourse to the criminal law, with liability being either of the traditional variety with the prosecution having to prove that

the accused did the prohibited act—'*actus reus*'—*and* did so with '*mens rea*', *i.e.* intentionally—knowingly with desire—or recklessly—heedless of the consequences—*or* of the 'strict liability' variety where knowledge of the wrongfulness of the act need not be proved;

(x) control may be imposed as a blanket prohibition on polluting emissions coupled with a system of permits allowing permission to pollute up to a preset level, with prosecution following transgression of the preset limit.

No one type of standard or mode of regulation has a monopoly. There is a range of regulation from strict liability criminal offences through differentiated standards applicable to varied circumstances, to highly discretionary modes of regulation where much depends upon individual value judgements with the court being the final arbiter. The choice of style of regulation *and* the thoroughness with which it is implemented vary according to the mix of economic, social, political, and ethical considerations surrounding the creation of any given regulatory measure and its application in particular circumstances. The historic British preference was for a highly discretionary style of regulation with exhortation, persuasion, and co-operation being more evident than the command and control techniques more favoured by mainland European nations.

The drawbacks of legal techniques of environmental regulation include difficulties of detection, proof, and complexity and uncertainty in the process of standard setting.

While the toxic qualities of particular pollutants can be scientifically determined, and thus an *estimate* made of likely lethal levels in given situations following experimentation on cell cultures, animals, and epidemiological studies, such estimates involve *extrapolations* from data and observed uncertainties. Figures thus obtained have to be divided by a safety factor which brings levels of exposure at which no adverse effects are observed down to a lower 'limit value'. Such a safety factor is an evaluative measure determined by a mix of issues—varieties of sensitivity of exposure, ranges of pathological conditions, the quality of data relied on—and political and ethical issues will be influential also. An example concerns the setting of vehicle emission limits. EC practice has tended to lower emissions limits *following* improvements in pollution reduction technology, and using as test parameters low average speeds starting with cold engines, arguably less than optimal for measuring NO_x emissions which increase with higher speeds. US practice on the other hand has tended to follow the 'as low as technically achievable'—ALATA—principle which forces the pace of technical improvement, and which uses much higher average vehicle speeds as test parameters.

Such complexities coupled with somewhat haphazard prosecution systems which can result in a range of penalties from nil upwards, fuel economists' arguments that criminal regulatory standards are too blunt as instruments of pollution control, while pollution taxes and levies would be more effective. Such arguments have not found general favour with government and public opinion on either side of the Atlantic. For all its drawbacks an increasingly sanctions-based regulatory approach is likely both as regards standard setting and implementation, if only in response to growing public environmental concern.

6 ILLUSTRATIVE PARTICULAR STYLES OF REGULATION

6.1 Planning and Hazardous Substances Controls

Many industrial developments occurred before the advent of planning control in the UK—effectively 1947. However, *new* chemical installations, for example, will be operational development and will require planning permission under the Town and Country Planning Act—TCPA—1990. A *somewhat* similar system exists in the USA in the form of 'zoning'. The District Council for an area has discretion whether or not to grant permission. Before doing so it will consult widely with, for example, NRA and HMIP, as well as taking into account the various plans that have been made concerning the land and their district. The public have no general *right* to be consulted, though this does exist in relation to certain 'bad neighbour' cases listed in the General Development Order—GDO—SI 1988/1813—a statutory instrument made under the TCPA. Chemical installations are not *per se* 'bad neighbours', but any building constructed to a height exceeding 20 m is and so most large plants will be caught. In other cases falling within regulations contained in SI 1988/1199 a process of environmental assessment—EA—will have to be undertaken as part of the decision making process. This is mandatory in some cases, *e.g.* oil refineries, asbestos works, and integrated chemical installations, while in others it is only necessary if the project will have 'significant' environmental effect, *e.g.* certain chemical plants such as pesticide producers. EA is needed because of the requirements of EC law—Directive 85/337/EEC. It enables a wider range of issues to be taken into account than is usual on a planning application, normally land use considerations only. The number of projects subject to EA in any given year will not be great. The USA has a similar process known as Environmental Impact Assessment—EIA.

A refusal of planning permissions entitles the disappointed developer to appeal to the Secretary of State. Conditions may be imposed on a grant of planning permission, again subject to a right of appeal. Conditions must serve land use purposes, must be reasonable *per se*, and must relate to the development permitted on the land. Unauthorized development can be required to stop by means of an enforcement notice, and if this is ignored criminal proceedings can be taken.

The Planning (Hazardous Substances) Control Act 1990 provides for a system of Hazardous Substances Consents operating parallel to planning procedures in relation to the presence on land of specified materials, sometimes in specified quantities. The basic requirement is that hazardous substance related activities will have to be specifically consented, generally by local authorities—District Councils and London Boroughs—who will have power to enforce relevant provisions, which extend to regulate hazardous substances on, over, or under land. The need for consent will arise once the aggregate quantity of a hazardous substance, as specified in regulations to be made by the Secretary of State, exceeds a specified 'controlled quantity'—again as specified by regulations. Thereafter the system of obtaining consent is similar to planning control: applications for consent have to be made in due form, with appropriate publicity, and have to be determined taking into account land use issues, and advice received from the Health and Safety Executive, another 'quango'. Consent may be given, with or without conditions, or refused; consents given may be revoked, subject to the payment of compensation. A disappointed applicant may appeal to the Secretary of State. Contravention of hazardous substances control constitutes an offence. It should also be noted that under the Health and Safety at Work Act 1974 various regulations have been made which supplement the foregoing controls. The Notification of Installations Handling Hazardous Substances Regulations—NIHHS—SI 1982/1357 require the Health and Safety Executive to be informed of developments involving the presence in, over, or under land of hazardous substances in specified quantities. The Executive are then in a consultative capacity with regard to planning issues concerning relevant land and can advise planning authorities to, for example, refuse any consents applied for. The Control of Industrial Major Accidents Regulations—CIMAH—SI 1984/1902 further require measures to prevent, or limit the effect of, major accidents arising from specified industrial activities where specified substances are in use or storage.

6.2 Integrated Pollution Control

IPC exists under the Environmental Protection Act—EPA—1990 and is a scheme whereby the Secretary of State is to plan for the reduction of specified pollutants which HMIP and District Councils will then apply on a process-by-process basis by way of specific authorizations. The EPA requires the three environmental media, land, air, and water, to be protected against polluting releases from any process of any substance capable of causing harm—including damage to the health of organisms or interference with ecological systems—to humankind or other living organisms. The Secretary of State has power to make regulations to prescribe processes needing authorization for their emissions.[9] These processes are contained in lists 'A' and 'B' in regulations, SI 1991/472. 'A' processes, the most complex, are overseen by HMIP who regulate *all* emissions to *all* the environmental media. 'B' processes are subject to local authority regulation in respect of *atmospheric* emissions. The 'A' processes include major thermal electricity power generation plants of 50+ MW capacity. This implements obligations under EC Directives 84/360 and 88/609 which impose requirements for staged reductions in SO_2 and NO_x emissions between 1993 and 2003. The following are generally subject to regulation by either HMIP or local authorities: fuel and power production; metal production and processing; the minerals and chemicals industries, waste disposal, and recycling, and other industries such as paper and dyestuffs manufacturing.

[9] Section 2, The Environmental Protection (Prescribed Processes and Substances) Regulations, SI 1991/472.

No-one may carry on a prescribed process without an authorization operated in accordance with any conditions imposed—these being subject to periodic review.[10] Applications for authorization may be refused and relevant authorities must keep public registers of information related to authorizations.

At the heart of the authorization procedure is the BATNEEC requirement—*i.e.* that authorities must use the '*best available techniques not entailing excessive costs*' to prevent emissions of prescribed substances or otherwise to reduce them to a minimum and render them harmless.[11] This phrase has found its way into UK law from International Law—The Geneva Convention of 1979, see above, via EC Directives, and thus represents an example of vertical integration of national, world regional, and international legal systems, a phenomenon still uncommon. BATNEEC has two elements, BAT and NEEC. BAT clearly contemplates more than just equipment, but relates to how it is used and how staff are trained and deployed. It also contemplates rising standards of emission control as techniques adapt to new state-of-the-art technologies. NEEC on the other hand seems to suggest that state-of-the-art measures need not be adopted if they would be excessively expensive and might only produce marginal decreases in emissions of a few percentage points—an application of cost benefit analysis theory. An individual BATNEEC authorization will probably be expressed in terms of performance rather than requiring specific technologies. Guidance notes on the IPC and BATNEEC systems will be issued by HMIP, see for example IPR 1/1 on large boilers and furnaces issued in March 1991. To reinforce the BATNEEC principle the Secretary of State has powers to make regulations concerning process releases of substances, setting standard limits to their amounts, or amounts at given times, their concentrations, and characteristics.[12]

With regard to 'A' processes where releases may take place into more than one environmental medium HMIP is to apply the BATNEEC principle according to the requirements of 'the best practical environmental option'—BPEO.[13] This is designed to ensure that the environmental media are considered together and that the mode of release is chosen which minimizes pollution of the environment as a whole.

Authorizations may be varied, revoked and enforced by the appropriate authority, as may failures to obtain authorizations; the ultimate sanction is a criminal prosecution.[14] Where a prosecution lies for failure to adopt BATNEEC the normal presumption that the prosecution must prove its case is reversed and it is for the accused to prove there was no better 'BATNEEC' available.[15] Further where an offence is committed by a company and it is proved that a director or other senior officer of the company connived in the commission of the offence, that individual will be criminally liable along with the company.[16]

[10] Section 6, EPA 1990.
[11] Section 7, EPA 1990.
[12] Section 3, EPA 1990.
[13] Section 7(7), EPA 1990.
[14] Section 23, EPA 1990.
[15] Section 25, EPA 1990.
[16] Section 157, EPC 1990.

This system of regulation largely supersedes those previously existing. It is much more sanctions based than its predecessors, clearly moving away from more discretionary systems they employed. Even so the 1990 Act appears less stringent than its transatlantic counterparts, particularly with regard to atmospheric pollution. The current US law can be traced back to 1967 followed by The Clean Air Act Amendments of 1970, 1977, and 1990. The general tendency in US practice has been for the law to become more proactive in attempting to force technological innovation—particularly with regard to vehicle emissions—and for it to become stricter and prescriptive, and more expensive and cumbersome to implement. As an example of this one may cite the move from the 'reasonably available control technology'—RACT—standards of the older legislation to the 1990 requirements for 'maximum achievable control technology'—MACT—standards in respect of some 250 emission sources of 189 hazardous air pollutants—HAPs. A further feature of US law has been to move away from state enforcement of the law to a mixture of federal enforcement through the NEPA and state enforcement supported by Federal fiscal inducements. Enforcement has not been all that it might have been and criticism has centred on:

(a) the over-regulatory, sanctions based approach;
(b) an over-reliance on technological answers to pollution problems;
(c) under-resourced enforcement agencies—the evil of over regulation and under-enforcement.[17]

The latest amendments to the law strengthen the enforcement powers of state and federal agencies, and allow fines of up to $1 M to be imposed, together with terms of imprisonment for 1–15 years for guilty individuals, which includes guilty employees, supervisors, managers, and corporate officials. Even so NEPA's ability to meet its new commitments has been called in question while new annual pollution compliance costs have been estimated at between $25 and $50 billion.[18] The parallels with UK issues can easily be seen.

6.3 Water

Many major polluters of water will be subject to IPC regulation by HMIP as 'A' processes under the Act of 1990, nevertheless NRA has major powers under the Water Resources Act 1991 over emissions from other processes and activities—*e.g.* agriculture—to control pollution of waters inland, around coasts, and in estuaries. These powers include achieving national water quality standards and objectives, powers of control by means of fixed discharge consents, and to prosecute those who, without consent or IPC authorization, cause or permit poisonous, noxious, or polluting matter to enter relevant waters. With regard to IPC authorizations the 1990 Act makes general provision for NRA to direct HMIP in relation to releases of emissions which will prejudice relevant water

[17] D. McGrory, 'Air Pollution Legislation in the United States and the Community', *European Law Review*, 1990, **15**, 298.
[18] Engelgau, 'Written on the Wind', *Resources*, No. 1, 1991, **13**, p. 7.

quality objectives. It will be noted, however, that the 1991 Act, like that of 1990, is not a pollution *eradication* measure. Provision is made for 'polluting permission' to be given by way of consent. Essentially both pieces of legislation illustrate the point made earlier—the conditions for eradication of pollution do not exist, and therefore the lesser form of pollution control has to be used.

7 CONCLUSION

Pollution regulation, from both legal and economic standpoints, is at an early stage of development. Scientific evidence of the existence of pollution is not enough to trigger an automatic regulatory response, and a whole complex of economic, social, political, administrative, and ethical issues has to be examined before a decision is made. This process will be reflected in the type and content of regulatory measures. The key to understanding pollution regulation is to realize that it operates within a regularly changing ethical context. It is the marriage of constants and variables that does much to give the subject its interest—and its irritations.

8 FURTHER READING

In addition to the references, reference may also be made to the following:

S. Ball and S. Bell, 'Environmental Law', Blackstone Press, London, 1991.
F. Cairncross, 'Costing the Earth', Business Books Ltd., London, 1991.
N. Haigh, 'EEC Environmental Law and Britain' (2nd Edn. Rev.) Longman, London, 1989.
N. H. Highton and M. G. Webb, 'Sulfur Dioxide from Electricity Generation: Policy Options for Pollution Control', *Energy Policy*, 1980, **8**, 61.
N. H. Highton and M. G. Webb, 'Pollution Abatement Costs in the Electricity Supply Industry in England and Wales', *J. Ind. Econ.*, 1980, **30**, 49.
D. J. Hughes, 'Environmental Law' (2nd Edn.), Butterworth, London, due 1992.
G. M. Richardson, A. I. Ogus, and P. Burrows, 'Policing Pollution', Oxford University Press, Oxford, 1983.
D. Pearce and A. Markandya, 'The Benefits of Environmental Policy', OECD, Paris, 1989.
D. Pearce and R. K. Turner, 'Economics of Natural Resources and the Environment', Harvester Wheatsheaf, Hemel Hempstead, 1990.
R. C. Schwing, B. W. Southworth, C. R. von Buseck, and C. J. Jackson, 'Benefit–cost Analysis of Automotive Emission Reductions', *J. Environ. Manage.*, 1980, **7**, 44.

Subject Index